Stem Cell-based Biosystems

Editors

Yi-Chen Ethan Li

Department of Chemical Engineering
Feng Chia University
Taiwan

I-Chi Lee

Department of Biomedical Engineering and Environmental Sciences
National Tsing Hua University
Taiwan

CRC Press
Taylor & Francis Group
Boca Raton London New York

CRC Press is an imprint of the
Taylor & Francis Group, an **informa** business
A SCIENCE PUBLISHERS BOOK

Book cover design: I-Chi Lee, Yi-Chen Ethan Li

First edition published 2024
by CRC Press
2385 NW Executive Center Drive, Suite 320, Boca Raton FL 33431

and by CRC Press
4 Park Square, Milton Park, Abingdon, Oxon, OX14 4RN

Library of Congress Cataloging-in-Publication Data (applied for)

ISBN: 978-0-367-65545-7 (hbk)
ISBN: 978-0-367-65546-4 (pbk)
ISBN: 978-1-003-13004-8 (ebk)

DOI: 10.1201/9781003130048

Typeset in Times New Roman
by Radiant Productions

Preface

In the past twenty years, there has been significant progress in stem cell science and technologies. With the deciphering of the biological properties of stem cells, scientists have been able to combined them with various technologies and utilize them for versatile applications in biomedical research. Recently, the emergence of stem cell therapy has led to an expectation of rapid growth in the stem cell-related market over the next decade, demonstrating the enormous economic value of related requirements. Simultaneously, these needs have also stimulated the development of the stem cell industry chain, which suggests that stem cell research will be expanded to an industry level. Therefore, stem cell-based biosystems have gained increasing attention.

This book presents the recent advances in stem cell-based biosystems, covering a broad spectrum of topics and state-of-the-art concepts in the field of stem cell application. The book is divided into three parts comprising ten chapters. The first part (Chapters 1–4) includes methods for isolating and culturing stem cells from different sources, recent technologies for regulating stem cell stemness through biomaterials, and common systems for culturing stem cells on a large scale. The second part (Chapters 5–8) discusses advanced technologies for biomimetic in vitro biosystems. In developing biomimetic *in vitro* biosystems, biomaterials play an indispensable role. Therefore, advanced technologies for two-dimensional (2D) and three-dimensional (3D) biomaterials for fabricating stem cell-based biomimetic systems are introduced. Moreover, there are a large number of new drug or drug repurposing testing on living biomimetic 2D/3D biosystems in tissue engineering, regenerative medicine, and precision medicine field. Therefore, stem cell-based organ-on-chips for drug screening have also been introduced to enhance the prediction precision of testing effectiveness. Furthermore, biomaterials combined with nanotechnology as drug carriers for controlled release to regulate stem cell behavior is discussed for precision medicine applications in the second part.

The third part (Chapters 9–10) presents the evaluation of stem cell-based biosystems for applications from *in vitro/ex vivo* to clinical therapy approaches. In-depth discussions of the standard testing method of *ex vivo* and animal models for biomaterials and the guidelines and case studies of stem cell therapy for clinical applications are included. These novel and ongoing technologies provide new insights and possibilities for fabricating stem cell-based biosystems. Readers will receive a unique perspective on the challenges and emerging opportunities in the applications of stem cell-based biosystems.

Yi-Che Ethan Li
Feng Chia University

I-Chi Lee
National Tsing Hua University

Contents

Preface iii

1. **Traditional Isolation and Culture Methods of Stem Cells** 1
Min-Huey Chen, Nai-Chen Cheng, Wen-Yen Huang, Sung-Jan Lin, Chia-Ning Shen, I-Chi Lee and *Yi-Chen Ethan Li*

2. **Biomaterial Effects on Isolation, Proliferation, and Regulation of Stem Cells** 32
Nai-Chen Cheng, Min-Huey Chen, Chia-Ning Shen, Wen-Yen Huang, Sung-Jan Lin, I-Chi Lee and *Yi-Chen Ethan Li*

3. **Scale-Up of Stem Cells** 63
Yi-Chen Ethan Li, Nai-Chen Cheng, Min-Huey Chen, Wen-Yen Huang, Sung-Jan Lin, Chia-Ning Shen and *I-Chi Lee*

4. **The Effects of Biomaterial Properties on the Behaviors of Stem Cells** 74
I-Chi Lee, Nai-Chen Cheng, Chia-Ning Shen, Wen-Yen Huang, Sung-Jan Lin, Min-Huey Chen and *Yi-Chen Ethan Li*

5. **The 2D Membrane Technology for Fabrication of 2D Stem Cell Culture System** 95
Nai-Chen Cheng, Wen-Yen Huang, Sung-Jan Lin, Min-Huey Chen, I-Chi Lee, Chia-Ning Shen and *Yi-Chen Ethan Li*

6. **The 3D Culture Technology for Fabrication of Stem Cell-based Biosystem** 113
Yi-Chen Ethan Li, Nai-Chen Cheng, I-Chi Lee, Wen-Yen Huang, Sung-Jan Lin, Chia-Ning Shen and *Min-Huey Chen*

7. ***In vitro* Biosystems—Stem Cell-based Organ on Chips** 175
I-Chi Lee, Nai-Chen Cheng, Chia-Ning Shen, Min-Huey Chen, Wen-Yen Huang, Sung-Jan Lin and *Yi-Chen Ethan Li*

8. **The Nanotechnology for Stem Cells** 205
Yi-Chen Ethan Li, I-Chi Lee, Nai-Chen Cheng, Wen-Yen Huang, Sung-Jan Lin, Chia-Ning Shen and *Min-Huey Chen*

9. Biomaterials use in *Ex vivo* Testing and Animal Model **216**
Chia-Ning Shen, Wen-Yen Huang, Sung-Jan Lin, Nai-Chen Cheng, Min-Huey Chen, I-Chi Lee and *Yi-Chen Ethan Li*

10. Guidance & Case Study for Stem Cell Therapy **233**
Wen-Yen Huang, Sung-Jan Lin, Chia-Ning Shen, Nai-Chen Cheng, Min-Huey Chen, I-Chi Lee and *Yi-Chen Ethan Li*

Index **251**

1

Traditional Isolation and Culture Methods of Stem Cells

Min-Huey Chen,[1] *Nai-Chen Cheng,*[2] *Wen-Yen Huang,*[3] *Sung-Jan Lin,*[3] *Chia-Ning Shen,*[4] *I-Chi Lee*[5,*] and *Yi-Chen Ethan Li*[6,*]

Since 1998, human stem cells have been successfully isolated and cultured into cell lines, a development which triggered great excitement in the medical community. With the development of genomic information, the application of stem cells, as well as the research of biology and the development of material technology, has brought about immense potential for the development of regenerative medicine. Currently, stem cells can be used to regenerate a variety of tissues, and the development of nanobiotechnology has also led to greater breakthroughs in regenerative medicine. Because stem cells are related to other cells in the development process and structure, the study of behaviors of stem cells can also be used as a reference for other tissue and organ regeneration application models.

Stem cells have the ability to divide infinitely and can produce specialized cells. There are totipotent stem cells, pluripotent stem cells, and multipotent stem cells.

When the sperm and egg are combined, a cell with the potential to develop into a complete organism is formed. This fertilized egg is called "totipotent". In the first hour of fertilization, the fertilized egg can split into two identical cells. That is to say,

[1] Graduate Institute of Clinical Dentistry, School of Dentistry, National Taiwan University, Taiwan.
[2] Department of Surgery, National Taiwan University Hospital, Taiwan.
[3] Department of Biomedical Engineering, College of Medicine and College of Engineering, National Taiwan University, Taiwan.
[4] Genomics Research Center, Academia Sinica, Taiwan.
[5] Department of Biomedical Engineering and Environmental Sciences, National Tsing Hua University, Taiwan.
[6] Department of Chemical Engineering, Feng Chia University, Taiwan.
* Corresponding authors: iclee@mx.nthu.edu.tw; yicli@fcu.edu.tw

if any one of the two cells is placed in the uterus, they will both have the ability to develop into a fetus. In fact, the formation of twins means when both the cells are placed in the uterus, they will each have the ability to develop into a fetus. That is, after separation, they each form an independent individual with the same gene.

Next, pluripotent stem cells are introduced. Four days after the egg is fertilized, the cell undergoes several divisions, begins to be specialized, and aggregates to form a hollow blastocyst. The outer cells of the blastocyst continue to form the placenta and various supporting tissues required for the development of the fetus in the uterus, and the inner cells of the blastocyst will eventually form various tissues of the human body. Although the inner cells can form various types of cells in the human body, these inner cells cannot form the placenta or the supporting tissues needed for the development of the uterus; these inner cells cannot form an organism. We call these inner cells "pluripotent"—that is, they can form many types of cells but not all cell types are required for fetal development. In addition, these plruipoten stem cells have no fully effective ability to form the whole fetus.

Pluripotent stem cells can be further specialized into multipotent stem cells. These multipotent stem cells can be re-formed into cells with special functions. For example, blood stem cells can produce red blood cells, white blood cells, and platelets, while skin stem cells can form various types of skin cells; these more specialized stem cells are called "multipotent". Multipotent stem cells exist in children and adults and play very important roles. Take blood stem cells as an example; blood stem cells exist in the bone marrow cavity, and a small amount is also present in blood circulation. Through the blood stem cells, our blood cells can be constantly renewed. Blood stem cells are very important for maintaining life. As mentioned earlier, multipotent stem cells can be found in some adult tissues. In fact, these stem cells can supply the cells that are gradually destroyed and lost in our body, such as the previously mentioned hematopoietic stem cells. Multipotent stem cells have not been completely discovered in all types of adult tissues, but research in this area is still increasing. For example, stem cells were once thought not to exist in the adult nervous system. However, there have been reports showing that neural stem cells can be successfully isolated from the nervous system of rats and mice, and human neural stem cells have also been isolated from fetal tissues. In addition, cells that may resemble neural stem cells have also been isolated from brain tissue that was surgically removed from epilepsy patients. As for whether adult multipotent stem cells have the same potential as pluripotent stem cells, there is still little evidence that multipotent stem cells in mammals, such as blood stem cells, can change the process of differentiation to produce skin cells, liver cells, or cells other than any blood cell type. However, researchers have gradually questioned these viewpoints gathered from experiments on animal. In these experiments, it has been confirmed that some adult stem cells were previously thought to develop into only one type of specialized cells, but in fact, they can develop into other kinds of specialized cells. For example, research conducted on mice pointed out that placing neural stem cells in the bone marrow cavity can produce different types of blood cells. In addition, experiments conducted on rats have confirmed that stem cells in the bone marrow can produce hepatocytes. These exciting findings point out that even when stem cells have begun to specialize in certain special environments, they are more flexible than originally thought. Previous studies have pointed out that adult stem cells have great potential for research and development of cell therapies [1]. For example, transplantation with adult stem cells has many advantages. If stem cells can be isolated from patients to guide their isolation and differentiation and then transplanted back into the patient's body, these cells should not cause rejection. In addition, the use of adult stem cells for cell therapy will inevitably reduce or even avoid the moral controversy that may be caused by the use of cells isolated from human embryos or fetuses.

Nevertheless, the application of adult stem cells has many limitations in practice. Firstly, not all tissues in the body have been isolated from adult stem cells. Furthermore, the amount of adult stem cells is extremely limited, and it is not easy to isolate and purify, and their number is decreased with age. Any attempt to treat the patient with stem cells from the patient's body requires the stem cells to be isolated from the patient's body and cultivated to a sufficient amount for treatment. For some urgent diseases, there may not be too much time to cultivate enough cells for treatment. In some diseases caused by genetic defects, genetic problems may still exist in the stem cells of patients, and the cells taken from these patients may not be suitable for transplantation. There is evidence that stem cells taken from adults may not have the same proliferation potential as young cells, and adult stem cells may experience errors due to various factors in daily life, including sunlight, toxic substances, and repeated DNA replication. Various effects may cause more abnormalities in DNA, and these latent shortcomings will limit the application of adult stem cells.

In order to determine the source of many specialized cells and tissues in the body and to develop new treatments, it is important to study the development potential of adult stem cells and compare them with pluripotent stem cells. Stem cells can be used in new therapies for the most severe diseases and can be guided to be differentiated into specific cells to promote the regeneration of tissues and organs [2]. Therefore, it is necessary to actively work on various related research. Scientists need to find the best source of these cells, and once identified, they can then develop new cell therapies. The development of stem cell lines, including pluripotent stem cells and multipotent stem cells, can be used to promote the regeneration of tissues and organs and is an important cell source for the development of regenerative medicine.

Embryonic stem cells

Embryonic stem cells (ESCs) are pluripotent or totipotent stem cells. This type of stem cell possesses self-renewal and unlimited proliferative abilities. Under adequate conditions, ESCs are able to differentiate into all cells of three embryonic germ layers and germ cells in the human body. Therefore, this type of stem cell has substantial potential in scientific research or future reading materials, pharmacology, and even clinical applications. In terms of ESCs, the cultivation of ESCs provides a platform to understand the formation of body cells and further investigate the mechanism and controllable factors for regulating ESCs. Therefore, the study of ESCs can be useful for knowledge enhancement in the field of biology. In addition, establishing an ESC culture model can also be used to progress the effects of pharmaceutical molecules and toxicants on the development of the human body, and even be used for drug repurposing. Moreover, if we can develop an efficient culture method for regulating specific differentiation of ESCs such as cardiac cells, liver cells, and pancreas cells, it might contribute to more versatile applications and benefit the transplantation of ESC-derived somatic cells for disease therapy. In fact, mice ESCs were isolated and established in the early 1980s. Increasing evidence confirmed how valuable ESCs are. Then, after nearly 20 years, Thomson and co-workers first isolated human ESCs and established the protocol to culture human ESCs [3]. Many *in vitro* protocols have been developed to culture ESCs. Using Thomson's isolation principle, Chen and co-workers established three human ESC cell lines with Taiwanese ancestry. In their study, they prepared different culture media for the derivation and maintenance of human ESC line. The first medium for ESC derivation consisted of a high glucose DMEM medium with 20% FBS, 2 mM glutamine, 0.1 mM 2-mercaptoethanol, 1% ITS, 1% non-essential amino acid, and 2000 IU/ml human LIF. The second medium for maintenance of human ESCs was the same as the first medium but used 10 ng/ml bFGF instead of 2000 IU/ml human LIF. Subsequently, the feeder layer was prepared by using primary murine embryonic fibroblasts for two different uses. For example, if the feeder layer is for the derivation of hESC lines, the fibroblast cell number is around 60,000 cells/cm^2, and the cell number for maintenance would be used around 30,000 cells/cm^2. After preparing the medium and feeder layer, the blastocysts with good morphology were selected according to a grading system published in Gradner's study [4]. Subsequently, the chosen blastocysts were treated with 0.5% proteinase for 4 min to remove zona pellucia and cultured with the first medium for 30 mins. Then, a commercial diluted rabbit anti-human serum (1:3 to 1:20) was used to treat blastocysts for 30 mins, washed by th first medium , and further treated with a diluted guineapig complement (1:8 to 1:50) for 15 mins. After treatment, the first medium was used to wash the blastocysts, and the dead trophectoderm cells were removed by pipetting the blastocysts. Following this, the inner cell mass was seeded on the feeder cells and incubated for two days. Notably, the inner cell mass can rapidly attach to the feeder layer within two days. Then the culture medium was changed daily, and the expanded inner cell mass was isolated from the outgrowing differentiated cells, exposed with 10 mg/ml dispase, collected, and reseeded on a fresh feeder layer. After another 7–14 days of incubation, the human ESC-like colonies formed were cut into small pieces and transferred to another fresh feeder layer with the second medium for maintenance. After another 7–14 days, the resulting hESC-like colonies were transferred in cut pieces into fresh feeder layers with the second medium (ES-M medium). Then the best human ESC colony was kept according to its morphology and size and was routinely transferred at a seven-day interval to establish a human ESC cell line. In addition, Gearhart and co-workers isolated human embryonic germ cells from postfertilization human embryos [5]. In the study, the mesenteries and gonadal rides tissues were obtained from 5 to 9-week-old postfertilization embryos and mechanically excised. Subsequently, the excised tissues were incubated in 0.25% trypsin for 5–10 mins or a solution consisting of 0.002% DNase type I, 0.1% collagenase type IV, and 0.01% hyaluronidase type V for two hours at 37°C. Afterwards, the cells from the isolated mesenteries and gonadal rides tissues were cultured on a SIM mouse embryo-derived thioguanine and ouabain resistant feeder layer, which is pre-mitotically inactivated by the treatment of 50 Gy γ-radiation. The cells were incubated using the DMEM medium with 15% FBS, 2 mM glutamine, 0.1 mM 2-mercaptoethanol, 10 μM forskolin, 1 ng/ml of human recombinant bFGF, 1000 units/ml of human recombinant LIF until confluence. After confluence, 0.25% trypsin or 0.05% trypsin/0.53 mMEDTA were used to treat resuspending cells for 5–10 mins, and were routinely subcultured every seven days. Moreover, recently, many studies tried to establish a whole human-based system without any feeder layers from animal sources to culture human ESCs [6, 7]. Therefore, several feeder layers, such as the human foreskin feeder, and human placental fibroblast, have been confirmed to maintain the growth of human ESCs. In addition, another method using animal serum-free medium, called condition medium, was also developed for culturing human ESCs [8, 9]. For example, a TeSR1 medium containing transforming growth factor-beta (TGF-β), γ-aminobutyric acid (GABA), pipecolic acid, bFGF, and lithium chloride can achieve the same effect as the mouse embryonic fibroblast condition medium [8]. Moreover, combining human extracellular matrices (ECM) such as collagen IV, fibronectin, laminin, or vitronectin can fabricate a culture system with feeder layers for culturing human ESCs. According to these studies, the success of human isolation and cultivation brings a promise to develop further various applications in tissue engineering, drug screening/repurposing, and regenerative medicine. However, using human ESCs still poses ethical issues in clinical applications. Therefore, other types of pluripotent and multipotent stem cells from adult tissues or reprogramming processes have been introduced in the following paragraph.

Induced pluripotent stem cells

Induced pluripotent stem cells (iPSCs) are stem cells generated by the co-expression of designated pluripotency-associated proteins in somatic cells and have the pluripotency to differentiate into three germ layers including ectoderm, mesoderm, and endoderm. Human iPSC technology has evolved rapidly since its inception in 2007 and significant progress has been made in the field of stem cell biology and regenerative medicine. Human iPSCs have been extensively used for disease modelling, drug discovery, and cell therapeutic development and have numerous applications in both basic research and the clinical setting. iPSC presents multiple advantages over standard cellular screens and has been applied to generate a 'disease in dish' model where it can be used to test drug efficacy and cytoxicities. The benefits include their human origin, ability to be differentiated into any desired cell type, ease of access, expansivity, and the possibility to generate personalized treatment utilizing patient-specific iPSCs. Furthermore, recent advances in genome editing technologies, particularly CRISPR-Cas9, have facilitated the rapid development of genetically engineered human iPSC-based disease models. Patient-specific iPSCs can also be utilized to generate physiologically relevant cellular platforms with 3D structures such as organoids. Indeed, novel pathogenic processes have been identified and new drugs discovered through iPSC screens. Together, the integration of human iPSC technology with recent advances in gene editing and 3D organoids makes iPSC-based systems a key component of medical research going forward.

Embryonic stem cells (ESCs) can give rise to specialized and terminally differentiated cells with low developmental potential. Terminal differentiations were once considered an irreversible process. However, the discovery that these terminally differentiated cells can be reprogrammed to a pluripotent state completely altered this prospective and the field of developmental biology. First conducted in Shinya Yamanaka's laboratory, the use of retrovirus to express transcription factors Oct4, Sox2. Klf4 and c-myc (OSKM) in somatic cells resulted in the generation of iPSCs with similar pluripotent properties as ESCs [10]. These transcription factors' cumulative activity was demonstrated to be both necessary and sufficient for converting somatic cells into iPSCs [11]. The path to iPSC reprogramming involves the loss of somatic gene expression and the progressive activation of pluripotency-related transcription factors. OSK, in particular, serves to activate the genome-wide reorganization enhancer by direct DNA binding and the manipulation of chromatin opening [12]. Although revolutionary, the use of OSKM alone for iPSC reprogramming can be an inefficient and stochastic [13]. Only a small percentage of initial somatic cells can reach the pluripotent state due to the resistance of distinct transcription factors to OSKM induction [14]. Therefore, in addition to OSKM, numerous other factors that enhance the reprogramming process have been discovered including Nanog, Esrrb, Lin28, Glis1, Nr5a2, Utf1, all4, and Foxh1 [15–22]. Introduction of reprogramming factors can be integrative or non-integrative. Integrative approaches, in which DNA sequences encoding reprogramming factors are incorporated into the genome of somatic cells, and non-integrative approaches, which do not lead to permanent genome integration have been used to generate iPSCs [23]. Retro- and lenti-viral vectors were used in the early days of iPSC production to introduce OSKM into somatic cells [23]. Although robust, the incorporation of viral particles into the host genome poses significant risk of insertional mutagenesis and malignancy, making the iPSC generated via these methods unsuitable for clinical applications [24]. Furthermore, the presence of viral components may also elicit an immunogenic response [25]. Non-integrative methods on the other hand have the advantage of producing "safe" iPSCs with a lower risk of acquiring secondary disease-causing mutations, making them more suitable for cell-based therapies [26]. Non-integrative approaches can be viral-based (adenovirus, the adeno-associated virus vector, and Sendai virus) [27]. DNA-based (plasmid, minicircle and episomal) [28]. or RNA-based (synthetic mRNAs and self-replicating RNAs) [29]. Viral-based vectors can be used to deliver reprogramming factors with limited genomic integration. However, viral vector selection is crucial as it can result in vastly different reprogramming efficiency. Adenoviral vectors exhibit a board range of cellular tropism and can carry large cargo with insertion beyond 8kb [30]. This allows the generation of iPSC from numerous cell types with the use of polycistronic construct design expressing multiple reprogramming factors under the control of a common promoter. The major limitation of adenoviral vector is the transient nature of factor expression (3–8 days) due to rapid clearance by dividing cells. This results in the requirement for a high viral titre and very low reprogramming efficiency.

Additionally, adenoviral vectors can induce immunogenic response in the host. Compared to adenovirus, adeno-associated virus (AAV) maybe more suitable for clinical applications due to low immunogenicity and stable transgene expression [31].

Limitation of AAV vectors includes restricted cargo capacity (~ 5kb), low reprogramming efficiency and up to 10% chance of genomic integration. Currently, sendai virus (SeV) has proven to be one of the most used viral-based method for iPSC reprogramming. SeV enables higher transduction efficiency and more stable transgene expression compared to both adenoviral and AAV vectors, leading to higher reprogramming efficiency. Moreover, due to its ability replicate exclusively within the cytoplasm, chances of genome integration are significantly reduced. Complete removal of residue SeV vectors can be achieved through multiple passages (> 10 passages). Indeed, human iPSCs has been generated using SeV from multiple cell sources including fibroblasts, cord blood CD34+, PBMC and skeletal myoblasts [32, 33]. DNA-based reprogramming methods enable the generation of relatively safe iPSCs with minimal possibility of genome integration. Plasmid or minicircle vectors, that is, plasmid that lacks bacterial backbone, can be designed to be polycistronic, encoding multiple reprogramming factors on a single vector. Both are easily synthesized and can be used to generate non-integrating iPSCs. However, due to the lack of self-replication and the requirement of multiple transfections, the reprogramming efficiency of these two methods is significantly lower compared to the use of viral-based methods described in the previous paragraph. To resolve issues faced by plasmid or mincicle vectors, episomal vectors can be used. Episomal vectors can self-replicate as extrachromosomal elements without genome integration. This enables high reprogramming factor expression with only a single transfection necessary. Multiple studies have successfully generated iPSCs from human fibroblasts and PBMCs using episomal vectors encoding transcription factors OSKM along with LIN28 and p53 suppression [34]. Indeed, reprogramming efficiency of episomal vectors can be comparable to viral-based methods. Compared to SeV, the loss of transgene expressing episome can occur within a shorter time frame, approximately two weeks. RNA-based reprogramming methods can be used to generate iPSC with virtually no genome integration possibilities. Conventional mRNA poses major disadvantages for iPSC reprogramming, including instability, transfection difficulties and an immunogenic response. Therefore, synthetic mRNA as well as self-replicating mRNA (srRNA) has been developed for the purpose of iPSC reprogramming. Synthetic mRNAs are designed with enhanced 5' and 3' UTRs, 5' cap, modified nucleotides and poly-A tail to enhance stability and reduce the immunogenic response. Using synthetic mRNA, iPSC has been successfully reprogrammed from skin fibroblasts, mesenchymal stromal cells, and amniotic fluid stem cells [35]. Despite advancement in synthetic mRNA stability, multiple daily mRNA transfections are still required [29]. Thus, srRNA was developed that features the RNA replication complex, which can extend the duration of the reprogramming factor expression to the extent where only a single transfection is required (Yoshioka et al. 2013). Two major strategies of mRNA delivery can be used, including cationic lipid and electroporation. Cationic lipids are formulated with positively charged liposomes that can interact with negatively charged RNA to form a RNA-liposome complex, which can fuse with the cell membrane, providing a simple solution to the delivery of synthetic mRNA into somatic cells [36]. Compared to cationic lipids, electroporation can be more complex to perform, requiring an instrument to generate an electrical pulse to disturb the phospholipid bilayer of the cell membrane. However, when optimized, electroporation can be highly efficient at delivering mRNA into difficult to transfect cell types [37].

Bone marrow derived mesenchymal stem cells

There are at least two types of stem cells in bone marrow: one is hematopoietic stem cells, and the other is non-hematopoietic stem cells. These non-hematopoietic stem cells are called mesenchymal stem cells (MSCs). These mesenchymal stem cells have been proven to be able to be differentiated into osteoblasts, adipocytes, chondrocytes, myocytes, and neurons, etc. Therefore, these cells have been considered to have great potential on the gene therapy of diseases. Previous studies have found at least two different forms of mesenchymal stem cells: one is a cone-shaped larger cell, and the other is a very small, rapidly regenerating cell. In a typical cell therapy method, the use of differentiated cells for transplantation, the clinical and technical problem lies in the problem of cultivating and amplifying these differentiated cells. Moreover, induction of cells for differentiation *in vitro* usually requires the cultivation of cells many times to increase the number of cells; these processes may cause cells to lose their phenotype [38]. In adults, multipotent stem cells can be applied to maintain connective tissues, including bones, cartilage, muscles, ligaments, and other tissues. These multipotent stem cells can be differentiated into various stem cell lines, namely osteoblasts, chondrocytes, adipocytes, and muscle cells, and are referred to as “mesenchymal stem cells” (MSCs) [38, 39].

Mesenchymal stem cells (MSCs) have been successfully derived from many species of organisms, including humans, rats, mice, dogs, and rabbits [40–46]. The multiple differentiation potentials of mesenchymal stem cells make them play a very important role in cell therapy. Cell therapy mainly relies on the transplantation of isolated cells to repair tissue defects. If articular cartilage is injured and does not penetrate the bone tissue under the cartilage, there will be no repair phenomenon. Therefore, it can be seen that the bone marrow component is important for the repair of the articular cartilage [47]. The invalid repair of cartilage is because the reparability of the cartilage itself is not enough, and the chondrogenic cells of the bone marrow cannot reach the expected repair site. Conversely, if the articular cartilage injury extends beyond the subchondral bone, the repair process can ensure that mesenchymal stem cells can be transferred from the bone marrow to the injured site for cartilage differentiation. However, the analysis of animal and human biopsies can show the facts that the main synthesis of tissue repair is fibrocartilage instead of the real articular cartilage [48].

The cells derived from bone marrow and periosteal cells are related to the repair of fractured bones [49]. Repairing fractures requires local cells to move from the bone marrow space into the fracture sites through endochondral ossification and the formation of bone eggplant. These cells differentiate into chondrocytes and finally into hypertrophic chondrocytes. If there is sufficient mechanical stability at the fracture site, there will be blood vessel infiltration, and the newly formed cartilage will have the effect of internal cartilage ossification. This process will cause the formation of ossification at the fracture site. In order to improve the repair of bone and cartilage, various biological transplantation methods, including transplantation using progenitor cells, have been developed. For example, osteochondral stromal cells have been used for transplantation, the whole bone marrow has been used to treat bone defects [50, 51], and the perichondral membrane and periosteum have been used to treat cartilage defects [52, 53]. The mesenchymal stem/progenitor cells separated from bone marrow and peripheral cartilage have been a large amount of incubation and transplanted into cartilage defects [22–24] and bone defects [44]. Furthermore, many studies have pointed out that biological factors such as bone morphogenetic proteins can guide the formation of bone in the bone area and other areas in animal models. Many studies involve the use of biological factors to promote the repair of cartilage defects [50, 54]. Although these transplants have been reported to promote the restoration of bone and cartilage, many areas still need to be studied, especially in cartilage repair, and a clinically applicable treatment method still needs to be developed.

The basic strategy of these treatments depends on the cartilage differentiation of mesenchymal stem cells in the transplant or due to the biological activation factors of the transplant. However, little is known about the differentiation and chondrogenesis of these cells. Cartilage regeneration *in vitro* has been successfully achieved in avian and embryonic mammalian cells and cell lines, and more valuable information has been obtained from these studies [55]. However, till now, there has been no *in vitro* system that can promote mesenchymal stem cells from cartilage and can be used to study the differentiation process directly. Researchers had once proposed a culture system that can promote cartilage differentiation *in vitro* from the progenitor cells derived from rabbit bone marrow. Johnstone et al. have also found that adding tissue growth hormone (TGF-beta) can promote the differentiation of progenitor cells from the bone marrow into chondrocytes [56]. Human mesenchymal stem cells have been isolated and mixed with non-osteogenic human fibroblasts. It was found that after deliberately mixing 25–50% of non-mesenchymal progenitor cells in this way, it will not cause a significant impact on the bone production of mesenchymal stem cells [57]. This result also makes the possible future use of mesenchymal stem cells in clinical applications. The feasibility of transplantation is improved because mesenchymal stem cells come from the connective tissue in the body, and some cells from other circulatory systems might be infiltrated, which may dilute the original concentration of human mesenchymal stem cells. Mesenchymal stem cells have the ability to multiply continuously without losing their multipotent properties [39]. There are two different strategies that can be used to achieve the roles of mesenchymal stem cells for tissue repair. The first method is to culture and increase the number of undifferentiated mesenchymal stem cells and then induce them into appropriate cell lines under a special environment and transplant them for tissue regeneration. Another method is to transplant a large number of cultured stem cells and directly induce them to differentiate into specific cell lines *in vivo* and then transplant them to promote the healing process. Regardless of the strategy to be adopted, a full understanding of the replication and differentiation of mesenchymal stem cells is necessary.

In recent years, researchers have been interested in the study of using mesenchymal stem cells for tissue repair because mesenchymal stem cells are widely available, easy to obtain, and can be multiplied in cell culture. Mesenchymal stem cells can provide a source of cells needed to repair skeletal muscle tissue. It is necessary to establish an *in vitro* model to study the differentiation mechanism of these progenitor cells. At present, the author has established a reliable and reproducible cell culture model by using mesenchymal stem cells in an *in vitro* model to regenerate cartilage. Human bone marrow stem cells were separated and centrifuged to obtain adsorbent cells, which can then be removed with trypsin. TGF and dexamethasone were used to differentiate chondrocytes. Using Alcian blue to stain cartilage-specific proteoglycans (proteoglycans) can quantify cartilage differentiation, and specific oligonucleotide primers can be used to distinguish the first, second, and tenth type of collagen fiber mRNAs in RT-PCR.

Adipose-derived stem cells

Adipose-derived stem cells (ASCs) are multipotent precursor cells from adipose stroma, which have a potential to differentiate into various cell types, such as bone, cartilage, fat, muscle, and even non-mesenchymal cell lineages, such as endothelial cells, and neurons [58]. Adipose tissue is now among the top choices of stem cell sources because of its accessibility, abundance, and less painful collection procedure when compared to other sources [59]. The increased incidence of obesity in particular makes the fat tissue readily accessible in the majority of cases [60]. ASCs can be maintained and expanded in culture for a long time without losing their differentiation capacity, leading to large cell quantities [59]. Therefore, ASCs are increasingly used in cell therapy and are considered highly promising for their future application in regenerative medicine [61].

There are mainly two types of adipose tissue: white adipose tissue and brown adipose tissue [62]. ASCs within brown adipose tissue depots exhibit different characteristics comparing to those isolated from white adipose tissue [63]. White adipose tissue is composed of subcutaneous white adipose tissue and visceral white adipose tissue [64]. ASCs are usually isolated from white adipose tissue, and ASCs from white fat tissue of various anatomical areas also exhibit different characteristics [65]. For example, ASCs from subcutaneous and visceral depots display intrinsic differences *in vitro*, such as proliferation and differentiation potentials [66]. Comparison studies demonstrated that stem cells from subcutaneous fat are more adipogenic and osteogenic upon induction than ASCs derived from visceral fat [67]. Compared to the abdomen, waist, and inner knee, the thigh seems to be a favorable donor site in terms of the viability and yield of ASCs [68]. Current methods for harvesting fat tissue for ASC isolation are mainly liposuction and direct excision. A direct excision surgery can yield a piece of adipose tissue, which needs to be minced into tiny particles to proceed with the ASC isolation procedures [69]. In contrast, liposuction involves using a cannula and negative pressure to suck out fat in small pieces, so further mincing is usually not required. Therefore, liposuction is generally preferred as a safe, well-tolerated, minimally invasive procedure to allow a high yield of stem cells from the aspirated fat [70]. There are numerous types of liposuction, including conventional, ultrasound-assisted, power-assisted, and laser-assisted liposuction. Among them, Coleman's aspiration technique is a commonly used method for the collection of subcutaneous adipose tissue. It relies on the gentle negative pressure created within a syringe, thus decreasing trauma to the aspirated fat [71]. The viability, yield, proliferative index, and stemness of ASCs can be influenced by different harvesting procedures [70]. Bajek et al. compared the biological properties of ASCs harvested through the following three approaches: surgical resection, power-assisted liposuction, and laser-assisted liposuction, and they recommended power-assisted liposuction because of the high proliferation potential and low senescence of the isolated ASCs [72]. However, another study indicated that the viable ASC yield from liposuction was significantly lower than that from lipectomy, while the apoptosis of cells from liposuction was significantly higher than from lipectomy, thus suggesting the ASCs from lipectomy have better biological characteristics [73]. Hence, several approaches to harvest fat tissue for ASC isolation have been well established, but conflicting data regarding their efficacy and the quality of isolated ASCs merits further study [70]. Next, the ASC isolation method was first proposed by Zuk et al. in 2001. Fat samples were washed extensively with equal volumes of phosphate-buffered saline (PBS) and digested at 37°C for 30 min with collagenase. The digested samples were neutralized with medium and centrifuged to obtain high-density pellets. After several steps of washing, cell pellets were dispersed and incubated. After cell attachment, the plates were washed extensively with PBS to remove residual non-adherent red blood cells, and the remaining cells are considered to be ASCs [74]. To date, many modified ASC isolation protocols have been proposed, but current literature has not concluded a standardized method to isolate ASCs for clinical application [63]. The most commonly employed method involves enzymatic digestion. Besides the most commonly used collagenase, other enzyme preparations used to achieve this fractionation include dispase and trypsin [75].

Since the clinical use of proteolytic enzymes may alter the physical properties and affect cell viability [76], the isolation of ASCs using non-enzymatic methods have also been advocated [70]. These non-enzymatic methods claimed to have the advantages of time-saving [77], or the isolated cells displayed a distinct and potentially favorable immunophenotype relative to the collagenase digestion [78]. However, the non-enzymatic methods generally require large amounts of lipoaspirate and have lower efficiency in cell recovery when compared to enzymatic methods [79]. Moreover, quantification of cells isolated after liposuction at the same harvesting site from the same patient can vary greatly depending on variation of the isolation protocol and the method of quantification [80]. As ASC has emerged as a new tool in the field of regenerative cytotherapy, a critical challenge to overcome in a standardized production process is to provide a sufficient amount of ASCs with preserved stemness and proliferation capability. Following the development history of the cultivation of multiple cell types, the *ex vivo* ASC culture started as two-dimensional (2D) culture. The only variables were culture medium, dishes, and the cultural environment. Isolated ASCs are typically expanded in monolayer culture on standard tissue culture plastics [81]. Classically, Dulbecco's Modified Eagle's medium (DMEM) containing 10% fetal bovine serum (FBS) has been used as culture medium for ASCs [82]. A study showed that among the different media formulations, modified Eagle medium alpha (α-MEM) supported a significantly faster cell expansion than the other basal media while still maintaining the full differentiation potential of ASCs [83]. In addition, DMEM/F12 has also been used for ASC expansion *in vitro* [84]. Most scientists added different growth factors as supplements. Baer et al. compared two laboratory-made media and three commercially available media with the standard medium and discovered that the supplement of growth factors, such as epidermal growth factor (EGF), basic fibroblast growth factor (bFGF), platelet-derived growth factor (PDGF), and growth factor-containing serum, promoted the cellular expansion [85]. Different cell morphology, surface CD markers, and pluripotency gene expression may be found after cultivation in various media.

Oral tissue-derived stem cells

Teeth are far more complicated than generally known, because the tooth itself is a complete organ. The embryonic development process and principles of tooth development are similar to many other organs in the body. It is through the embryonic epithelial cells of the ectoderm. The embryo epithelium interacts with mesenchymal cells in the mesoderm. Therefore, in addition to solving the problem of missing teeth, the study of tooth regeneration has a more important key significance: that is to lead the regeneration of organs to a more advanced level. Scientists have recognized that the process of life growth in nature has its consistency and commonality, and "following the principles of nature" is the wisest method. All research on regenerative medicine is actually exploring the process of life growth in nature and simulating it. Only the conditions required for its growth can be successful; since the development of teeth and other organs all begin at the embryonic stage, once the regeneration of the teeth can be successful, it means that the regeneration of other organs has a chance to succeed, and the regeneration of teeth can become another research model of organ regeneration. If researchers can make new teeth, they will be able to make even larger organs, leading medical treatment to the new century of regenerative medicine. In addition, the number of teeth is relatively large and easy to obtain, and they will not cause immediate harm to life. Therefore, scientists have more room for discussion and research. This is why tooth regeneration has become a research focus of great interest to scientists. There have also been many breakthroughs and advances in the research on regeneration in the past few years.

Mandible bone marrow stem cells are mesenchymal stem cells. Regarding the origin of embryonic development, the stem cells contained in the oral and maxillofacial tissues are unique and worthy of use. Judging from the process of embryonic development, the oral and maxillary face is the earliest part of the development, and there are many potential stem cells in it. Even the bone marrow stem cells of the mandible are derived from the neural crest. As they are different from the source of bone marrow stem cells from other long bones of the limbs, mandibular bone marrow stem cells have more differentiation potential and can also guide differentiation into bone, cartilage, blood vessels, and nerve cells. Each tooth in the mouth is an organ derived from the interaction of epithelial cells and mesenchymal tissue during the embryonic period. The source of mandible bone marrow stem cells is also from the neural crest. It is different from the bone marrow stem cells of other limbs and has more differentiation potential. Mandible bone marrow stem cells have greater potential than ordinary bone marrow stem cells; therefore, mandibular bone marrow stem cells can treat periodontal disease and promote bone regeneration. In addition, they can be guided into cartilage and bone tissue. Recently, several previous studies showed the effects of using mandible bone marrow. Moreover, oral-maxillo-facial has a great influence on the appearance of a person. Many oral cancer patients have a very special potential for mandibular bone marrow stem cells after undergoing surgical treatment. Therefore, researchers are committed to using the mandible. Bone marrow stem cells are placed on the scaffold to regenerate the mandible, and the dual cell guidance method is used to culture. The stem cells respectively guide cartilage and bone cells to establish temporomandibular joint regeneration to treat patients with temporomandibular joint damage. At the same time, stem cells and biomaterials can also be used to reconstruct the jawbone.

Studies have also pointed out that the occurrence of cancer may be related to cancer stem cells. If cancer stem cells can be found and treated for them, there will be a greater breakthrough in the treatment of oral cancer. Recently, several previous studies showed the effects of using mandible bone marrow stem cells on salivary gland regeneration and nerve regeneration. Many patients who suffer from oral cancer receive radiotherapy, which causes salivary gland insufficiency. At present, scholars are actively studying the methods of salivary gland regeneration. Studies have confirmed that bone marrow stem cells can be transdifferentiated into salivary gland cells [86, 87]. Moreover, animal experiments have also proved that the use of bone marrow stem cells for cell therapy can repair the salivary glands that are damaged by radiation [87]. The use of mandibular bone marrow stem cells may also be helpful in the future to treat patients with insufficiency of salivary glands or those with impaired salivary gland function due to oral cancer receiving radiotherapy. In addition, if the mandibular bone marrow stem cells can be guided to differentiate into nerve cells, it may be possible to treat patients with trigeminal neuropathy.

In addition to mandible bone marrow stem cells, various stem cells found in teeth have considerable potential. At present, various stem cells have been discovered in teeth, including stem cells from human exfoliated deciduous teeth (SHED), permanent teeth pulp stem cells, incompletely developed root apical stem cells (stem cells from apical papilla (SCAP)), periodontal stem cells, tooth germ cells, dental pulp stem cells, and so on [88–91]. For example, an earlier study provides a conventional protocol to isolate and culture tooth germ cells to obtain the oral tissue-derived stem cells [92]. An explant outgrowth technique without collagenase treatment was used to isolate a mandibular molar tooth germ from 4-day-old rats. Then, PBS was used to wash the tooth germ, and the tissues were cut into small fragments to release tooth germ cells. Subsequently, the cells and small fragments were collected through a centrifugal process for 5 mins. Afterwards, the pellet consisting of tissue fragments and cells was cultured with DMEM medium in the presence of 20% fetal calf serum containing endothelial cell growth supplement. After 6 to 8 days of incubation, tooth germ cells could migrate out from excised tissue fragments and achieve a confluence morphology. Then the remained tissue fragments were removed and placed in another culture plate to harvest more tooth germ cells. After 30 days of subculture, the procedure could yield around 1×10^8 cells for experiments. In addition, the other study also introduces a method to isolate and culture human dental pulp stem cells [93]. Similarly, human third molars were isolated from donors, and the molars were washed by using sterile PBS. Then, the molars were destroyed, and isolated the pulp tissues from the molars. The isolated pulp tissues were further excised into small fragments and cultured with DMEM medium containing 10% FBS. Subsequently, the dental pulp stem cells could migrate out, show a spindle shape morphology before confluence, and then the cells would undergo a subculture process once the cell achieves confluence. In general, the presence of stem cells from human exfoliated deciduous teeth (SHED) in the pulp of deciduous teeth has considerable potential. It can be guided to differentiate into bone, cartilage, nerves, etc., and may be used in related bone regeneration and nerve regeneration. Therefore, there are currently deciduous tooth banks that can retain fallen deciduous teeth and separate deciduous tooth pulp stem cells [94]. As a possible future application to make up for the regret of not leaving cord blood. Stem cells can also be found in the dental pulp of permanent teeth and have been proven to be transdifferentiated into odontoblasts and promote dentinogenesis. They can be used to treat dental caries and as capping cell therapy. In addition, dental pulp stem cells of permanent teeth have been proven to promote neural differentiation and can also be used to promote optic nerve regeneration [95, 96]. Studies have found that two-thirds of the root tip of the tooth has quite special stem cells from the apical papilla, which can promote tooth root development, dentin, and periodontal formation; so it can be used to treat tooth fracture or root apical hypoplasia [97]. If combined with periodontal stem cells, it may also be used to treat the repair of tooth root replantation in person that was involved in a car accident. Researchers have found that a combination of hydroxyapatite and root tip stem cell acts as the middle tube of the tooth root, places periodontal stem cells around it, and then implants in the mouth of the pig, which can form a root with the periodontal ligament. The manufacture of dental crowns is good news for patients with missing teeth. Periodontal stem cells can promote periodontal regeneration, treat periodontal disease, and can be used for the periodontal of the extracted and impacted wisdom teeth in the future. Tooth stem cells can all be induced into inducible and multifunctional stem cells, which also brings opportunities for more applications.

Neural stem/precursor cells

In the human body, the nervous system includes two major systems in the body. One is the central nervous system (CNS) which contains the spinal cord, medulla oblongata, pons, cerebellum, midbrain, thalamus, and cerebral cortex brain; the other one is the peripheral nervous system (PNS), which primarily consists of nerves and ganglions. The cells in CNS are two principal types: neurons and glial cells. The typical neuron consists of a cell body (soma), dendrites, and an axon. The cell body serves not only as a metabolic center but also as an integrator for all incoming information that is received by the dendrites. Dendrites are thin, branched processes that extend from the cytoplasm of the cell body, which provides a receptive area that transmits electrical impulses to the cell body. The axon could transmit electrical signals and impulses to other neurons or brain regions. The major function of the axon is that conduct impulses away from the cell body. The length of axons is from a millimeter long to a meter or more. Additionally, although many can regenerate a severed portion or sprout small new branches under certain conditions, most neurons cannot divide by mitosis. However, most of the cells in the nervous system are glia. Although they lack action potentials, as do neurons, they provide neurons with structural support and maintain an appropriate microenvironment essential for neuronal survival and function. Dissimilar to neurons, glial cells retain limited mitotic abilities. The major types of glial cells in the CNS are astrocytes, oligodendrocytes, and microglia. The astrocytes in the grey matter are called protoplasmic astrocytes. In contrast, mainly astrocytes in the white matter are fibrous astrocytes. During development, astrocytes provide a guide or framework for neuronal migration. In response to injury, proliferation or hypertrophy of astrocytes can result in the formation of astrocytic scars and then modulate the neurite outgrowth of neurons. The function of oligodendrocytes is responsible for the formation of myelin, a flattened sheet-like process that repeatedly wraps around the axon of the neuron. It speeds up the propagation of an electric signal from one neuron to another. Another type of glia is microglia cells, the immune effector cells of the CNS, and the predominant cells involved in CNS inflammation. Upon injury, activated microglia cells migrate to the site of damage and then proliferate and phagocytose cell debris [98]. In addition, in the nervous systems, neural stem/precursor cells (NSPCs) possess the ability to regenerate functional neural cells for treating neural diseases or injuries [99]. Okano's team and Weiss's team have successfully isolated NSPCs from lateral ventricle tissues and spinal tissues from mammals [100, 101]. Currently, neural stem cells can be isolated from human, mice, and rat tissues, such as 3- to 4-month-old human embryonic striatum through induction of labor with the water bag method, or 14- to 16-day-old embryonic brain or cortex tissues of mice/rats. Here, we will briefly describe a protocol for isolating NSPCs in laboratories. NSPCs were obtained from pregnant Wistar rat embryos on days 14–15 [102]. Briefly, embryonic rat cerebral cortices were dissected, cut into small pieces, and mechanically triturated in cold HBSS. The dissociated cells were collected by centrifugation and resuspended in the DMEM/F12 medium containing DMEM-F12 and N2 supplement in the absence of serum [102]. A hemocytometer counted the number of live cells by trypan blue exclusion assay. Cerebral cortical NSPCs were purified and cultured in culture plates at a density of 50,000 cells/cm^2 in the above culture medium in the presence of bFGF at a concentration of 20 ng/ml. After 1–3 days of *in vitro* incubation, cells underwent cell division, and the proliferating cells formed clusters of cells, termed neurospheres, which were suspended in the medium. Subsequently, adherent cells were discarded and suspended neurospheres were collected by centrifugation, mechanically dissociated, and subcultured in a culture plate in the fresh culture medium containing the same concentration of bFGF. These cells grew into new spheres in the subsequent 2–3 days. The procedure of subculture was repeated again to achieve purified cortical NSPCs and to produce more neurospheres ~ 100–150 μm in size. Although many protocols provide the methods to isolate NSPCs from embryonic tissues, the source and ethical issues limit the use of NSPCs in regenerative medicine. Therefore, isolating NSPCs and then establishing NSPCs cell lines enable overcoming the limited source of native tissues and avoiding ethical issues. In Table 1-1, several NSPC cell lines are presented, which can be helpful for scientists to investigate the cell-cell and cell-matrix interactions of NSPCs.

Table 1-1. The immortalized NSPCs cell line for biomedical applications.

Cell line	Species	Source*	Immortalized gene	Differentiation potential
HiB5	Rat	E16 Hippocampus	tsA58/U19T-ag	Multipotency
RN33B	Rat	E13 medullary raphe tissues	tsA58T-a	neuron
ST14A, ST-79-13A, ST-86	Rat	E14 Striatum	tsA58/U19T-ag	Multipotency
C17-2, C27-3	Mice	PN cerebellum EGL	v-myc	Multipotency
RN46A	Rat	E13 medullary raphe tissues	tsA58T-ag	neuron
CSM14.1.4	Rat	Ventral tegmental area	tsA58T-ag	unknown

*E: Embryonic; PN: Postnatal; EGL: External granular layer

Next, intrinsic programs and complex cell interactions determine the behavior of NSPCs. To regulate proliferation and differentiation of NSPCs, the proliferation and cell fate determination in the developing brain are regulated extrinsically by complex interactions between relatively large numbers of growth factors and neurotransmitters. For example, Choi et al. showed that combinations of bFGF and insulin-like growth factor-I (IGF-I) had an additive effect on neural differentiation [103]. Kokuzawa et al. reported that the addition of the hepatocyte growth factor (HGF) to a medium containing bFGF or EGF increased the promoting effect of HGF on the proliferation and differentiation of NSPC [104]. Takahashi et al. reported that retinoic acid and neurotrophins collaborate to regulate the neurogenesis of adult-derived neural stem cells [105]. Okabe et al. also reported that nestin-positive cells could be induced to differentiate into functional post-mitotic neurons by a polyornithine/laminin-coated substrate and DMEM/F12 medium containing the modified N3 supplement, bFGF, and laminin (mN3L medium) [106]. Shah and Anderson showed that BMP2 could promote neuronal differentiation of neural crest stem cells [107]. Pasternak et al. showed that mouse embryonic stem cells cultured in media with insulin, transferrin, selenium, and fibronectin preferentially supported the development of neural cells [108]. Based on the evidence from experiments, there are many strategies available to isolate and modulate the behavior of neural stem and progenitor cells (NSPCs). This represents a significant advance over the earlier theory that the central nervous system lacked any postnatal regenerative capacity, and it holds promise for treating neural tissues damaged by injury or disease. In the following chapters, we will introduce more applications of NSPCs in establishing stem cell-based systems for tissue engineering and regenerative medicine.

Hair follicle stem cells

The skin provides first-line protection for our body against environmental insults, such as chemical irritation, physical injury, pathogens, etc. Structurally, skin comprises the epidermis, dermis, subcutaneous tissue, and appendages [109]. Among the skin appendages, the hair follicle (HF) is a tiny organ that contributes to several important biological functions, such as thermoregulation, insulation, sensation, and camouflage. Human hair, especially scalp hair, has critical ornamental functions essential for social communication [110]. The hair follicle is a dynamic organ that undergoes cyclic involution and regeneration throughout life. The hair cycle consists of anagen (regeneration), catagen (destruction), and telogen (resting). Prolonged arrest in telogen, disruption of anagen, or loss of HFs can lead to unwanted alopecia that often negatively impacts the quality of life of individuals by inducing psychosocial distress or compromising the sense of wellbeing. Therefore, treatments that can alleviate hair loss or promote HF regeneration are of high clinical significance. The discovery of HF stem cells (HFSCs) in 1990 has advanced the current understanding of HF biology as well as the pathology of alopecia [111]. Despite such advances, their translation into effective clinical therapies for various types of alopecia is still an unmet medical challenge, especially when HFs are lost in cases of scarring alopecia [112]. Hair transplantation redistributes remaining HFs to regions of hair loss, and no new HFs are generated. With the advance in regenerative biology, scientists in the hair research field have started to employ tissue engineering technology for HF neogenesis. Various cell culture techniques, such as three-dimensional cell culture or organoid culture, have been developed to generate HF organoids or hair germ-like microtissues by combining epithelial and mesenchymal cells for HF regeneration. Here, we give an overview of HF biology and HF stem cells, summarize the current progress of bioengineering for HF regeneration, and discuss the future direction.

The HF is an ectodermal organ composed of epithelial and mesenchymal components. The epithelial component contains HFSCs that can give rise to various differentiated cell types. The mesenchymal part includes specialized fibroblasts of dermal papilla (DP) and the dermal sheath (DS). Similar to other epithelial organs, the morphogenesis of HFs relies on a dynamic, not yet well-understood epithelial-mesenchymal interaction (EMI) [113–115], which is driven and controlled by a series of precisely choreographed reciprocal interaction, for example, secreted growth factors, differentially expressed growth factor receptors and transcription factors, adhesion molecules, changes in the extracellular matrix, and cell rearrangement. Disturbance of these signaling pathways and events disrupts HF morphogenesis. A number of molecular signals have been identified to be involved in the EMI during HF morphogenesis, among which are Wnt signaling, bone morphogenic protein (BMP) signaling, hedgehog signaling, fibroblast growth factor (FGF) signaling, Notch signaling, epidermal growth factor (EGF) pathway signaling, platelet-derived growth factor (PDGF) signaling, transforming growth factor (TGF) signaling, etc. [113, 115–117].

After HF morphogenesis, HF starts cycling by initiating the first catagen. During hair cycles, an upper permanent segment, spanning from the follicular infundibulum to the bulge, is always preserved, whereas the segment below exhibits cyclic alteration [118–121]. Similar to HF morphogenesis, distinctive molecular signaling pathways are involved in the delicate regulation of different phases of the HF cycle [122]. Within HFs, the DP is a signaling center that secretes stimulating morphogens to trigger the activation of HFSCs to initiate HF regeneration from telogen [113, 115–117]. Its inductive and instructive influences on the HF epithelium, such as activating HFSCs to kickstart anagen, organizing the ordered differentiation of matrix cells for hair shaft production in anagen, and guiding the epithelial involution in catagen, have been revealed experimentally [123–125]. For example, DP also promotes catagen progression by inducing epithelial cell death through TGFβ1 signaling [126, 127]. Defects of specific signaling pathways in DPs may lead to premature anagen disruption, failure to cycle, and even loss of HFs [128, 129]. Activation of HFSCs powered HF regeneration from telogen to anagen [111, 118, 119]. HFSCs were generated early during HF morphogenesis from the peripheral part of basal placode cells and are later deployed to their niche, the bulge area. The bulge is a localized thickened epithelial structure at the lowermost part of the isthmus and encircles the hair shaft right below the sebaceous gland [130–132]. In 1876, Unna first postulated that the bulge is the area where the club hair continues to grow ("Haarbeet"; "hair bed") [133, 134]. In 1902, this structure was renamed by Stöhr as "Wulst" ("bulge") [133, 135]. Recent studies show that cells in the secondary hair germ of a telogen HF are another HFSC population that is primed to be activated [118, 119]. Hair follicle regeneration from telogen to anagen is orchestrated by the sequential activation of the two stem cell populations: primed HFSCs in the secondary hair germ are activated first, followed by the activation of relatively quiescent HFSCs in the bulge [118, 119, 136, 137] (Figure 1-1). During telogen to anagen transition, the lower segment of HFs expands dramatically into a long cylinder with an enlarged hair bulb in the proximal end that penetrates deeper into the dermis [138, 139].

Based on the cell dynamics and morphological alterations, anagen is subdivided into six sub-stages, from anagen I to VI [121, 136]. In anagen I and II, the activation of primed HFSC in the secondary hair germ produces transit-amplifying cells that

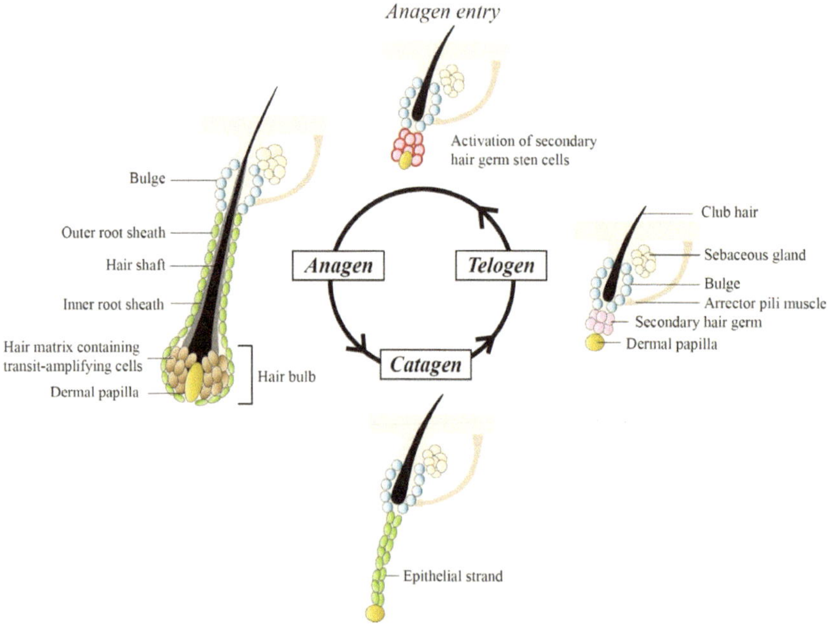

Figure 1-1. The hair cycle and structures of HFs in different hair cycle stages. (Reprinted with permission from [110] Copyright (2020) Springer Nature.)

contribute to the formation of the initial hair bulb and differentiate into most of the inner layer cells of the impermanent lower segment [118, 119, 136, 140, 141] (Figure 1-1). Quiescent bulge HFSCs are activated later in anagen III, and their descendants regenerate the uppermost outer root sheath of the HF lower segment [119, 140, 142]. Compared with the extensive proliferation of primed HFSCs, shorter activation during relatively limited proliferation of bulge HFSCs renders them label-retaining [111, 118, 119, 121, 143, 144]. In anagen III to VI, the anagen HF continues to grow downward and penetrates into adipose tissue, and eventually forms the mature anagen HF in anagen VI.

In anagen VI, the highly proliferative matrix cells in the hair bulb continuously proliferate to support the formation of internal seven concentric layers and the continued elongation of the hair shaft [120, 121, 145, 146]. With a cell division rate as high as 10.5 to 12.5 hours/cycle, matrix cells are one of the most actively dividing cells in the body [147, 148]. The duration of anagen VI varies among HFs, leading to variation in the final length of the produced hair shaft. In the meantime, bulge HFSCs are maintained in quiescent and do not contribute to hair growth in the full anagen [118, 119, 142]. The reciprocal interaction between HF epithelium and DP is indispensable for anagen initiation and progression [114, 122, 149, 150]. Disruption of their interaction can lead to alopecia [151, 152]. Catagen is a relatively transient phase between anagen and telogen driven by epithelial apoptosis [126, 153, 154, 121, 137, 155]. During catagen, it is believed that cell apoptosis with the removal of dead cells also drives the cell rearrangement of the lower segment, which progressively shrinks toward the bulge [118, 126, 137, 154, 155]. The structure of the upper permanent segment is preserved in the telogen HF, while the lower segment eventually shrinks to a short structure of the secondary hair germ, which is connected to the bulge from below [120]. The length of telogen HF is dramatically reduced compared with an anagen HF (Figure 1-1). In the telogen HF, cells in the bulge and the SHG remain relatively inactive in cell division. Compared with the classification of anagen sub-stages, rigorous criteria for characterizing each sub-stage of catagen are lacking. There is a lack of molecular markers that can signal the beginning of catagen and each sub-stage of catagen. Judging from the extent of apoptosis and structural involution, catagen is generally classified into eight sub-stages from catagen I to VIII [154]. Conventionally, the beginning of catagen is considered when cell proliferation in the hair bulb halts at the end of anagen VI with the appearance of TUNEL+ apoptotic cells in the hair bulb [136, 137, 154, 155]. The epithelial strand is gradually shortened due to the apoptosis of epithelial cells in the lower segment. Live imaging unveiled that the dying epithelial cells are phagocytosed by adjacent keratinocytes [126]. Compared with interfollicular epithelium that removes apoptotic keratinocytes by direct transepithelial elimination [156], engulfing dead cells by follicular cells seems to be an evolutional novelty of HFs adapted for its characteristic closed structure. In parallel with epithelial death and removal, the dermal sheath cells also undergo concomitant apoptosis in the proximal end of HF during catagen [157]. Lineage tracing demonstrated that some progenitor cells in the dermal sheath survive catagen and repopulate DP cells during telogen [157].

HFSC activation is subject to non-cell-autonomous regulation from the niche, which is composed of diverse cell types, including DP cells, immune cells, adipose tissue, lymphatic vessels, vascular vessels, and nerves [125, 158–170]. Followed by the advance of functional identifications of various cell types in the HFSC niche, the sophisticated interaction between HFSCs and each niche cell type has been progressively unveiled. HFSCs simultaneously receive both activating and suppressive signals from different cells within the niche. The behavior of HFSCs, either to remain quiescent or to become activated, depends on the summation of both activating and suppressive signals [171, 172]. The two major counteracting signals are Wnt/β-catenin and BMP signaling pathways [161, 171, 173]. The competitive balance of Wnt/β-catenin and BMP signaling determines the final behavior of HFSCs. Higher Wnt/β-catenin signaling promotes HFSC activation and elicits HF regeneration, while higher BMP signaling keeps HFSCs in quiescence [161, 171–175]. Moreover, the TGF-β2, Foxp1, oncostatin M, and adrenaline signaling pathways have also been shown to regulate HFSC activity [125, 162, 171, 176, 177]. The DP is a cluster of specialized fibroblasts. These cells play pivotal roles in regulating not only embryonic HF morphogenesis but also postnatal hair cycling by serving as an instructive niche for HFSCs and progenitors in HF regeneration [178]. Interaction between DP and HFSCs is essential to initiate a new anagen. Ablation of DP impairs HF regeneration from telogen. During the telogen-to-anagen transition, DP cells secret signaling ligands, such as TGF-β and FGF7, to active HFSC [119, 125]. In mature anagen, the DP is enveloped by the HF matrix containing hyperproliferative transit-amplifying progenitors in the hair bulb, where DP regulates the proliferation and differentiation of transit-amplifying progenitors for the production of inner root sheath and hair shafts [178]. Continued active Wnt/β-catenin signaling in DP is essential for anagen progression and maintenance [129]. Inactivating Wnt/β-catenin signaling in DP suppresses HFSC activation and inhibits anagen entry [129]. During catagen, DP also regulates apoptosis-driven HF regression through TGF-β signaling [126]. In addition to its physiological function, our study suggested that when HFs are damaged by chemo- and radiotherapy, DP also instructs ectopic progenitor-mediated anagen repair to avoid catagen entry [179]. These results demonstrate the key role of DP in the regulation of the hair cycle.

The cells in DP are also dynamic during different hair cycle phases. Recently, it was shown that some of the DP cells leave the HF during catagen [180]. DP cells exhibit little, if not at all, proliferative activity *in vivo*. Therefore, the lost DP cells must be replenished for the maintenance of HF growth. It was speculated that a distinct precursor of DP may exist in the surrounding dermal sheath (DS) [181]. Lineage tracing showed that a subset of DS cells, referred to as DS precursors, are capable of self-renewing and acting as a reservoir that can replenish DP cells over consecutive hair cycles [157, 182]. Depleting these DS precursors results in retarded hair regeneration and altered hair type specification, suggesting their role in maintaining DP functions [157]. DS also has a smooth muscle-like contractile function that contributes to HF regression in catagen. The contraction of DS helps to pull DP upward toward HFSCs to maintain their close contact that is required for proper epithelial-mesenchymal interaction for HF regeneration [183]. These studies also support the notion that pathology that impairs DP functions may result in unwanted hair loss.

Dysfunction of DP is a key factor in hair disorders such as androgenetic alopecia (AGA). AGA, or male pattern alopecia, is the most common hair disease with prominent premature HF aging. Male pattern alopecia shows accelerated HF atrophy with hair shaft miniaturization, impaired anagen initiation, shortened anagen duration and hair length, and finally, possible loss of the entire HF, resulting in diminished hair amount [151, 184]. This disease has a genetic predisposition that affects DP functions. In the balding area, DP has highly increased expression of type II 5a-reductase, an enzyme that converts testosterone into highly potent dihydrotestosterone through 5α-reduction of testosterone [185, 186]. Although the mechanisms are not fully clarified yet, dihydrotestosterone seems to induce premature senescence in DP due to persistent androgen receptor activation. Prolonged exposure to dihydrotestosterone compromises DP functions, leading to deteriorating hair growth and prolonged telogen [187, 188]. DP cells isolated from the balding scalp of AGA patients show prominent signs of senescence, such as loss of proliferative potential, an increase of senescent molecular markers, increased expression of inflammatory cytokines, etc. [189, 190]. Compared with non-balding DP cells, the balding DP cells not only lose the ability to promote HFSC proliferation but also produce inhibitory factors that suppress HFSCs and disrupt keratinocyte proliferation [152, 191–193]. Wnt signaling is essential for anagen initiation and maintenance. The activation of Wnt signaling is inhibited in balding DP through its overexpression of Dkk1, a negative regulator of Wnt signaling [174, 175, 192]. It has been shown that TGF-β1 signaling in DP regulates anagen-to-catagen transition [126, 194]. The production of TGF-β1 is also increased in balding DP and inhibits keratinocyte proliferation [193]. Furthermore, compared with non-balding DP, balding DP secrets a higher level of proinflammatory cytokine of IL-6 that not only inhibits anagen entry by suppressing keratinocyte proliferation but also disrupts normal anagen progression through interrupting the interaction between DP and progenitor cells [152, 191, 195]. Treatment by targeting restoration and/or preservation of normal functions of DP cells has been employed clinically. 5α-reductase represents an important example. 5α-reductase inhibitors, including finasteride and dutasteride, are both used clinically for the treatment of AGA. Finasteride mainly inhibits type-II 5α-reductases, the main 5α-reductase subtype in DP, while dutasteride suppresses both type-I and type-II 5α-reductases [196, 197]. Long-term treatment with finasteride and dutasteride promotes hair growth in patients with AGA [198, 199]. Another FDA-approved medication for treating baldness is topical minoxidil [200, 201]. Minoxidil is a potassium channel opener originally designed for hypertension treatment [202]. Though the mechanisms are still not well characterized, it is speculated to promote hair growth through their effects on blood vessels or potassium channels [203]. Since AGA shows a progressive nature, medical treatment is not effective in patients with severe hair loss. Hair transplantation surgery can be considered in patients with advanced AGA. Hair transplantation works by redistributing remaining HFs from the occipital scalp to the hairless area. It does not increase the number of HFs. Therefore, in patients with extensive hair loss, insufficient healthy HFs can limit its effect. Furthermore, hair transplantation itself does not prevent continued HF miniaturization and hair loss. If finasteride or dutasteride is not continuously used following hair transplantation, patients might need follow-up transplantations afterwards [204]. In addition to the DP, dysfunction of HF stem cells (HFSCs) can also lead to alopecia. It has been speculated that the decrease of HF size or miniaturization of HFs and hair shafts in AGA is related to the loss of HFSCs. It has been revealed that the HFSCs are still retained in miniaturized HFs in the balding scalp of AGA, but their activation is compromised [205]. The behavior of HFSCs is regulated by both intrinsic factors and by the extrinsic microenvironment or the niche. In early anagen, Foxc1 is upregulated right after HFSC activation by suppressing continued HFSC activation by activating the nuclear factor of activated T-cell c1 (NFATc1) and bone morphogenic protein (BMP) signaling, the two main mechanisms that contribute to the quiescence of HFSCs [161, 206]. The HFSCs in aged HFs have an increased expression of NFATc1, which may prolong the telogen phase of aged HFs by suppressing HFSC activation [207]. Skin aging can alter the normal extracellular environment due to the loss of key extracellular matrix proteins. Extrinsically, the aging change of HFSCs is triggered by DNA damage-induced proteolysis of type 17 collagen, a key extracellular matrix component that governs HFSC maintenance. Loss of the extracellular matrix niche of type 17 collagen leads to the elimination of HFSCs through terminal epidermal differentiation [151]. The decrease and finally complete loss of HFSCs can lead to the loss of the entire HF. The intrinsic abnormality of stress-induced asymmetric cell division in aged HFSCs also drives the destabilization of type 17 collagen, which also promotes terminal differentiation of HFSCs and accelerates their exhaustion. Stabilization of type 17 collagen rescues HFSC exhaustion [208]. Apart from transepidermal elimination of HFSCs, a recent study using live image and lineage tracing technologies showed another way of HFSCs during aging. Aged HFSCs escape from the epithelial component to the dermis, leading to a loss of HFSCs. Interestingly, the aged HFSCs display reduced expression of cell adhesion and extracellular matrix proteins under the control of Foxc1 and NFATc1. Ablation of Foxc1 and NFATc1 recapitulate phenotypes of HF aging. This study provides a new perspective on the functions of Foxc1 and NFATc1 in controlling the loss of HFSCs during aging [209].

When HFs are lost, inducing HF neogenesis through bioengineering is of promise for the treatment of alopecia. First demonstrated in rodents, a freshly isolated DP is capable of inducing HF neogenesis when it is properly placed in contact with the epidermis [210–213]. Since the treatment of alopecia usually needs to replenish thousands of lost HFs, DP cells should be expanded. Additionally, their HF inductive property should be preserved during their expansion. However, in the conventional two-dimensional monolayer culture condition, DP cells gradually lose their hair-inductive capability with decreased expression of signature genes after a few passages [214, 215]. Inamatsu et al. successfully preserved hair-inductive ability in rat DP cells after 70 passages by culturing them in the conditioned medium harvested from skin keratinocytes [216]. Kishimoto et al. and Rendl et al. retained the hair inductivity of cultured mouse DP cells by the treatment of Wnt3a and BMP6, respectively [215, 217]. By a similar approach for human DP cell expansion, Qiao et al. showed that the hair-inductive potential of two-dimension-cultured human DP cells could be maintained when they are cultured in the conditioned medium harvested from newborn foreskin keratinocytes [218, 219]. In terms of HF induction, key differences exist between rodent and human DP cells. Cultured human DP cells are less efficient in inducing new HFs, and the therapeutic effects of two-dimension-cultured human DP cells on inducing new HFs have not been demonstrated clinically [220, 221].

References

[1] Baksh, D., L. Song, and R.S. Tuan. Adult mesenchymal stem cells: characterization, differentiation, and application in cell and gene therapy. J Cell Mol Med, 2004. 8(3): 301–16.
[2] Tuan, R.S., G. Boland, and R. Tuli. Adult mesenchymal stem cells and cell-based tissue engineering. Arthritis Res Ther, 2003. 5(1): 32–45.
[3] Thomson, J.A. et al. Embryonic stem cell lines derived from human blastocysts. Science, 1998. 282(5391): 1145–7.
[4] Gardner, D.K. and W.B. Schoolcraft. Culture and transfer of human blastocysts. Curr Opin Obstet Gynecol, 1999. 11(3): 307–11.
[5] Shamblott, M.J. et al. Derivation of pluripotent stem cells from cultured human primordial germ cells. Proc Natl Acad Sci U S A, 1998. 95(23): 13726–31.
[6] Inzunza, J. et al. Derivation of human embryonic stem cell lines in serum replacement medium using postnatal human fibroblasts as feeder cells. Stem Cells, 2005. 23(4): 544–9.
[7] Amit, M. et al. Human feeder layers for human embryonic stem cells. Biol Reprod, 2003. 68(6): 2150–6.
[8] Ludwig, T.E. et al. Derivation of human embryonic stem cells in defined conditions. Nat Biotechnol, 2006. 24(2): 185–7.
[9] Sato, N. et al. Maintenance of pluripotency in human and mouse embryonic stem cells through activation of Wnt signaling by a pharmacological GSK-3-specific inhibitor. Nat Med, 2004. 10(1): 55–63.
[10] Takahashi, K. et al. Induction of pluripotent stem cells from adult human fibroblasts by defined factors. Cell, 2007. 131(5): 861–72.
[11] Inoue, H. et al. iPS cells: a game changer for future medicine. EMBO J, 2014. 33(5): 409–17.
[12] Li, D. et al. Chromatin accessibility dynamics during iPSC reprogramming. Cell Stem Cell, 2017. 21(6): 819–833 e6.
[13] Hu, K. All roads lead to induced pluripotent stem cells: the technologies of iPSC generation. Stem Cells Dev, 2014. 23(12): 1285–300.
[14] Cevallos, R.R. et al. Human transcription factors responsive to initial reprogramming predominantly undergo legitimate reprogramming during fibroblast conversion to iPSCs. Sci Rep, 2020. 10(1): 19710.
[15] Theunissen, T.W. et al. Nanog overcomes reprogramming barriers and induces pluripotency in minimal conditions. Curr Biol, 2011. 21(1): 65–71.
[16] Zhang, X. et al. Esrrb activates Oct4 transcription and sustains self-renewal and pluripotency in embryonic stem cells. J Biol Chem, 2008. 283(51): 35825–33.
[17] Zhang, J. et al. LIN28 Regulates stem cell metabolism and conversion to primed pluripotency. Cell Stem Cell, 2016. 19(1): 66–80.
[18] Li, L. et al. Glis1 facilitates induction of pluripotency via an epigenome-metabolome-epigenome signalling cascade. Nat Metab, 2020. 2(9): 882–892.
[19] Heng, J.C. et al. The nuclear receptor Nr5a2 can replace Oct4 in the reprogramming of murine somatic cells to pluripotent cells. Cell Stem Cell, 2010. 6(2): 167–74.
[20] Bao, Q. et al. Utf1 contributes to intergenerational epigenetic inheritance of pluripotency. Sci Rep, 2017. 7(1): 14612.
[21] Zhang, J. et al. Sall4 modulates embryonic stem cell pluripotency and early embryonic development by the transcriptional regulation of Pou5f1. Nat Cell Biol, 2006. 8(10): 1114–23.
[22] Wang, L. et al. FOXH1 is regulated by NANOG and LIN28 for early-stage reprogramming. Sci Rep, 2019. 9(1): 16443.
[23] Liu, G. et al. Advances in pluripotent stem cells: history, mechanisms, technologies, and applications. Stem Cell Rev Rep, 2020. 16(1): 3–32.

[24] Okita, K., T. Ichisaka, and S. Yamanaka. Generation of germline-competent induced pluripotent stem cells. Nature, 2007. 448(7151): 313–7.
[25] Bessis, N., F.J. GarciaCozar, and M.C. Boissier. Immune responses to gene therapy vectors: influence on vector function and effector mechanisms. Gene Ther, 2004. 11(Suppl 1): S10–7.
[26] Schlaeger, T.M. et al. A comparison of non-integrating reprogramming methods. Nat Biotechnol, 2015. 33(1): 58–63.
[27] Lieu, P.T. Reprogramming of human fibroblasts with non-integrating RNA virus on feeder-free or xeno-free conditions. Methods Mol Biol, 2015. 1330: 47–54.
[28] Wang, A.Y.L., and C.Y.Y. Loh. Episomal induced pluripotent stem cells: functional and potential therapeutic applications. Cell Transplant, 2019. 28(1 suppl): 112S–131S.
[29] Steinle, H. et al. Generation of iPSCs by nonintegrative RNA-based reprogramming techniques: benefits of self-replicating RNA versus Synthetic mRNA. Stem Cells Int, 2019. p. 7641767.
[30] Zhou, W., and C.R. Freed. Adenoviral gene delivery can reprogram human fibroblasts to induced pluripotent stem cells. Stem Cells, 2009. 27(11): 2667–74.
[31] Senis, E. et al. AAV vector-mediated in vivo reprogramming into pluripotency. Nat Commun, 2018. 9(1): 2651.
[32] Churko, J.M., P.W. Burridge, and J.C. Wu. Generation of human iPSCs from human peripheral blood mononuclear cells using non-integrative Sendai virus in chemically defined conditions. Methods Mol Biol, 2013. 1036: 81–8.
[33] Okumura, T. et al. Robust and highly efficient hiPSC generation from patient non-mobilized peripheral blood-derived CD34(+) cells using the auto-erasable Sendai virus vector. Stem Cell Res Ther, 2019. 10(1): 185.
[34] Lee, M. et al. Efficient exogenous DNA-free reprogramming with suicide gene vectors. Exp Mol Med, 2019. 51(7): 1–12.
[35] Warren, L., and C. Lin. mRNA-Based Genetic Reprogramming. Mol Ther, 2019. 27(4): 729–734.
[36] Hou, X., et al. Lipid nanoparticles for mRNA delivery. Nat Rev Mater, 2021. 6(12): 1078–1094.
[37] Campillo-Davo, D. et al. The ins and outs of messenger RNA electroporation for physical gene delivery in immune cell-based therapy. Pharmaceutics, 2021. 13(3).
[38] Bonaventure, J. et al. Reexpression of cartilage-specific genes by dedifferentiated human articular chondrocytes cultured in alginate beads. Exp Cell Res, 1994. 212(1): 97–104.
[39] Caplan, A.I. Mesenchymal stem cells. J Orthop Res, 1991. 9(5): 641–50.
[40] Dennis, J.E., and A.I. Caplan. Porous ceramic vehicles for rat-marrow-derived (Rattus norvegicus) osteogenic cell delivery: effects of pre-treatment with fibronectin or laminin. J Oral Implantol, 1993. 19(2): 106–15; discussion 136–7.
[41] Dennis, J.E., and A.I. Caplan. Differentiation potential of conditionally immortalized mesenchymal progenitor cells from adult marrow of a H-2Kb-tsA58 transgenic mouse. J Cell Physiol, 1996. 167(3): 523–38.
[42] Dennis, J.E. et al. Osteogenesis in marrow-derived mesenchymal cell porous ceramic composites transplanted subcutaneously: effect of fibronectin and laminin on cell retention and rate of osteogenic expression. Cell Transplant, 1992. 1(1): 23–32.
[43] Haynesworth, S.E. et al. Characterization of cells with osteogenic potential from human marrow. Bone, 1992. 13(1): 81–8.
[44] Kadiyala, S. et al. Culture expanded canine mesenchymal stem cells possess osteochondrogenic potential in vivo and in vitro. Cell Transplant, 1997. 6(2): 125–34.
[45] Wang, W.G. et al. In vitro chondrogenesis of human bone marrow-derived mesenchymal progenitor cells in monolayer culture: activation by transfection with TGF-beta2. Tissue Cell, 2003. 35(1): 69–77.
[46] Wakitani, S. et al. Repair of rabbit articular surfaces with allograft chondrocytes embedded in collagen gel. J Bone Joint Surg Br, 1989. 71(1): 74–80.
[47] Hunziker, E.B., and L.C. Rosenberg. Repair of partial-thickness defects in articular cartilage: cell recruitment from the synovial membrane. J Bone Joint Surg Am, 1996. 78(5): 721–33.
[48] Shapiro, F., S. Koide, and M.J. Glimcher, Cell origin and differentiation in the repair of full-thickness defects of articular cartilage. J Bone Joint Surg Am, 1993. 75(4): 532–53.

[49] O'Driscoll, S.W. and R.B. Salter. The repair of major osteochondral defects in joint surfaces by neochondrogenesis with autogenous osteoperiosteal grafts stimulated by continuous passive motion. An experimental investigation in the rabbit. Clin Orthop Relat Res, 1986(208): 131–40.
[50] Paley, D. et al. Percutaneous bone marrow grafting of fractures and bony defects. An experimental study in rabbits. Clin Orthop Relat Res, 1986(208): 300–12.
[51] Werntz, J.R. et al. Qualitative and quantitative analysis of orthotopic bone regeneration by marrow. J Orthop Res, 1996. 14(1): 85–93.
[52] Coutts, R.D. et al. Rib perichondrial autografts in full-thickness articular cartilage defects in rabbits. Clin Orthop Relat Res, 1992(275): 263–73.
[53] Kreder, H.J. et al. Biologic resurfacing of a major joint defect with cryopreserved allogeneic periosteum under the influence of continuous passive motion in a rabbit model. Clin Orthop Relat Res, 1994(300): 288–96.
[54] Otsuka, Y. et al. Requirement of fibroblast growth factor signaling for regeneration of epiphyseal morphology in rabbit full-thickness defects of articular cartilage. Dev Growth Differ, 1997. 39(2): 143–56.
[55] Denker, A.E., S.B. Nicoll, and R.S. Tuan. Formation of cartilage-like spheroids by micromass cultures of murine C3H10T1/2 cells upon treatment with transforming growth factor-beta 1. Differentiation, 1995. 59(1): 25–34.
[56] Mohammad Ali Khalilifar, Mohamadreza Baghaban Eslaminejad, Mohammad Ghasemzadeh, Samaneh Hosseini and Hossein Baharvand. *In vitro* and *in vivo* comparison of different types of rabbit mesenchymal stem cells for cartilage repair. Cell J., 2019. 21(2): 150–160.
[57] Lennon, D.P. et al. Dilution of human mesenchymal stem cells with dermal fibroblasts and the effects on in vitro and in vivo osteochondrogenesis. Dev Dyn, 2000. 219(1): 50–62.
[58] Zuk, P.A. et al. Human adipose tissue is a source of multipotent stem cells. Molecular Biology of the Cell, 2002. 13(12): 4279–4295.
[59] Mazini, L. et al. Regenerative capacity of adipose derived stem cells (ADSCs), Comparison with Mesenchymal Stem Cells (MSCs). 2019. 20(10): 2523.
[60] Bajek, A. et al. Adipose-derived stem cells as a tool in cell-based therapies. Archivum Immunologiae et Therapiae Experimentalis, 2016. 64(6): 443–454.
[61] Bacakova, L. et al. Stem cells: their source, potency and use in regenerative therapies with focus on adipose-derived stem cells – a review. Biotechnology Advances, 2018. 36(4): 1111–1126.
[62] Tsuji, W., J.P. Rubin and K.G. Marra. Adipose-derived stem cells: Implications in tissue regeneration. World J Stem Cells, 2014. 6(3): 312–21.
[63] Raposio, E. and N.J.C.P.i.S.C.B. Bertozzi. Isolation of ready-to-use adipose-derived stem cell (ASC) Pellet for Clinical Applications and a Comparative Overview of Alternate Methods for ASC Isolation. 2017. 41(1): 1F. 17.1-1F. 17.12.
[64] Torres, N., A.E. Vargas-Castillo, and A.R. Tovar. Adipose tissue: white adipose tissue structure and function. pp. 35–42. In: Caballero, B., P.M. Finglas, and F. Toldrá (Eds.). Encyclopedia of Food and Health. 2016, Academic Press: Oxford.
[65] Si, Z. et al. Adipose-derived stem cells: Sources, potency, and implications for regenerative therapies. Biomedicine & Pharmacotherapy, 2019. 114: 108765.
[66] Tang, Y. et al. A comparative assessment of adipose-derived stem cells from subcutaneous and visceral fat as a potential cell source for knee osteoarthritis treatment. Journal of Cellular and Molecular Medicine, 2017. 21(9): 2153–2162.
[67] Kim, B. et al. Gene expression profiles of human subcutaneous and visceral adipose-derived stem cells. Cell Biochemistry and Function, 2016. 34(8): 563–571.
[68] Tsekouras, A. et al. Comparison of the viability and yield of adipose-derived stem cells (ASCs) from different donor areas. In Vivo, 2017. 31(6): 1229.
[69] Zhang, J. et al. Adipose-derived stem cells: current applications and future directions in the regeneration of multiple tissues. Stem Cells International, 2020. p. 8810813.
[70] Palumbo, P. et al. Methods of isolation, characterization and expansion of human adipose-derived stem cells (ASCs): An Overview. Int J Mol Sci, 2018. 19(7).
[71] Pu, L.L.Q. et al. Autologous fat grafts harvested and refined by the Coleman technique: a comparative study. Plast Reconstr Surg, 2008. 122(3): 932–937.

[72] Bajek, A. et al. Does the harvesting technique affect the properties of adipose-derived stem cells?-The comparative biological characterization. J Cell Biochem, 2017. 118(5): 1097–1107.
[73] Bian, Y. et al. A comparative study on the biological characteristics of human adipose-derived stem cells from lipectomy and liposuction. PLOS ONE, 2016. 11(9): e0162343.
[74] Zuk, P.A. et al. Multilineage cells from human adipose tissue: implications for cell-based therapies. Tissue Engineering, 2001. 7(2): 211–228.
[75] Banyard, D.A. et al. Implications for human adipose-derived stem cells in plastic surgery. Journal of Cellular and Molecular Medicine, 2015. 19(1): 21–30.
[76] Olenczak, J.B. et al. Effects of collagenase digestion and stromal vascular fraction supplementation on volume retention of fat grafts. Annals of Plastic Surgery, 2017. 78(6S).
[77] Francis, M.P. et al. Isolating adipose-derived mesenchymal stem cells from lipoaspirate blood and saline fraction. Organogenesis, 2010. 6(1): 11–14.
[78] Shah, F.S. et al. A non-enzymatic method for isolating human adipose tissue-derived stromal stem cells. Cytotherapy, 2013. 15(8): 979–85.
[79] Bellei, B. et al. Maximizing non-enzymatic methods for harvesting adipose-derived stem from lipoaspirate: technical considerations and clinical implications for regenerative surgery. Sci Rep, 2017. 7(1): 10015.
[80] Prantl, L. et al. Adipose tissue-derived stem cell yield depends on isolation protocol and cell counting method. Cells, 2021. 10(5).
[81] Mizuno, H., M. Tobita, and A.C. Uysal. Concise review: Adipose-derived stem cells as a novel tool for future regenerative medicine. STEM CELLS, 2012. 30(5): 804–810.
[82] Nakashima, Y. et al. A Liquid Chromatography with Tandem Mass Spectrometry-Based Proteomic Analysis of Cells Cultured in DMEM 10% FBS and Chemically Defined Medium Using Human Adipose-Derived Mesenchymal Stem Cells. 2018. 19(7): 2042.
[83] Lund, P. et al. Effect of growth media and serum replacements on the proliferation and differentiation of adipose-derived stem cells. Cytotherapy, 2009. 11(2): 189–197.
[84] Lin, Y. et al. Multilineage differentiation of adipose-derived stromal cells from GFP transgenic mice. Molecular and Cellular Biochemistry, 2006. 285(1): 69–78.
[85] Baer, P.C. et al. Human adipose-derived mesenchymal stem cells in vitro: evaluation of an optimal expansion medium preserving stemness. Cytotherapy, 2010. 12(1): 96–106.
[86] Lin, C.-Y. et al. Transdifferentiation of bone marrow stem cells into acinar cells using a double chamber system. J Formos Med Assoc, 2007. 106(1): 1–7.
[87] Lin, C.Y. et al. Cell therapy for salivary gland regeneration. J Dent Res, 2011. 90(3): 341–6.
[88] Miura, M. et al. SHED: stem cells from human exfoliated deciduous teeth. Proc Natl Acad Sci U S A, 2003. 100(10): 5807–12.
[89] Gronthos, S. et al. Postnatal human dental pulp stem cells (DPSCs) in vitro and in vivo. Proc Natl Acad Sci U S A, 2000. 97(25): 13625–30.
[90] Sonoyama, W. et al. Mesenchymal stem cell-mediated functional tooth regeneration in swine. PLoS One, 2006. 1: e79.
[91] Seo, B.M. et al. Investigation of multipotent postnatal stem cells from human periodontal ligament. Lancet, 2004. 364(9429): 149–55.
[92] Chen, R.S. et al. Cell-surface interactions of rat tooth germ cells on various biomaterials. J Biomed Mater Res A, 2007. 83(1): 241–8.
[93] Hsiao, D. et al. Characterization of designed directional polylactic acid 3D scaffolds for neural differentiation of human dental pulp stem cells. J Formos Med Assoc, 2020. 119(1 Pt 2): 268–275.
[94] Kuo, T.F. et al. Bone marrow combined with dental bud cells promotes tooth regeneration in miniature pig model. Artif Organs, 2011. 35(2): 113–21.
[95] Arthur, A. et al. Adult human dental pulp stem cells differentiate toward functionally active neurons under appropriate environmental cues. Stem Cells, 2008. 26(7): 1787–95.
[96] Mead, B. et al. Intravitreally transplanted dental pulp stem cells promote neuroprotection and axon regeneration of retinal ganglion cells after optic nerve injury. Invest Ophthalmol Vis Sci, 2013. 54(12): 7544–56.
[97] Banchs, F., and M. Trope. Revascularization of immature permanent teeth with apical periodontitis: new treatment protocol? J Endod, 2004. 30(4): 196–200.
[98] Joshi, D., and M. Behari. Neuronal stem cells. Neurol India, 2003. 51(3): 323–8.

[99] McKay, R. Stem cells in the central nervous system. Science, 1997. 276(5309): 66–71.
[100] Pincus, D.W. et al. Fibroblast growth factor-2/brain-derived neurotrophic factor-associated maturation of new neurons generated from adult human subependymal cells. Ann Neurol, 1998. 43(5): 576–85.
[101] Weiss, S. et al. Multipotent CNS stem cells are present in the adult mammalian spinal cord and ventricular neuroaxis. J Neurosci, 1996. 16(23): 7599–609.
[102] Tsai, R.Y. and R.D. McKay. Cell contact regulates fate choice by cortical stem cells. J Neurosci, 2000. 20(10): 3725–35.
[103] Choi, K.C. et al. Effect of single growth factor and growth factor combinations on differentiation of neural stem cells. J Korean Neurosurg Soc, 2008. 44(6): 375–81.
[104] Kokuzawa, J. et al. Hepatocyte growth factor promotes proliferation and neuronal differentiation of neural stem cells from mouse embryos. Mol Cell Neurosci, 2003. 24(1): 190–7.
[105] Takahashi, J., T.D. Palmer, and F.H. Gage. Retinoic acid and neurotrophins collaborate to regulate neurogenesis in adult-derived neural stem cell cultures. J Neurobiol, 1999. 38(1): 65–81.
[106] Okabe, S. et al. Development of neuronal precursor cells and functional postmitotic neurons from embryonic stem cells in vitro. Mech Dev, 1996. 59(1): 89–102.
[107] Shah, N.M., and D.J. Anderson. Integration of multiple instructive cues by neural crest stem cells reveals cell-intrinsic biases in relative growth factor responsiveness. Proc Natl Acad Sci U S A, 1997. 94(21): 11369–74.
[108] Pachernik, J. et al. Neural differentiation of mouse embryonic stem cells grown in monolayer. Reprod Nutr Dev, 2002. 42(4): 317-26.
[109] Driskell, R.R. et al. Defining dermal adipose tissue. Exp Dermatol, 2014. 23(9): 629–31.
[110] Chen, C.L. et al. Functional complexity of hair follicle stem cell niche and therapeutic targeting of niche dysfunction for hair regeneration. J Biomed Sci, 2020. 27(1): 43.
[111] Cotsarelis, G., T.T. Sun, and R.M. Lavker. Label-retaining cells reside in the bulge area of pilosebaceous unit: implications for follicular stem cells, hair cycle, and skin carcinogenesis. Cell, 1990. 61(7): 1329–37.
[112] Castro, A.R., and E. Logarinho. Tissue engineering strategies for human hair follicle regeneration: How far from a hairy goal? Stem Cells Transl Med, 2020. 9(3): 342–350.
[113] Pinkus, H. Embryology of hair. pp. 1–32. In: Montagna, W., and R. Ellis (Eds.). The Biology of Hair Growth. 1958, Academic Press: New York.
[114] Philpott, M., and R. Paus. Principles of hair follicle morphogenesis. pp. 75–110. In: Chuong, C. (Ed.). Molecular Basis of Epithelial Appendage Morphogenesis. 1998, R.G. Landers Company: Austin, Texas, U.S.A.
[115] Millar, S.E. Molecular mechanisms regulating hair follicle development. J Invest Dermatol, 2002. 118(2): 216–25.
[116] Botchkarev, V.A., and J. Kishimoto. Molecular control of epithelial-mesenchymal interactions during hair follicle cycling. J Investig Dermatol Symp Proc, 2003. 8(1): 46–55.
[117] Plikus, M., J. Sundberg, and C. Chuong. Mouse skin ectodermal organs. pp. 691–730. In: Fox, J. et al. (Eds.). The Mouse in Biomedical Research, 2007. Academic Press: Burlington, MA, USA.
[118] Hsu, Y.C., H.A. Pasolli, and E. Fuchs. Dynamics between stem cells, niche, and progeny in the hair follicle. Cell, 2011. 144(1): 92–105.
[119] Greco, V. et al. A two-step mechanism for stem cell activation during hair regeneration. Cell Stem Cell, 2009. 4(2): 155–69.
[120] Paus, R., and G. Cotsarelis. The biology of hair follicles. N Engl J Med, 1999. 341(7): 491–7.
[121] Muller-Rover, S. et al. A comprehensive guide for the accurate classification of murine hair follicles in distinct hair cycle stages. J Invest Dermatol, 2001. 117(1): 3–15.
[122] Chen, C.C., and C.M. Chuong. Multi-layered environmental regulation on the homeostasis of stem cells: the saga of hair growth and alopecia. J Dermatol Sci, 2012. 66(1): 3–11.
[123] Jahoda, C.A., and A.J. Reynolds. Dermal-epidermal interactions--follicle-derived cell populations in the study of hair-growth mechanisms. Journal of Investigative Dermatology, 1993. 101(1 Suppl): 33S–38S.
[124] Kollar, E.J. The induction of hair follicles by embryonic dermal papillae. Journal of Investigative Dermatology, 1970. 55(6): 374–8.

[125] Oshimori, N., and E. Fuchs. Paracrine TGF-beta signaling counterbalances BMP-mediated repression in hair follicle stem cell activation. Cell Stem Cell, 2012. 10(1): 63–75.
[126] Mesa, K.R. et al. Niche-induced cell death and epithelial phagocytosis regulate hair follicle stem cell pool. Nature, 2015. 522(7554): 94–7.
[127] Foitzik, K. et al. Control of murine hair follicle regression (catagen) by TGF-beta1 in vivo. The FASEB Journal, 2000. 14(5): 752–760.
[128] Michael Rendl, Lisa Polak and Elaine Fuchs. BMP signaling in dermal papilla cells is required for their hair follicle-inductive properties. Genes Dev., 2008. 15: 543–57.
[129] Enshell-Seijffers, D. et al. beta-catenin activity in the dermal papilla regulates morphogenesis and regeneration of hair. Dev Cell, 2010. 18(4): 633–42.
[130] Morita, R. et al. Tracing the origin of hair follicle stem cells. Nature, 2021. 594(7864): 547–552.
[131] Ouspenskaia, T. et al. WNT-SHH antagonism specifies and expands stem cells prior to niche formation. Cell, 2016. 164(1-2): 156–169.
[132] Xu, Z. et al. Embryonic attenuated Wnt/beta-catenin signaling defines niche location and long-term stem cell fate in hair follicle. Elife, 2015. 4: e10567.
[133] Ackerman, A.B., P.A. De Viragh, and N. Chongchitnant. Anatomic, histologic and biologic aspects of hair follicles and hairs. pp. 35–102. In: Neoplasms with Follicular Differentiation. 1993, Lea & Febiger: Philadelphia/London.
[134] Unna, P. Beiträge zur Histologie und Entwiekelungsgeschichte der menschlichen Oberhaut und ihrer Anhangsgebilde. Arch Mikr Anat, 1876. 12: 29.
[135] Stöhr, P. Entwicklungsgeschichte des menschlichen Wollhaares. Anat. Hefte, 1903. 23: 65.
[136] Chase, H.B., R. Rauch, and V.W. Smith. Critical stages of hair development and pigmentation in the mouse. Physiol Zool, 1951. 24(1): 1–8.
[137] Chase, H.B. Growth of the hair. Physiol Rev, 1954. 34(1): 113–26.
[138] Slominski, A. et al. Hair follicle pigmentation. J Invest Dermatol, 2005. 124(1): 13–21.
[139] Stenn, K., S. Parimoo, and S. Prouty. Growth of the hair follicle: a cycling and regenerating biological system. pp. 111–130. In: Chuong, C. (Ed.). Molecular Basis of Epithelial Appendage Morphogenesis. 1998, R.G. Landers Company: Austin, Texas, U.S.A.
[140] Rompolas, P. et al. Live imaging of stem cell and progeny behaviour in physiological hair-follicle regeneration. Nature, 2012. 487(7408): 496–9.
[141] Legue, E., and J.F. Nicolas. Hair follicle renewal: organization of stem cells in the matrix and the role of stereotyped lineages and behaviors. Development, 2005. 132(18): 4143–54.
[142] Hsu, Y.C., L. Li, and E. Fuchs. Transit-amplifying cells orchestrate stem cell activity and tissue regeneration. Cell, 2014. 157(4): 935–49.
[143] Tumbar, T. et al. Defining the epithelial stem cell niche in skin. Science, 2004. 303(5656): 359–63.
[144] Zhang, Y.V. et al. Distinct self-renewal and differentiation phases in the niche of infrequently dividing hair follicle stem cells. Cell Stem Cell, 2009. 5(3): 267–78.
[145] Sequeira, I., and J.F. Nicolas. Redefining the structure of the hair follicle by 3D clonal analysis. Development, 2012. 139(20): 3741–51.
[146] Legue, E., I. Sequeira, and J.F. Nicolas. Hair follicle renewal: authentic morphogenesis that depends on a complex progression of stem cell lineages. Development, 2010. 137(4): 569–77.
[147] Cattaneo, S.M., H. Quastler, and F.G. Sherman. Proliferative cycle in the growing hair follicle of the mouse. Nature, 1961. 190: 923–4.
[148] Vanscott, E.J., T.M. Ekel, and R. Auerbach. Determinants of rate and kinetics of cell division in scalp hair. J Invest Dermatol, 1963. 41: 269–73.
[149] Blanpain, C., and E. Fuchs. Epidermal homeostasis: a balancing act of stem cells in the skin. Nat Rev Mol Cell Biol, 2009. 10(3): 207–17.
[150] Goldstein, J., and V. Horsley. Home sweet home: skin stem cell niches. Cell Mol Life Sci, 2012. 69(15): 2573–82.
[151] Matsumura, H. et al. Hair follicle aging is driven by transepidermal elimination of stem cells via COL17A1 proteolysis. Science, 2016. 351(6273): aad4395.
[152] Huang, W.Y. et al. Stress-induced premature senescence of dermal papilla cells compromises hair follicle epithelial-mesenchymal interaction. J Dermatol Sci, 2017. 86(2): 114–122.
[153] Stenn, K.S., and R. Paus. Controls of hair follicle cycling. Physiol Rev, 2001. 81(1): 449–494.

[154] Lindner, G. et al. Analysis of apoptosis during hair follicle regression (catagen). Am J Pathol, 1997. 151(6): 1601–17.
[155] Dry, F.W. The coat of the mouse (Mus musculus). Journal of Genetics, 1926. 16(3): 54.
[156] Murphy, G. et al. The molecular determinants of sunburn cell formation. Exp Dermatol, 2001. 10(3): 155–60.
[157] Rahmani, W. et al. Hair follicle dermal stem cells regenerate the dermal sheath, repopulate the dermal papilla, and modulate hair type. Dev Cell, 2014. 31(5): 543–58.
[158] Pena-Jimenez, D. et al. Lymphatic vessels interact dynamically with the hair follicle stem cell niche during skin regeneration in vivo. EMBO J, 2019. 38(19): e101688.
[159] Brownell, I. et al. Nerve-derived sonic hedgehog defines a niche for hair follicle stem cells capable of becoming epidermal stem cells. Cell Stem Cell, 2011. 8(5): 552–565.
[160] Festa, E. et al. Adipocyte lineage cells contribute to the skin stem cell niche to drive hair cycling. CELL, 2011. 146(5): 761–771.
[161] Plikus, M.V. et al. Cyclic dermal BMP signalling regulates stem cell activation during hair regeneration. Nature, 2008. 451(7176): 340–4.
[162] Wang, E.C.E. et al. A Subset of TREM2(+) dermal macrophages secretes oncostatin M to maintain hair follicle stem cell quiescence and inhibit hair growth. Cell Stem Cell, 2019. 24(4): 654–669 e6.
[163] Castellana, D., R. Paus, and M. Perez-Moreno. Macrophages contribute to the cyclic activation of adult hair follicle stem cells. PLoS Biol, 2014. 12(12): e1002002.
[164] Ali, N. et al. Regulatory T cells in skin facilitate epithelial stem cell differentiation. Cell, 2017. 169(6): pp. 1119–1129 e11.
[165] Jahoda, C.A., K.A. Horne, and R.F. Oliver. Induction of hair growth by implantation of cultured dermal papilla cells. Nature, 1984. 311(5986): 560–2.
[166] Paus, R. et al. Mast cell involvement in murine hair growth. Dev Biol, 1994. 163(1): 230–40.
[167] Botchkarev, V.A. et al. Hair cycle-dependent changes in adrenergic skin innervation, and hair growth modulation by adrenergic drugs. J Invest Dermatol, 1999. 113(6): 878–87.
[168] Fan, S.M.-Y. et al. External light activates hair follicle stem cells through eyes via an ipRGC-SCN-sympathetic neural pathway. Proceedings of the National Academy of Sciences of the United States of America, 2018. 115(29): E6880–E6889.
[169] Chen, C.C. et al. Organ-level quorum sensing directs regeneration in hair stem cell populations. Cell, 2015. 161(2): 277–90.
[170] Yu, Z. et al. Hoxc-dependent mesenchymal niche heterogeneity drives regional hair follicle regeneration. Cell Stem Cell, 2018. 23(4): 487–500 e6.
[171] Plikus, M.V. New activators and inhibitors in the hair cycle clock: targeting stem cells' state of competence. J Invest Dermatol, 2012. 132(5): 1321–4.
[172] Murray, P.J. et al. Modelling hair follicle growth dynamics as an excitable medium. PLoS Comput Biol, 2012. 8(12): e1002804.
[173] Plikus, M.V. et al. Self-organizing and stochastic behaviors during the regeneration of hair stem cells. Science, 2011. 332(6029): 586–9.
[174] Myung, P.S. et al. Epithelial Wnt ligand secretion is required for adult hair follicle growth and regeneration. J Invest Dermatol, 2013. 133(1): 31–41.
[175] Choi, Y.S. et al. Distinct functions for Wnt/beta-catenin in hair follicle stem cell proliferation and survival and interfollicular epidermal homeostasis. Cell Stem Cell, 2013. 13(6): 720–33.
[176] Leishman, E. et al. Foxp1 maintains hair follicle stem cell quiescence through regulation of Fgf18. Development, 2013. 140(18): 3809–18.
[177] Choi, S. et al. Corticosterone inhibits GAS6 to govern hair follicle stem-cell quiescence. Nature, 2021. 592(7854): 428–432.
[178] Morgan, B.A. The dermal papilla: an instructive niche for epithelial stem and progenitor cells in development and regeneration of the hair follicle. Cold Spring Harb Perspect Med, 2014. 4(7): a015180.
[179] Huang, W.Y. et al. Mobilizing transit-amplifying cell-derived ectopic progenitors prevents hair loss from chemotherapy or radiation therapy. Cancer Res, 2017. 77(22): 6083–6096.
[180] Aamar, E. et al. Hair-follicle mesenchymal stem cell activity during homeostasis and wound healing. J Invest Dermatol, 2021. 141(12): 2797–2807 e6.

[181] Jahoda, C.A. Cell movement in the hair follicle dermis - more than a two-way street? J Invest Dermatol, 2003. 121(6): ix–xi.
[182] Chi, W.Y., D. Enshell-Seijffers, and B.A. Morgan. De novo production of dermal papilla cells during the anagen phase of the hair cycle. J Invest Dermatol, 2010. 130(11): 2664–6.
[183] Heitman, N. et al. Dermal sheath contraction powers stem cell niche relocation during hair cycle regression. Science, 2020. 367(6474): 161–166.
[184] Courtois, M. et al. Ageing and hair cycles. Br J Dermatol, 1995. 132(1): 86–93.
[185] Hibberts, N.A., A.E. Howell, and V.A. Randall. Balding hair follicle dermal papilla cells contain higher levels of androgen receptors than those from non-balding scalp. J Endocrinol, 1998. 156(1): 59–65.
[186] Chen, W., C.C. Zouboulis, and C.E. Orfanos. The 5 alpha-reductase system and its inhibitors. Recent development and its perspective in treating androgen-dependent skin disorders. Dermatology, 1996. 193(3): 177–84.
[187] Thornton, M.J. et al. Differences in testosterone metabolism by beard and scalp hair follicle dermal papilla cells. Clin Endocrinol (Oxf), 1993. 39(6): 633–9.
[188] Randall, V.A. et al. The hair follicle: a paradoxical androgen target organ. Horm Res, 2000. 54(5-6): 243–50.
[189] Bahta, A.W. et al. Premature senescence of balding dermal papilla cells in vitro is associated with p16(INK4a) expression. J Invest Dermatol, 2008. 128(5): 1088–94.
[190] Randall, V.A., N.A. Hibberts, and K. Hamada. A comparison of the culture and growth of dermal papilla cells from hair follicles from non-balding and balding (androgenetic alopecia) scalp. Br J Dermatol, 1996. 134(3): 437–44.
[191] Kwack, M.H. et al. Dihydrotestosterone-inducible IL-6 inhibits elongation of human hair shafts by suppressing matrix cell proliferation and promotes regression of hair follicles in mice. J Invest Dermatol, 2012. 132(1): 43–9.
[192] Kwack, M.H. et al. Dihydrotestosterone-inducible dickkopf 1 from balding dermal papilla cells causes apoptosis in follicular keratinocytes. J Invest Dermatol, 2008. 128(2): 262–9.
[193] Inui, S. et al. Identification of androgen-inducible TGF-beta1 derived from dermal papilla cells as a key mediator in androgenetic alopecia. J Investig Dermatol Symp Proc, 2003. 8(1): 69–71.
[194] Paus, R. et al. Transforming growth factor-beta receptor type I and type II expression during murine hair follicle development and cycling. J Invest Dermatol, 1997. 109(4): 518–26.
[195] Turksen, K. et al. Interleukin 6: insights to its function in skin by overexpression in transgenic mice. Proc Natl Acad Sci U S A, 1992. 89(11): 5068–72.
[196] McConnell, J.D. et al. Finasteride, an inhibitor of 5 alpha-reductase, suppresses prostatic dihydrotestosterone in men with benign prostatic hyperplasia. J Clin Endocrinol Metab, 1992. 74(3): 505–8.
[197] Bramson, H.N. et al. Unique preclinical characteristics of GG745, a potent dual inhibitor of 5AR. J Pharmacol Exp Ther, 1997. 282(3): 1496–502.
[198] Kaufman, K.D. et al. Finasteride in the treatment of men with androgenetic alopecia. Finasteride Male Pattern Hair Loss Study Group. J Am Acad Dermatol, 1998. 39(4 Pt 1): 578–89.
[199] Olsen, E.A. et al. The importance of dual 5alpha-reductase inhibition in the treatment of male pattern hair loss: results of a randomized placebo-controlled study of dutasteride versus finasteride. J Am Acad Dermatol, 2006. 55(6): 1014–23.
[200] Rossi, A. et al. Minoxidil use in dermatology, side effects and recent patents. Recent Pat Inflamm Allergy Drug Discov, 2012. 6(2): 130–6.
[201] De Villez, R.L. Topical minoxidil therapy in hereditary androgenetic alopecia. Arch Dermatol, 1985. 121(2): 197–202.
[202] Zappacosta, A.R. Reversal of baldness in patient receiving minoxidil for hypertension. N Engl J Med, 1980. 303(25): 1480–1.
[203] Headington, J.T. Hair follicle biology and topical minoxidil: possible mechanisms of action. Dermatologica, 1987. 175 Suppl 2: 19–22.
[204] Jimenez, F. et al. Hair transplantation: Basic overview. J Am Acad Dermatol, 2021. 85(4): 803–814.

[205] Garza, L.A. et al. Bald scalp in men with androgenetic alopecia retains hair follicle stem cells but lacks CD200-rich and CD34-positive hair follicle progenitor cells. J Clin Invest, 2011. 121(2): 613–22.
[206] Wang, L. et al. Foxcl reinforces quiescence in self-renewing hair follicle stem cells. Science, 2016. 351(6273): 613–7.
[207] Keyes, B.E. et al. Nfatcl orchestrates aging in hair follicle stem cells. Proc Natl Acad Sci USA, 2013. 110(51): E4950–9.
[208] Matsumura, H. et al. Distinct types of stem cell divisions determine organ regeneration and aging in hair follicles. Nature Aging, 2021. 1(2): 190–204.
[209] Zhang, C. et al. Escape of hair follicle stem cells causes stem cell exhaustion during aging. Nature Aging, 2021. 1(10): 889–903.
[210] Oliver, R.F. The experimental induction of whisker growth in the hooded rat by implantation of dermal papillae. J Embryol Exp Morphol, 1967. 18(1): 43–51.
[211] Oliver, R.F. Ectopic regeneration of whiskers in the hooded rat from implanted lengths of vibrissa follicle wall. J Embryol Exp Morphol, 1967. 17(1): 27–34.
[212] Cohen, J. The transplantation of individual rat and guineapig whisker papillae. J Embryol Exp Morphol, 1961. 9: 117–27.
[213] Oliver, R.F. The induction of hair follicle formation in the adult hooded rat by vibrissa dermal papillae. J Embryol Exp Morphol, 1970. 23(1): 219–36.
[214] Kishimoto, J. et al. Selective activation of the versican promoter by epithelial- mesenchymal interactions during hair follicle development. Proc Natl Acad Sci U S A, 1999. 96(13): 7336–41.
[215] Kishimoto, J., R.E. Burgeson, and B.A. Morgan. Wnt signaling maintains the hair-inducing activity of the dermal papilla. Genes Dev, 2000. 14(10): 1181–5.
[216] Inamatsu, M. et al. Establishment of rat dermal papilla cell lines that sustain the potency to induce hair follicles from afollicular skin. J Invest Dermatol, 1998. 111(5): 767–75.
[217] Rendl, M., L. Polak, and E. Fuchs. BMP signaling in dermal papilla cells is required for their hair follicle-inductive properties. Genes Dev, 2008. 22(4): 543–57.
[218] Qiao, J., E. Philips, and J. Teumer. A graft model for hair development. Exp Dermatol, 2008. 17(6): 512–8.
[219] Qiao, J. et al. Hair follicle neogenesis induced by cultured human scalp dermal papilla cells. Regen Med, 2009. 4(5): 667–76.
[220] Higgins, C.A. et al. Microenvironmental reprogramming by three-dimensional culture enables dermal papilla cells to induce de novo human hair-follicle growth. Proc Natl Acad Sci U S A, 2013. 110(49): 19679–88.
[221] Hu, S. et al. Dermal exosomes containing miR-218-5p promote hair regeneration by regulating beta-catenin signaling. Sci Adv, 2020. 6(30): eaba1685.

2

Biomaterial Effects on Isolation, Proliferation, and Regulation of Stem Cells

Nai-Chen Cheng,[1] *Min-Huey Chen,*[2] *Chia-Ning Shen,*[3] *Wen-Yen Huang,*[4] *Sung-Jan Lin,*[4] *I-Chi Lee*[5,*] and *Yi-Chen Ethan Li*[6,*]

Tissue engineering is a multidisciplinary field comprising biology and engineering technology which creates artificial tissues/organs or *in vitro* biomimetic models. In terms of tissue engineering, to investigate cell-cell, cell-matrix, and cell-signal interactions in the *in vivo* physiological environment, how to combine cell, biomaterials, and signal factors plays an dispensable role in regulation the proliferation, differentiation, and even isolation of cells or stem cells. Especially, biomaterials enable to provide a cell-anchored substrate and create a tissue-like *in vitro* environment for survival and growth of cells. For example, the influence of surface hydrophilicity on stem cells' adhesion, motility, cytoskeletal organization, and differentiation is significant. In a study, hydrogels were synthesized by copolymerizing acrylamide with amino acid monomers at 7:1, 6:2, 5:3, and 3:5 mol ratios. The variable lengths of amino acid side chains (from C1 to C10) altered the hydrophobicity of matrix, while the functional group, charge density, surface

[1] Department of Surgery, National Taiwan University Hospital, Taiwan.
[2] Graduate Institute of Clinical Dentistry, School of Dentistry, National Taiwan University, Taiwan.
[3] Genomics Research Center, Academia Sinica, Taiwan.
[4] Department of Biomedical Engineering, College of Medicine and College of Engineering, National Taiwan University, Taiwan.
[5] Department of Biomedical Engineering and Environmental Sciences, National Tsing Hua University, Taiwan.
[6] Department of Chemical Engineering, Feng Chia University, Taiwan.
* Corresponding authors: iclee@mx.nthu.edu.tw; yicli@fcu.edu.tw

roughness, and bulk mechanical properties remained the same. The water contact angle increased as amino acid side chains became longer, ranging from about 27° to about 85°. Bone marrow-derived MSCs were cultured on these hydrogels, and showed better adhesion and spreading on C5 hydrogels (contact angle at about 58°). These results can be explained by the hydrophobicity-driven conformational changes of the pendant side chains at the interface, leading to differential binding of proteins. The amino acid side chains need to be long enough to access protein binding sites, while longer chains collapse due to increased hydrophobicity. C5 hydrogels have a moderate length of side chains and moderate hydrophilicity, which results in better protein adsorption ability, and further promotes cell adhesion. Moreover, it was observed that MSCs cultured on C5 hydrogels could differentiate into osteogenic and myogenic lineage with osteogenic and myogenic medium, while those cultured on C3 hydrogels failed to differentiate. Taken together, the surface biomaterial property of the 2D membrane, such as hydrophilicity, has striking influences on the stem cell behaviors [1]. In addition, the stiffness and elasticity of the stem cell niches also play important roles in determining cell behaviors. For example, during the initial week in culture, it is possible to reprogram the differentiation with the supplement of soluble induction factors. However, after several weeks, matrix elasticity seems to determine the cells to commit to certain specific lineages [2]. In this series of study, polyacrylamide gels were developed as elasticity-modifiable substrates and coated them with type 1 collagen for favorable adhesiveness to MSCs. The stiffness could be modified with different acrylamide/bis-acrylamide ratios. Naive MSCs showed extreme sensitivity to substrate elasticity. The MSCs on soft substrates that mimic brain tissue (0.1 ~ 1 kPa) adhered, spread, exhibited an increasingly branched, filopodia-rich morphology, and showed the most significant expression of neurogenic transcripts. Stiffer matrices that mimic muscle (8 ~ 17 kPa) baked spindle-shaped cells that expressed increased myogenic messages, and the stiffest matrices that mimic collagenous bone (25 ~ 40 kPa) grew polygonal MSCs with osteogenesis. The upregulated non-muscle myosin II (NMM II) activity was necessary for MSCs to respond to stiffer matrices and commit to specific lineages [3]. Moreover, it is well known that the surface charge or functional groups provide versatile applications to regulate cell behaviors. For example, Change and co-workers developed chitosan and poly(vinyl alcohol) (PVA) hybrid substrates for use in cell fractions [4]. In their design, chitosan polymers with the primary amine of glucosamine functional group endow chitosan polymers with a pH-sensitive property. Then, through blending chitosan and PVA polymers, the hybrid polymeric membrane offers a rapid cell detachment rate, and more than 99% of cells could detach from the hybrid membrane within 60 mins by controlling the adsorption of ECM protein at different pH values. According to the above studies, it can be understand that through controlling the properties of biomaterials such as wettability (i.e., hydrophilic or hydrophobic ability), mechanical properties, and surface characterizations, the various cell-matrix interactions can be applied for isolation, proliferation, or differentiation of stem cells. Therefore, in this chapter, we will focus on the case of biomaterials for regulating cell behaviors and the detail biomaterials properties will be discussed in Chapter 4.

The biomaterial effects on ESC

In pluripotent ESC culture, a long-term culture is required to induce the expansion of ESCs for harvesting enough cell numbers for experiments. In general, although people believe ESC with an unlimited division ability, the average doubling time of human ESCs requires more than 36 hours, which may cause a critical limitation for culture ESC *in vitro* [5]. Until now, the conventional approach for expanding human ESCs only uses a feeder layer. However, the method of using feeder layers may be concerned about the xeno-contamination. Recently, many reports indicated that using feeder-layer free methods to culture ESCs can improve the abovementioned scenario and maintain stemness and pluripotency [6, 7]. Biomaterials used as an alternative bring a promise to culture, expand ESCs, and further overcome the challenges to the ES culture technologies in the clinic, which are concerns about reducing the risk of xeno-contamination occurrence and supplying the uniform cellular quality and quantity. For example, an earlier study shows that using alkali treatment to cleave biodegradable polymeric backbones such as PLGA or PLLA presents more hydroxyl and carboxyl function group and increase hydrophilicity, allowing for promoting the proliferation of ESCs [8]. In addition, another study also reported that encapsulating ESC in a 3D hydrogel can force ESC aggregation, further form an Embryoid bodies (EBs), and enhance cell viability, growth, and pluripotency [9]. Therefore, the information also indicated that ESCs could sense the culture microenvironment properties, such as elasticity or ECM components, to adapt their fate and behaviors related to the properties of the biomaterial [10]. In addition to proliferation, Wang and co-workers also investigate the structural effects of biomaterials on the differentiation of ESCs. It has been confirmed that the pluripotency of ESC-derived EB organoids are a valuable aspect of their applications in tissue engineering and regenerative medicine [11]. Therefore, Wang and co-workers fabricated an inverted colloidal crystal scaffold with well-controlled pore sizes (i.e., 90, 170, 210, 280, and 400 μm) to obtain EB-like spheroids [12]. These studies have demonstrated that properly designed scaffolds can generate EB-like spheroids with varying sizes. Furthermore, the hepatic differentiation behavior of the EB-like spheroids can be observed when their size is around 200 μm after being harvested from the scaffolds. Moreover, the results also appeared that the nutrients delivered in an EB-like spheroid might become an obstacle to physical restriction when the pore size is over 270 μm. The information implied that the biomaterial effect significantly influences the growth, maintenance, and pluripotency of ESCs.

The biomaterial effects on iPSCs

In addition, iPSCs have become an essential tool for the investigation of organ development in recent years, allowing for the modelling of disease and drug efficacy and brought renewed hope for cell therapy. Generation and maintaining iPSCs of the highest quality are crucial. Genome stability, pluripotency, self-renewal capacity, and differentiation potential are all factors that must be considered prior to the application of iPSCs. For research into organ development, failure to maintain the quality of iPSCs can reduce their capacity to generate differentiated cells. For disease modelling, poor quality iPSCs can skew the disease-causing outcome, whilst the use of iPSC in the clinical setting requires the highest degree of consistency and pass rigorous testing.

Advancements in culture media, growth factors, and chemical compound selection has provided a framework for iPSC-related work. These alone, however, are insufficient to generate and maintain iPSCs. The niche in which the iPSC is cultured is also critical. The goal is to mimic the *in vivo* environment in which nature has evolved to sustain pluripotency as well as differentiation into appropriate adult cells. This is where designing biomaterials that assist in the controlling iPSC fate can assist with the advancement of iPSCs. Biomaterials are coating or scaffolds that mimic nature niches [13]. As the field continues to advance, the production of biomaterials can be either natural or synthetic.

iPSCs have traditionally been cultivated on a feeder layer of mouse embryonic fibroblasts. This raises several concerns, including the difficulty of eliminating contaminated animal-derived fibroblasts. To overcome this issue, natural or synthetic biomaterials can be utilized. Naturally derived biomaterials include native ECM or polymers derived from decellularized tissues. For example, chitosan, silk, fibrin, gelatin, alginate, agarose, hyaluronic acid, and collagen are used in the production of scaffolds that supports iPSC attachment and growth. Scaffolds made from natural materials have lower or no immunogenicity, decent repeatability, and a higher differentiation rate due to the presence of native integrin-binding sites [14]. Other advantages of using natural materials in scaffold manufacturing include hydrophilicity and cell affinity. Matrigel, a basement-membrane matrix produced from murine Engelbreth-Holm Swarm sarcomas is commonly used to maintain feeder-free iPSC culture [15]. It contains laminin, collagen IV, entactin, and heparin sulfate proteoglycan prelecan. It also has a transforming growth factor (TGF), fibroblast growth factor (FGF), and enzyme metrix metalloproteinases (MMPs) that contribute to the ability of Matrigel to support iPSC culture. Indeed, iPSCs maintained on Matrigel has been shown to be maintained for > 100 population doubling with normal karyotype, proliferation as well as expression of pluripotent markers including OCT4, SSEA4, Tra-1-60, and Tra-1-81 [16]. Moreover, iPSC cultured on Matrigel maintains the ability to differentiate into three germ layers *in vivo* when transplanted into immunodeficient mice. Despite many successes, there are several disadvantages of Matrigel, including degradation under long-term storage and xenogenic contaminants. Additionally, complex and variable composition of Matrigel makes it difficult to optimize for specific cellular behaviors to achieve desired biological outcome [17]. Synthetic materials, on the other hand, can be controllable with specific biocompatibility, biodegradability, and physical properties [17]. Examples of synthetic materials include derivatives made from polyacrylamide (PAM) [17] or polyethylene glycol (PEG) [18]. Various modifications to synthetic scaffold have been made to mimic key cell-matrix interactions found in Matrigel that are necessary to maintain iPSC growth and pluripotency. These includes physical properties such as dimensionality, stiffness, and topography [17]. Alteration to physical properties can be achieved by altering hydrophilicity, thickness, and surface charge and these properties have been shown to be crucial in determining if long-term iPSC culture is attained. For example, by comparing a selection of > 80 polymers with varying composition and molecular weight, it was demonstrated that only selected polymers can support iPSC proliferation short-term and only one, that is, poly(methyl vinyl ether-alt-maleic hydride), which can sustain long-term culture to a similar degree as Matrigel [19]. Apart from physical properties, proper biochemical properties are also critical. Synthetic scaffold that supports growth factors stabilization, cell-matrix interactions, and creating spatial patterning are also key to determining if iPSC pluripotency can be maintained. For example, the generation of a heparin-mimetic scaffold that mimicked heparin sulfate proteoglycans allowed for the maintenance of iPSC culture for > 20 passages in chemically defined medium without significant alteration to the expression of pluripotent markers, OCT4, and NANOG [19]. It was suggested that the heparin-mimetic scaffold bound and protected basic fibroblast growth factor (bFGF), a key growth factor in sustaining iPSC culture, from proteolytic degradation.

Another key aspect of synthetic scaffold is to support iPSC adhesion and expansion. Binding of integrin receptors present on iPSC to Matrigel coated surfaces is crucial. Therefore, the development of different peptide combinations that mimics Matrigel is ongoing. Fibronectin-derived peptide Arg-Gly-Asp (RGD) that binds to integrins $\alpha v\beta 3$ and $\alpha v\beta 5$ or vitronectin-derived peptides are commonly used to promote cell adhesion, enabling long-term maintenance of iPSC pluripotency [20, 21]. Lastly, the generation of the 3D scaffold compared to 2D coating has been shown to further enhance programming efficiency of somatic cells into iPSCs. Indeed, the use of 3D RGD coated PEG hydrogel promoted mesenchymal-to-epithelial transition and epigenetic plasticity which translated into a threefold increase in iPSC reprogramming efficiency [22].

The biomaterial effects on MSCs

Adult stem cells are somatic stem cells and have been confirmed to exist in many tissues or organs. Compared with ESCs, even though ESCs have pluripotency, adult stem cells provide more safe treatment and are easier to use [23]. Adult stem cells could be obtained from mesenchymal, adipose, brain, skin, and other tissues in the human body. In terms of adult stem cells, MSCs have been shown their potential applications in tissue engineering and regenerative medicine because MSCs can differentiate into adipogenic, osteogenic, and chondrogenic tissues [24]. However, previous studies indicated that the average doubling time of MSCs canbe distribute in a range from 1.3 to 16 days [25, 26]. To rapidly and stably harvest MSCs for applications, Bertolo and co-workers provided a strategy for grafting a growth factor, bFGF, on a collagen microcarrier via an EDS/NHS crosslinked method [27]. Through the crosslink, it can reduce the degradation of collagen microcarrier and increase by the four times the cell numbers of MSCs via the regulation of bFGF compared with a non-crosslinked microcarrier in the absence and presence of bFGF.

In addition, to regulate the differentiation of MSCs, the physiochemical properties of the culture microenvironment also play an essential role in controlling the differentiation behaviors of MSCs. It is well known that the surface chemical features of biomaterials contribute a sufficient factor to modulate the MSC differentiation. For example, the methyl group-based polymeric and unmodified glass and surface properties support the survival of MSCs; amine and thiol-based surfaces offer osteogenic differentiation effects on MSCs; alcohol and carboxyl groups can enhance the MSCs with highly efficient chondrogenic differentiation [28]. In addition to the surface chemistry features, the physical properties, such as substrate stresses, also directly influence the differentiation behavior of MSCs [29]. For example, an earlier study shows that using the substrate with micropatterns to alter the cell shape into a spherical structure can enhance the expression of adipogenic differentiation; in contrast, inducing cells with a flat shape endows MSCs with preferential osteoblastic differentiation [30]. Based on this information, a wide range of natural, synthetic, and hybrid polymers, such as hyaluronan, cellulose, silk, PLGA, and others, have been used to design the cell culture systems for inducing the adipose, neural, chondrogenic, and osteogenic differentiations [31–36]. Previous studies have identified that contact angles and cellular roundness are associated with cellular motility, cytoskeletal organization, differentiation capability, and stemness. Researchers have designed different 2D micropattern islands to test the hypothesis that cell shapes (probably influenced by ECMs) could determine the fate of stem cells at the single-cell and molecular levels. For example, Connelly et al. developed a substrate with micropatterned adhesive isolated islands composed of type I collagen in different diameters and aspect ratios (major/minor axis ratios) [37]. They surrounded those islands with poly-oligo (ethylene glycol methacrylate) (POEGMA) brushes resistant to protein adsorption. They seeded human epidermal stem cells on the islands to investigate the signaling pathways of different responses to the various triggers of the local ECM contact. The keratinocytes that remained round on the small (20 μm in diameter) circular islands presented a higher Involucrin positive percentage (indicating differentiation) than those spreading on islands with a 50 μm diameter or aspect ratio of 8. Moreover, the actin cytoskeleton directed shape-induced differentiation by regulating serum response factor (SRF) transcriptional activity.

Similarly, McBeath et al. used variously sized fibronectin islands printed on polydimethylsiloxane (PDMS) substrates to alter the cell shapes to regulate human MSCs' commitment to different lineages [38, 39]. These MSCs were allowed to adhere, flatten, and spread, thus not shutting down the constitutively active RhoA, and underwent osteogenesis. Meanwhile, most of the MSCs that remained round on limited adhesive areas showed no RhoA expression, and they turned into adipocytes. Under transforming growth factor β (TGF-β) stimulation, those adherent and well-spread hMSCs showed upregulated SMC genes. In contrast, under the same TGFβ stimulation, those MSCs that remained round suppressed the constitutively active Rac1 and the N-Cadherin and finally underwent chondrogenesis.

Wang et al. chose photoreactive azidophenyl-derivatized poly(vinyl alcohol) (AzPhPVA) to prepare micropattern islands of different sizes, geometry (circle, triangle, square, pentagon, hexagon), and aspect ratios to control the morphogenesis of human MSCs [40]. The results showed increased nuclear activity and cellular elasticity (due to the cytoskeletal organization), and decreased stemness with a more extensive spreading area and aspect ratio. However, the geometry had no significant influence on the expression of stem cell markers. It is possible that the morphology during the stem cell cultivation was related to environmental stress and the corresponding cytoskeletal tension could affect both the differentiation capability and the stemness. Furthermore, other synthetic stimuli-responsive polymeric substrates also enable inducing MSCs to form tissue-like microstructures. For example, thermo-sensitive and pH-sensitive polymeric substrates provide surface features to adsorb ECM proteins on the surface for MSC anchoring. Subsequently, while MSCs achieved confluence, a stem cell sheet formed on an ECM-adsorbed substrate could be rapidly harvested without the treatment of enzymes by changing the temperature or pH value of the culture environment because the ECM detached from the substrates [41, 42]. Moreover, previous studies also show that using a thermo-sensitive methylcellulose hydrogel to fabricate a 3D MSC spheroid could be a stem cell-based inherent niche that activates the secretion of immunomodulatory factors and pro-regenerative paracrine molecules for enhancing therapeutic benefits of cell-based therapy [43].

The biomaterial effects on ASCs

Adipose-derived stem cells (ASCs) are mesenchymal stem cells (MSCs) identified within the subcutaneous fat tissue. The most apparent advantage of ASCs over other MSCs is their harvesting procedure is much less invasive. Both *in vitro* and *in vivo* ASC research have flourished, and most results are promising at present. The ASCs are known to involve the healing process in both cellular and secretory pathways. The ASCs proliferate and differentiate into different types of mature cells, and their secretome, including growth factors and cytokines, helps create a suitable microenvironment for healing [44]. Usually, cellular expansion is necessary to meet the therapeutic requirements, and *ex vivo* cultivation requires meticulous manipulation and standardization [45]. Since ASCs lose their surface pluripotency markers essential for cellular renewal and differentiation capabilities during traditional monolayer culture [46], researchers have endeavored to develop novel biomaterials that mimic the *in vivo* microenvironment to foster ASC stemness as well as other favorable characteristics.

SVF isolated from fat tissue is a cluster of heterogeneous cells with various morphology, proliferation rates, and differentiation efficacy [47]. Different cellular subtypes survived, proliferated, and differentiated diversely in the respective cultivation microenvironment. After serial passages of conventional monolayer culture, expression levels of mesenchymal markers, such as CD29, CD44, and CD90, increased and became more uniform. In contrast, mRNAs for several pluripotent genes, such as Nanog, Oct-4, and Rex-1, only existed in early passages [46]. Hence, biomaterials have been applied to isolate ASCs with better regenerative potential. In a study, ASCs were isolated from a primary fat tissue solution using: (1) conventional culture, (2) a membrane filtration method, (3) a membrane migration method where the primary cell solution was permeated through nylon mesh filter membranes with pore sizes ranging from 11 to 80 μm. ASCs isolated by the membrane migration method were found to have the highest MSC surface marker expression and efficient differentiation into osteoblasts. Osteogenic differentiation ability of ASCs and MSC surface marker expression were correlated, but osteogenic differentiation ability and pluripotent gene expression were not [48].

In another study, human ASCs cultured for five passages were filtered through nylon mesh filter membranes coated with and without extracellular matrix proteins to obtain the permeation solution. Subsequently, the culture media were filtered via the membranes to obtain the recovery solution. Then, the membranes were cultured in a cell culture medium to obtain the migrated cells from the membranes. The ASCs in the permeation solution, through any type of nylon mesh filter membrane having 11 and 20 μm pore sizes, had lower osteogenic differentiation ability than conventional ASCs cultured on tissue culture polystyrene (TCP) dishes for passage 5. However, the ASCs purified by the membrane migration method through nylon mesh filter membranes coated with recombinant vitronectin, which have 11 and 20 μm pore sizes, showed a higher proliferative activity as well as higher osteogenic differentiation potential than the conventional ASCs cultured on TCP dishes for passage 5 [49]. Therefore, the biomaterial membrane filtration and migration methods would be useful for cell sorting for ASCs with high proliferation and high osteogenic differentiation ability.

ASCs lose their surface pluripotency markers after several passages, leading to lower proliferative activity [46, 48]. To resolve this issue, investigators have tried different approaches to create a microenvironment that mimicked the target tissue's extracellular matrix (ECM) [50–52]. Yu et al. showed that promoting ECM synthesis via the ERK signaling pathway was the key for ASC stemness enhancement *in vitro* [53]. Studies have verified that engineered microenvironments benefit cell-cell interaction and cell-scaffold adhesion and further improve the ASC survival and stemness [51, 52]. Hence, researchers have tried stimulating endogenous ECM deposition or fabricating extrinsic scaffolds to build a similar microenvironment. For example, ECM derived from MSCs has recently been shown to be able to maintain the differentiation potential of MSCs during culture expansion and to restore the activities of aging MSCs, suggesting that MSC-derived ECM may be a suitable culture substrate to enhance the bioactivity of biomaterial scaffolds for MSCs [54]. Similarly, ECM deposited by ASCs also supported adhesion and proliferation of primary human fibroblasts and dermal microvascular endothelial cells, indicating their potential as platforms for wound-healing studies [55].

In addition to endogenous ECM deposited by ASCs [56], scientists have also tried to exogenously shape the microenvironment for ASC incubation. Modern biochemical technology facilitates the development of these ASC-compatible scaffolds (Table 2-1). Organic materials were the first choice because of their inherent biocompatibility. For example, hydrogels can encapsulate ASCs with enhanced cellular proliferation and differentiation capabilities in the semi-liquid phase. Hydrogels can also facilitate cellular transplantation and provide gradual ASC release during biodegradation by applying different cross-linking methods to modify these raw materials and create ideal characteristics. For example, Cheng et al. used temperature-sensitive chitosan/gelatin hydrogels to cultivate and deliver ASCs [50]. The hydrogel was hydrophilic and porous and mimicked the genuine ECM. The *in vitro* and *in vivo* results revealed significantly improved viability, increased VEGF concentration, and consequently, enhanced angiogenesis relative to the chitosan hydrogel.

Table 2-1. Examples of biomaterial effects on ASCs.

Study	Biomate-rials	Modified Steps	Modifications	Endpoint	Outcome
Monolayer Culture					
Lin H.R. et al. (2017) [48]	Nylon mesh filters (11, 20, 41, 60, and 80 μm)	Cell culture/Cell isolation	Membrane filtration (Permitting SVF through the filters and collecting the permeation solution and the recovery solution separately)/ Membrane migration (placing a filter on the SVF solution during the ASC culture and letting the cells migrate through it).	*In vitro* MSC surface marker expression, osteogenic differentiation capability, and pluripotent gene expression.	1. Compared to ASC isolation with conventional monolayer culture, the membrane filtration method featured much less time consumption and comparable osteogenic differentiation ability. The membrane migration method gave the highest purity of ASCs with the highest MSC surface marker expression osteogenic differentiation ability though taking more time. 2. The osteogenic differentiation rate of the cells increased with increasing MSC surface marker expression but decreased with increasing pluripotent gene expression.
Cell Spheroids					
Cheng N.C. et al. (2012) [57]	Chitosan films	Cell spheroid formation	Seeding the 3rd passage ASCs on chitosan films for seven days.	Nude mice, intramuscular injection of the ASC spheroid PBS suspension.	Modifiable cell spheroids were successfully formed with excellent cell viability, enhanced *in vitro* ASC pluripotency, robust endogenous ECM, and improved *in vivo* cellular survival rate after intramuscular injection.
Tsai C.C. et al. (2019) [58]	1. Agarose microwell plate 2. Ggelatin/mTG hydrogel	1. Cell spheroid formation 2. Cell spheroid encapsulation	Seeding the 3rd passage ASCs evenly onto microwells plate, then incubating them for three days for cell spheroid formation, then mixing them with gelatin/mTG solution, and then gelating the solution and cultivating them for seven days.	ICR mice, subcutaneous injection of the gelatin/ mTG hydrogel.	1. Standardized cell spheroids were obtained by seeding cells in the microwells plate. 2. The cell spheroids encapsulated in the gelatin/mTG hydrogel presented good viability, great stemness, injectability, and *in vivo* biocompatibility after ICR mice subcutaneous injection.

Lu T.-Y. et al. (2020) [59]	1. Agarose microwell plate 2. Gdelatin/mTG hydrogel	1. Cell spheroid formation 2. Cell spheroid encapsulation	(As above)	Wistar rat, burn wound, gelatin/mTG hydrogel with ASC spheroids fixed with a Tegaderm film.	*In vivo* burn wound healing experiments showed that the cell spheroid with hydrogel accelerated wound contraction with reduced scab development, faster tissue regeneration, thicker epidermises, and improved wound architecture.
Cell Sheets					
Yu J. et al. (2014) [53]	L-ascorbate 2-phosphate (A2-P)	Cell sheet formation	Adding A2-P to the culture medium of the 3rd passage ASCs for seven days.	Nude mice, healing-impaired full-thickness cutaneous wound, ASC sheet in 100 µL PBS covered with a Tegaderm film.	1. A2-P-treated ASCs successfully composited cell sheets with incremental ECM, grew faster, maintained their stemness, improved the *in vivo* cellular survival rate, and enhanced wound healing. 2. Partially through attenuating ERK signaling pathway, A2-P-induced collagen synthesis was required for stemness enhancement.
Yu J. et al. (2018) [67]	L-ascorbate 2-phosphate (A2-P)	Cell sheet formation	(As above)	(As above)	1. In the animal model experiment, ASC sheets reduced scar formation via releasing HGF and CTRP3, and consequently improved the quality of the regenerated skin and benefited wound healing. 2. More ASCs in sheets survived during the healing process in the animal model experiment, and all ASCs vanished in 28 days.
Scaffolds					
Cheng N.-C. et al. (2008) [60]	cECM-derived porous scaffold	Cell culture	Seeding the 4th passage ASCs on the cECM-derived porous scaffold and culturing them for six weeks.	*In vitro* chondrogenesis	A native cECM-derived porous scaffold could induce the chondrogenic differentiation of ASCs without exogenous growth factors.
Cheng N.C. et al. (2017) [50]	Temperature-sensitive chitosan/ glycerophosphate hydrogel (TCGH)	Cell encapsulation	Suspending the 3rd passage ASCs in the TCGH, and then gelating the hydrogel at 37°C.	C57/B6 mice, full-thickness cutaneous wound, TCGH with ASCs fixed with a Tegaderm film.	TCGH encapsulation provided an ECM-like microenvironment, enhanced *in vitro* viability of ASCs, increased VEGF release, allowing the gradual release of ASCs to facilitate tissue angiogenesis *in vivo*.

Table 2-1 contd. ...

...Table 2-1 contd.

Study	Biomate-rials	Modified Steps	Modifications	Endpoint	Outcome
Lin Y.H. et al. (2017) [51]	Keratin/ chitosan-azide (KE/CHI-AZ) UV-crosslinked composite film	Cell culture	Seeding the 3rd passage ASCs onto the KE/CHI-AZ composite film and culturing them for 14 days.	*In vitro* osteogenic differentiation capability.	The keratin/chitosan scaffolds presented long-term stability in PBS, elasticity, and capability in repeated water absorption, significantly improved the cellular viability, promoted cell-cell interaction, and enhanced ASC osteogenic differentiation into mature osteoblasts.
Tsai C.-Y. et al. (2017) [52]	Nano-grooved gelatin films cross-linked by EDC/NHS	Cell culture	Seeding the 3rd passage ASCs onto the gelatin films for 14 days with neural induction.	*In vitro* neurogenic differentiation capability.	The gelatin films presented natural-organ-like softness and hydrophilic properties, benefited cell adhesion, enhanced cellular alignment with the grooves, and promoted neural differentiation.
Tsai C.-C. et al. (2020) [61]	cECM-encapsulated gelatin/mTG Hydrogel	Cell encapsulation	Suspending and mixing the 1st passage ASCs with the cECM-encapsulated gelatin/mTG hydrogel, and then gelating it and incubating them for 14 days.	New Zealand white rabbits, full-thickness knee-joint osteochondral defect, intra-operative transplantation of the cECM-encapsulated gelatin/mTG hydrogel.	The cECM-encapsulated gelatin/mTG hydrogel showed little toxicity to the cells, enhanced ASC proliferation and chondrogenic differentiation potential, controlled *in vivo* inflammation, and facilitated the hyaline cartilage regeneration.
Yang B. et al. (2020) [62]	Conductive PEDOT/alginate (PEDOT/Alg) porous scaffold	Cell encapsulation	Seeding the 3rd passage ASCs on PEDOT/Alg scaffold, then starting electrical pulse stimulation 24 hours later.	*In vitro* cardiomyogenic differentiation capability.	1. Cross-linked PEDOT existed in the alginate matrix as particles, overcame its fragileness, and introduced excellent conductivity. 2. The PEDOT/Alg porous scaffold provided BADSCs with suitable attachment and benefited the proliferation and cardiomyogenic differentiation, especially under electrical stimulation.

SVF: stromal vascular fraction; ASCs: adipose-derived stem cells; mTG: microbial transglutaminase; HGF: Hepatocyte growth factor; CTRP3: C1q/TNF-Related Protein 3; cECM: cartilage extracellular matrix; EDC: N-(3-dimethylaminopropyl)-N0-ethylcarbodiimide hydrochloride; NHS: N-hydroxysuccinimide; PEDOT: poly(3,4-ethylenedioxythiophene); BADSCs: brown adipose-derived stem cells

Cheng et al. successfully formed ASC spheroids by seeding the ASCs on chitosan films. The spheroids contained robust endogenous ECM and upregulated the pluripotency genes of the ASCs. The ASCs also showed enhanced differentiation capabilities after spheroid formation, including increased transdifferentiation efficiency into neuron and hepatocyte-like cells [57]. However, despite their favorable characteristics, these ASCs spheroids were very motile and spontaneously aggregated into each other on chitosan films. The varied spheroid size was unfavorable for standardization and clinical applications. Hence, Tsai et al. later successfully fabricated an agarose microwell plate for standardized cell spheroid formation with uniform sizes [58]. The standardized ASC spheroids also showed enhanced stemness compared to ASCs in the monolayer culture condition. They also developed a hydrogel by cross-linking gelatin with microbial transglutaminase (mTG) to encapsulate the cell spheroids. The encapsulated ASC spheroids showed good differentiation capabilities, especially adipogenesis and chondrogenesis, compared to the cell suspension group. Furthermore, the gelatin/mTG-hydrogel-encapsulated ASC spheroids enhanced healing in a rat burn-wound healing experiment, showing accelerated wound contraction with reduced scab development, faster tissue regeneration, thicker epidermises, and improved wound architecture [59]. Moreover, a native porcine cartilage-derived porous ECM scaffold could induce the chondrogenic differentiation of ASCs without exogenous growth factors [60]. Encapsulation of both ASCs and porcine cartilage ECM in gelatin/mTG hydrogel can be used for chondrogenic differentiation. Transplantation of the hydrogels to the knee joint defects of rabbits regenerated a grossly smooth articular surface with histologically hyaline-like tissue in eight weeks [61].

Physical properties of the engineered biomaterial microenvironment also affect stem cell behaviors. Tsai et al. successfully fabricated nano-grooved gelatin films and found that engineered anisotropic topographical cues could facilitate the ASC growth and specific differentiation [52]. ASCs successfully grew, proliferated, and differentiated along with the specific orientation of the nano-grooves on the adhesive gelatin films, and the oriented growth facilitated neural differentiation of ASCs. Moreover, Yang et al. employed poly(3,4-ethylenedioxythiophene) (PEDOT) for the conductivity needed for electrical stimulation for cardiomyogenic differentiation [62]. The chemically cross-linked PEDOT existed in the alginate matrix as particles, overcame its fragileness, and provided good electrical conductivity. The PEDOT/alginate porous scaffold provided brown fat-derived ASCs with suitable attachment sites and benefited their proliferation and cardiomyogenic differentiation, especially under electrical stimulation.

Base on above physiochemical properties of biomaterial effects on regulation of ASCs, ASCs are also fabricated as microtissue-like structures through the biomaterial induction. Yamato et al. synthesized poly(N-isopropylacrylamide), whose surface is relatively hydrophobic at 37°C but hydrophilic below 32°C, to facilitate cellular delivery and recovery without enzyme digestion [63]. As the cellular adhesiveness is temperature-responsive and controllable, cells spontaneously detach and can be non-invasively and non-destructively harvested at low temperature in a contiguous cell sheet form with intact cell-cell junctions and deposited ECM on this temperature-responsive polymer. With this novel method, scientists have produced cell sheets of different cell types and applied them in various tissue reconstructions and engineering, including ocular surfaces, periodontal ligaments, cardiac patches, post-myocardial infarction (MI) cellular therapy [64], 3D vascularized cardiac tissue engineering [65], bladder augmentation, and cartilage repair and regeneration [66]. Approaching from a different direction, Yu et al. successfully composited cell sheets simply by adding L-ascorbate 2-phosphate (A2-P), a stable form of ascorbic acid, to the culture medium on a common plastic culture dish [53]. The cell sheets contained incremental ECM deposition with a robust collagen type I, fibronectin, and laminin network. The cell sheets accelerated the ASC growth rate while simultaneously maintained their stemness. By supplementing the collagen synthesis inhibitors in ASC culture with A2-P, the impaired expression of stemness markers indicated the requirement of A2-P-induced collagen synthesis for ASC stemness enhancement. In experiments of a murine-impaired cutaneous wound model, cell sheets facilitated the ASC transplantation, improved the survival rate of the transplanted ASCs, enhanced angiogenesis, and accelerated wound closure. Moreover, ASC sheets reduced scar formation via releasing the hepatocyte growth factor (HGF) and C1q/TNF-related protein 3 (CTRP3) [67]. Consequently, ASC sheets improved the regenerated skin quality and increased wound healing. Although increased ASC survival was noted in the initial phase of the wound-healing process, no transplanted ASC lasted for more than 28 days, which minimized the concern of long-term side effects of stem cell transplantation. Although some evidence revealed no transplanted ASC lasted more than 28 days *in situ* [67], the long-term concerns of allogeneic cellular transplantation were still not clarified. Therefore, the secretome of ASCs and their therapeutic potential have been extensively investigated. In addition to secreting a wide range of growth factors and cytokines, stem cells also release an abundance of extracellular vehicles (EVs), which could be readily isolated from the culture media by centrifugation and concentrated with sucrose solution [68]. The accessibility of potential mass production of ASC-EVs makes them a promising candidate for cell-free regenerative therapy, despite the obstacles that need to be clarified or resolved. The complete understanding of the underlying mechanism, therapeutic effectiveness and reproducibility, cost-effective production, the optimal and individualized dosing and regimen for different applications and different diseases, and biosafety must be addressed for ASC-EV therapy [68, 69].

Obtaining sufficient EVs for therapeutic purposes is also an important issue, and a recent publication revealed that 3D culture of MSCs markedly increased the amount of membrane-bound vesicles on the cell surface, leading to enhanced EV excretion [70]. Moreover, biodegradable or highly porous hydrogels have been utilized to load EVs and to deliver a sustained therapeutic effect to various tissues. The hydrogels can also prevent the EVs from being cleared prematurely and allow the delivery of a more localized and concentrated EV dosage by placing the hydrogel directly at or in the proximity of the target site [71].

Additionally, on the surface of 2D biomaterial membranes of low adhesiveness, many cell types tend to aggregate simultaneously and form multicellular spheroids, and most of the studies showed promisingly improved therapeutic potentials [72]. The spheroid formation is known to be mediated by both cell-cell and cell-matrix interactions. The cell-cell connection is formed by homophilic cadherin binding, while cell-matrix interactions rely on the binding of integrin and ECM. ASC spheroids could be fabricated by seeding the ASCs on chitosan films, and they exhibited multiple benefits that can increase their therapeutic potentials [57]. The spheroids contained robust endogenous ECM, like fibronectin and laminin, and upregulated the pluripotency genes of the ASCs, Sox-2, Oct-4, and Nanog. In a serum-deprivation condition, ASC spheroids showed superior cell survival than in the monolayer culture. The ASCs also showed improved cellular survival and good intramuscular injectability and biocompatibility. The spheroid form would temporarily inhibit the proliferation capability of ASCs. Still, after being seeded back to TCPC, those ASCs readily generated colony-forming units (CFUs) and exhibited even higher proliferation capability compared to the monolayer ASCs constantly cultured on TCPS. After spheroid formation, the ASC transdifferentiation capability into neurogenic and hepatogenic lineages was also enhanced. Compared with ASCs constantly cultured in monolayers, expanding them after short-term spheroid formation can achieve superior regenerative capacity [73]. Spheroid-derived ASCs exhibited higher expansion efficiency and lower senescence. They also showed higher expression of stemness markers, angiogenic growth factors, and CXCR4 (regulating migration, engraftment, and proliferation of stem cells). CXCR4 upregulation was associated with enhanced matrix metalloproteinases (MMP-9 and MMP-13) expression and chemotaxis in spheroid-derived ASCs. In a cutaneous wound model of nude mice, after intralesional injection of spheroid-derived ASCs, they observed faster healing, enhanced angiogenesis, more ASC engraftment, and ASC differentiation toward endothelial and epidermal lineages. These findings suggested that short-term spheroid formation of ASCs before monolayer expansion can enhance their regenerative capacity by promoting stemness, angiogenesis, and chemotaxis, thus increasing the therapeutic potential for tissue regeneration.

The biomaterial effects on DSCs

Recently, another type of adult stem cell, dental stem cell, greatly increased attention to developing various therapeutic strategies in regeneration medicine. Dental stem cells are multipotent stem cells that are easily isolated from oral tissues and have immuno-modularity and high flexibility [74]. In addition, compared with other adult stem cells, the procedure of dental stem cell isolation is a less invasive technique, and the source of waste oral tissues has no ethical issues. Therefore, through the combination of appropriate scaffolds and desirable cytokines, growth factors, and other biomolecules, great numbers of dental stem cell-based culture systems could be developed to regulate the proliferation, differentiation, and other functions of dental stem cells. In general, dental stem cells derived from many different sites in oral tissues (Figure 2-1) [75], such as exfoliated deciduous teeth, apical papilla, tooth germ, periodontal ligaments, and gingival tissues, that contain dental follicle precursor cells, dental pulp stem cells, periodontal ligament stem cells, tooth germ stem/progenitor cells, and gingival-derived stem cells. Recently, combining natural or synthetic biomaterials and dental stem cells have been widely developed as pulp-like tissues for inducing mineralization and dentin formation. For example, a collagen hydrogel containing bFGF, VEGF, and TGF-β provided an active surface

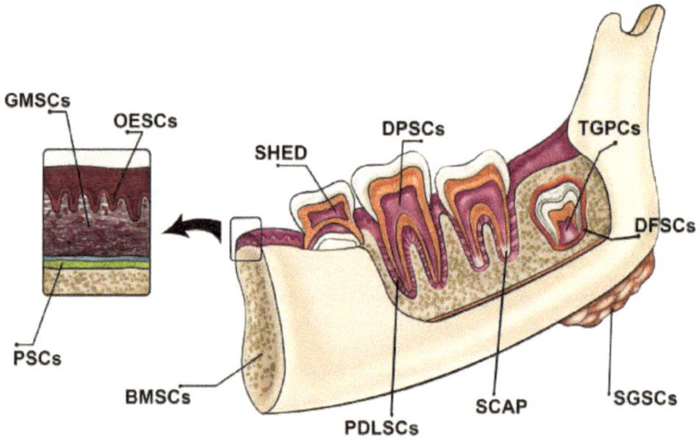

Figure 2-1. The source of dental stem cells in oral and maxillofacial tissues. (Reprinted with permission from [74] Copyright (2022) Springer Nature.)

to promote dental stem cells with endodontic differentiation [76]. In addition, dental stem cells have also demonstrated their potential applications for periodontal and alveolar bone regeneration. For example, dental stem cells seeded on a collagen sponge scaffold significantly enhance the periodontal tissue regeneration of patients with deep intraosseous defects caused by chronic periodontitis [77]. Furthermore, some structural and chemical properties of ceramic biomaterials are similar to the structure and composition of bone, so it has been confirmed that combining collagen, chitosan, and bioactive glass nanoparticle can induce osteogenesis and new bone formation of dental stem cells for repairing femoral bone defects [78]. Moreover, many studies have also shown that the surface properties and composition of biomaterials play a role in regulating dental stem cells' stemness [79, 80]. Therefore, Chen and co-workers used a hydrophilic poly(vinyl alcohol) to culture tooth germ cells and found that the ERK1/2 pathway of suspended tooth germ cell spheroids was inhibited and then promoted the osteogenesis of the suspension type of tooth germ cell spheroids [81, 82]. In addition to bone regeneration, due to the multipotency of dental stem cells, dental stem cells have also confirmed their applications in nerve regeneration by using different biomaterials [83]. An earlier study showed that a hybrid alginate/hyaluronic acid 3D scaffold with NGF could significantly induce human dental stem cells from periodontal gingival tissues to differentiate into neural tissues. Hsiao and co-workers also investigated the relationship between the printed fiber gap of 3D-printed PLA scaffolds and neural differentiation of human dental pulp stem cells [84]. The study found that human dental pulp stem cells showed a high expression of neuronal differentiation when the gap was around 150 mm to 200 mm. Furthermore, Carnevale also confirmed that using a human dental stem cell-laden collagen scaffold as an implant is useful to stimulate the axonal regeneration and reconnected nerve from both stumps in sciatic nerve defect mice [85]. Therefore, in the paragraph, the information showed that biomaterials provide various effects to regulate the behaviors of dental stem cells and offer a promising strategy to create versatile applications for dental stem cells.

The biomaterial effect on NSCs

Biomaterials also could affect the behaviors of cells via their chemical property, wettability (hydrophobicity/hydrophilicity), roughness, and rigidity [86]. Previous studies show that the change of substrates has a significant influence on the behaviors of neural stem cells; for example, tissue culture polystyrene (TCPS) is a non-adherent substrate for NSPCs to keep them in suspension and then form into neurospheres. NSPCs are considered anchorage-independent cells, which differentiate into different cell types when they attach to the substrate under appropriate conditions. For example, NSPCs will attach to poly-D-lysine or poly-L-ornithine-coated substrates and then differentiate into astrocytes, oligodendrocytes, and neurons [87, 88]. In addition, the factors responsible for NSPC differentiation into specific lineages include signals coming from various diffusible factors to either augment or inhibit the differentiation capabilities of NSPCs [89]. Traditionally, the serum would be added into the cell culture system to support the proliferation or survival of cultured cells. However, the serum is a complex mixture containing various components with different molecular weights. In terms of NSPCs, it has been shown that differentiation of NSPCs is influenced by numerous factors in the serum, including various growth factors, cytokines, neurotransmitters, proteins, and polyamines [90]. For example, Kaufman and Barrett have demonstrated that serum fractions with a molecular weight greater than 100 kDa were toxic to cultured neurons [91]. Therefore, an earlier study investigates the effects of combining biomaterials and serum components on the differentiation of NSPCs [92]. In the study, Li and co-workers used a centrifugal method to isolate the serum fractions with molecular weight less than 100 kDa (100 KD) and 50 kDa (50 KD), and between 50 kDa and 100 kDa (50 K–100 K). Afterward, NSPCs were cultured on four polymeric membranes with different wettability, polyvinylidene fluoride (PVDF), poly (ethylene-co-vinyl alcohol) (EVAL), and polyvinyl alcohol (PVA) in the medium containing the three isolated serum fractions. Compared with the PVDF and PVA polymers, EVAL polymer has a hydrophobic ethylene segment and hydrophilic vinyl alcohol segment that provides an adequate wettability between PVDF and PVA polymer with a better protein adsorption behavior [93]. Therefore, the results indicated that the combination of EVAL biomaterial and 100 KD medium provided a synergistic effect to dominate the neuronal differentiation of NSPCs. In addition, the abovementioned shows that NSPCs could undergo differentiation once NSPCs started to attach to the culture substrates. One study also provides a concept to prepare an NSPC-monolayer system that combines growth factor, ECM protein, and biomaterials [94]. Compared with the conventional medium treatment, the growth factor and ECM protein were pre-adsorbed on the 2D or 3D biomaterial surface, and then NSPCs were seeded on the biomaterials. Afterward, the NSPCs enables to attach to the biomaterials, show a high proliferation rate, form a stem cell monolayer, and have no differentiation phenomenon after three days of incubation. Interestingly, the stem cell monolayer could be further induced to differentiate into the astrocyte-rich or neuron-rich monolayer on the 2D or 3D biomaterial surfaces by treating the serum-contained medium or 100 kD fraction-contained medium, respectively. The study contributes an easy method to coat NSPCs for forming the NSPC-rich monolayer on the different biomaterial surfaces. In addition to the adsorption of growth factors or ECM protein on the substrate surfaces, grafting the biomolecules on biomaterial surfaces to form a bioactive culture substrate is a common strategy. For example, Li and co-workers immobilized a laminin-derived peptide, Gly-Tyr-Ile-Gly-Ser-Arg (GYIGSR), on the EVAL membrane surface [95]. YIGSR is a cell-binding domain in laminin protein, which can promote cell migration. Therefore, compared with only an EVAL membrane or laminin-coated surface, the GYIGSR immobilized EVAL membrane significantly promotes the migration and neuronal differentiation of NSPCs. Moreover, the GYIGSR-immobilized EVAL membrane is a permanent modification, offering a long-term culture surface to support differentiated neuron survival. Compared with the laminin-coated surface through physical adsorption, the neurons still attached to the modified EVAL membrane and expressed neuronal functional synaptic activity after 21 days of incubation. In addition to the immobilization of peptides, Ren and co-workers investigate the chemical functional group effect on the behavior of neural stem cells [95]. They used glass coverslips as a basic substrate and then grafted -OH, $-SO_3H$, -COOH, $-NH_2$, -SH, and $-CH_3$ functional groups on the glass coverslips. Regarding the effect on cell adhesion, the results show that the $-SO_3H$, and -COOH modified glass coverslips have similar wettability but have no significant influence on the cell morphology. Compared with the -SO3H groups, the -OH group possessed more hydrophilic properties, so the NSPCs showed a smaller contact area. Subsequently, the morphologies of NSPCs cultured on -NH2, -SH, and -CH3 are rounded shape cells, and only cells on -NH2 express a neurite-extending structure. Regarding the migration behavior, the -OH, -SO3H, -COOH functional groups provide NSPCs with a low migration ability, but in these three groups, NSPCs on the -SO3H-modified surface has a slight migration behavior compared to the other two groups. The -NH2 functional group endows NSPCs with the best migration behaviors compared to all functional groups. Furthermore, regarding the differentiation behavior of NSPCs, the -OH functional group modified surface offers a high hydrophilic affinity, which can prevent protein being adsorbed on the surface, leading to an inhibition effect on NSPCs. The -SO3H-modified surface can promote NSPCs differentiating into oligodendrocytes; the results also show that the -NH2-modified surface with a positive charge benefits the protein adsorption and then promotes neuronal differentiation of NSPCs. Therefore, the results provided information for designing the surfaces to culture NSPCs for different purposes. Next, exogenous electrical stimulation is a feasible, physical, and flexible technique that has been confirmed to enable significant enhancement of the proliferation, migration, and differentiation of neurons and NSPCs. Currently, electroactive biomaterials, such as conductive materials, have been demonstrated to deliver electrical signals to trigger the intrinsic electrical characteristics, electro-mechanical, and elector-chemical action in a neuron or NSPCs [96–98]. Recently, carbon nanotube has been proven to boost electrical signaling for neurons [99, 100] and even protect neurons from injury in ischemia stroke [101]; therefore, based on the excellent electrical properties of graphene [102], it might be suitable for neural tissues. Vis-à-vis its application in neuronal regeneration, Park et al. were the first to explore the fundamental effects of graphene-based material on the behaviors of hNSCs [103]. They cultured hNSCs on a 2D graphene film, the neuronal differentiation of hNSCs were enhanced and differentiated neurons showed excellent neural activity on the graphene electrode. Additionally, for nerve injury, although autologous nerve graft by suture is a "gold standard" for the small damaged nerve defect/gap in clinical therapy [104], a surgical tension effect on the nerve cable (inhibiting nerve regeneration) [105] and a slow axon regeneration rate (about 2–5 mm/day) [106] will cause obstacles for the regeneration of a long nerve defect/gap. It reveals that an appropriate supporting bridge is necessary to correctly guide axon growth to reconnect both damaged nerve ends.

The biomaterial effects on DP and HFSCs

Physiologically, the DP in the hair bulb grows as a three-dimensional dense aggregate [107–109]. It was demonstrated that culturing DP cells into a three-dimensional spheroidal microtissue that is similar to their natural intracellular organization is beneficial to preserving their hair-inductive capability [109]. Three-dimensional culture has been successfully developed by forcing cells to aggregate in a hanging drop method [110]. Higgins et al. created three-dimensional human DP spheres morphologically akin to intact DP *in vivo* by the hanging drop method [111]. The expression profiles of these hanging drop-generated three-dimensional spheres have many similarities to physiological intact DP [111]. For example, decreased expression of a smooth muscle actin with preservation of versican was observed in these hanging drop-generated DP spheres. However, due to the limited culture medium used in each hanging drop, this might not be an optimal method to generate DP aggregates.

With the advent of tissue engineering, various bioreactors have been tested and applied to the generation of three-dimensional microtissues of various cell types, such cardiomyocytes, cartilage cells, hepatocytes, etc. [112–115]. A bioreactor that is able to produce DP spheroidal microtissues on a large scale will be highly advantageous for future clinical applications. DP cells are usually expanded on an adhesive substratum and grow as two-dimensional monolayered cells. We used poly(ethylene-co-vinyl alcohol) (EVAL) as the culture substratum for DP cells. Compared with conventional culture substratum, EVAL is less adhesive to DP cells and also promotes DP cell migration. Due to the relatively high intercellular adhesiveness of DP cells, DP cells actively move, collide, and self-aggregate into spheroidal microtissues on the EVAL surface [116]. Further coating of the extracellular matrix protein fibronectin on EVAL increases the production of DP microtissues by increasing cell adhesion to the culture surface without reducing cell migration rate [117]. The DP microtissues formed on EVAL better preserve their molecular markers and are able to induce HF neogenesis when subcutaneously injected into the skin [116]. High-throughput production of DP microtissues can be achieved because hundreds of DP spheroid microtissues form spontaneously on the EVAL within several days after cell seeding. It has been shown that adipose tissue surrounding HFs plays a critical role in regulating hair growth [118, 119]. It was shown that when DP cells are co-cultured with adipose-derived stem cells on chitosan; these two types of cells spontaneously form spheroidal heterotypic microtissues. The formed heterotypic microtissues show a core-shell structure with an outer shell composed of adipose-derived stem cells and an inner core composed of DP cells. Compared with homotypic DP cell microtissues, the heterotypic microtissues show enhanced HF inductivity. The result indicates adipose-derived stem cells are capable of enhancing the HF-inductive capability of DP microtissues [120].

Since hair shows regional differences, such as hair thickness, spacing, alignment, and length in different body regions, these properties of regenerated new HFs need to be precisely controlled to mimic the natural hair patterns. Because up to thousands of new HFs are needed for a single patient, a well-controlled platform for generating a desirable pattern of regenerated HFs is needed for a favorable clinical outcome. In mouse pelage, the number of DP cells correlates with the hair size [121, 122]. In humans, the thickness of a hair shaft is correlated with the volume of DP, and a large DP contains more DP cells that support the growth of a larger HF with a thicker hair shaft [123]. By counting the cell numbers of DP in rat whiskers with different diameters, we also found a positive correlation of DP cell numbers with the diameter of whiskers [124].

In bioengineering, we speculated that by controlling the size of the transplanted DP microtissues, we might be able to control the thickness of the neogenetic hair. For this purpose, we need to develop a scalable method to produce DP microtissues of precisely controlled size. We developed a hydrophilic polyvinyl alcohol (PVA)-coated culture platform that can be automated for mass production of human and rat DP microtissues with wide-range controllable sizes and cell numbers. DP cells are unable to attach to PVA-coated surfaces. When DP cells are seeded in PVA-coated tubes, all the cells quickly aggregate into a single floating spheroid with progressive compaction during culture. The DP cells can be seeded into arrays of PVA-coated tubes for scalable production of DP microtissues of a controllable size. We used DP microtissues of various sizes to induce new HFs and found an interesting result. We found that although DP microtissues of a larger size have higher HF inductivity as compared with DP microtissues of a smaller size, the diameter of the regenerated hair shafts shows no significant difference between the two groups [124]. Therefore, size regulation of neogenetic HFs is more complicated than we have expected. From an engineering point of view, there are several possibilities that might account for this result. First: cultured DP cells preserve the "size memory" of the original HFs, and therefore the regenerated HFs always remodel back to the original sizes. Second, the size of regenerated HF is stochastic and thereby doesn't correlate with either the size of transplanted DP microtissue or the size of the original HFs where DP cells are derived. Third, the injected DP microtissues may reshape or resize *in vivo* during HF neogenesis. The regulation of morphogen signaling pathways, such as BMP, noggin, Wnt/b-catenin, etc., may also be involved in HF neogenesis *in vivo*. Understanding how these morphogens regulate the size of HFs during natural morphogenesis will help us to control the size of the regenerated HFs in the future.

Stem cell biology and developmental biology are two cornerstones of regenerative medicine. By combing the knowledge from stem cell biology and developmental biology, constructing a functional organoid for tissue regeneration has shown new promise in tissue engineering [125–128]. In embryonic development,

HF morphogenesis is driven by the reciprocal interaction of the epidermal keratinocytes and the dermal fibroblasts. To induce HF neogenesis by DP cells, DP microtissues should be transplanted delicately right below the epidermis to facilitate the interaction between DP and the epidermis. The other way for HF neogenesis is to generate HF organoids *in vitro* first and then transplant these organoid primordia to the skin for further maturation into functional HFs. To produce HF organoids *in vitro*, the epidermal keratinocytes are isolated and mixed to form structured microtissues [124, 129]. Indeed, a proof of concept in the generation of folliculoid microtissues or organoids *in vitro* has been demonstrated by two-step culture methods. Ihara et al. generated the organoid structures of HFs *in vitro* by co-culturing rat-dissociated epidermal cells and dermal cells into a two-step rotation and floatation method [130, 131]. Qiao et al. also demonstrated that dissociated embryonic dermal cells and hair peg keratinocytes could grow into HF-like "protohairs" in a two-step culture method [132].

Recently, several innovative biomaterial-facilitated approaches have been developed to generate HF organoids that can grow into functional HFs after transplantation [127, 133–137]. Havlickova et al. successfully generated human folliculoid microspheres by culturing dissociated adult outer root sheath cells and DP cells in extracellular matrix gel [137]. Researchers in Japan also successfully developed HF germ by co-culturing dissociated mouse embryonic epidermal cells with mouse embryonic mesenchymal cells or human DP cells in collagen microgel [127, 133, 134].

To control the space and alignment of HF germ, we developed programmable CO_2 laser-assisted surface microfabrication on PVA-coated glass chips. This method can deploy heterotypic cells of DP cells, keratinocytes, and fibroblasts into desired patterns for the controllable formation of folliculoid microtissues [138]. Abaci and colleagues reconstructed the physiological three-dimensional folliculoid human microtissues using 3D-printed molds. Importantly, enhanced vascularization in the microenvironment of bioengineered folliculoid microtissues can promote graft survival after transplantation [136]. These works provide new concepts to develop novel strategies for reconstituting an integrated microenvironment for bioengineered HFs.

Though the formation of individual epithelial organ germs *in vitro* has been achieved using dissociated embryonic cells [128, 130–132, 139, 140], the direct formation of individual HF organ germs through self-assembly by use of adult keratinocytes from hairless skin has not been demonstrated. It is believed that adult keratinocytes are intrinsically restricted to self-organize with mesenchymal cells into HF germs *in vitro*. On the contrary, the self-assembly of these two cell populations is also affected by *in vitro* culture conditions, such as the culture substratum, growth medium, and seeded cell numbers. Since thousands of new HFs are needed for the transplantation to the bald scalp of patients with alopecia, the generation of sufficient HF germs for clinical need remains a major challenge. To build a biomaterial-facilitated HF germ *in vitro*, heterotypic cells must be assembled. Several studies have been done in guiding heterotypic cells into repeated patterns of flat or monolayered cell domains on biomaterials through surface microfabrication [141–143]. For example, by selective patterning of adhesive extracellular matrix protein and sequential seeding of heterotypic cells, hepatocytes co-cultured with fibroblasts in repeated adjacent regions showed to enhance the function of hepatocytes [115]. Fundamental factors governing the patterning of heterotypic cells into compact multicellular organ primordia on substratum surface is not well understood yet.

We previously reported that EVAL, containing both hydrophilic and hydrophobic domains, is a unique polymer that can promote the self-assembly of DP cells into spheroidal microtissues that are able to induce HF neogenesis [144]. Furthermore, we also intended to introduce two different cell types containing not only DP cells but also keratinocytes to generate the HF germ. The introduction of heterotypic cell types can complicate the system [145]. Considering the simultaneous seeding of two cell types, there are many possibilities in the epithelial-mesenchymal interaction that the final morphologies or intercellular organizations can be patterned (Figure 2-2).

In the microtissues produced by the hanging drop method, DP cells are able to aggregate by eliminating the contact effect between substratum and cells; however, keratinocytes are unable to aggregate and show a random distribution pattern within DP aggregates (Figure 2-3). Compared to a random mixture of epithelial and mesenchymal cells, the pre-patterned compartmented distribution of epithelial cells and mesenchymal cells within an organ germ has been shown to promote epithelial-mesenchymal interaction [128, 139]. Therefore, we attempted to develop a one-step seeding approach on EVAL to massive production of HF germ for effective HF neogenesis. When simultaneously seeding DP cells and keratinocytes on the EVAL surface, DP cells can bring an effect to mobilize keratinocytes to aggregate. The DP cells and keratinocytes then progressively form the microtissues with a layered core-shell structure. This approach has unveiled an important feature of EVAL-facilitated microtissues that cells have non-randomly compartmented distribution in the microtissue: DP cells with higher adhesivity are preferentially located in the center, and keratinocytes with lower adhesivity are deployed to the periphery. The spontaneously formed layered structure observed on EVAL is similar to the natural three-dimensional organization of the hair bulb in HFs: a shell of keratinocytes surrounding the core of aggregated DP cells (Figure 2-3). Importantly, the respectively specific gene expression profiles in DP and keratinocytes are perfectly preserved in the DP-keratinocyte heterotypic microtissues, supporting the fact of hair neogenesis by microtissue transplantation (Figures 2-3C and 2-3D).

It has been shown that the close intercellular contact between DP cells is associated with preserving the HF induction ability [109, 146, 147]. Additionally, the close interaction between keratinocytes and DP is indispensable for maintaining the differentiation of keratinocytes to support hair growth. Hence, the formation of the layered hybrid DP-keratinocyte microtissues by our one-step approach is beneficial to facilitate the epithelial-mesenchymal interaction as well as HF neogenesis. Moreover, we also demonstrate that the differentiation fate of keratinocytes can be changed by close interaction with DP cells in the microtissues. It has been shown that DP cells are important in maintaining keratinocytes in a follicular fate [137]. Our results demonstrate that through the heterotypic cell interaction in the microtissues, the keratinocytes derived from adult hairless skin with an epidermal differentiation state can be reshaped towards follicular differentiation. This suggests that not only the intercellular organization is affected by other types of cells, but

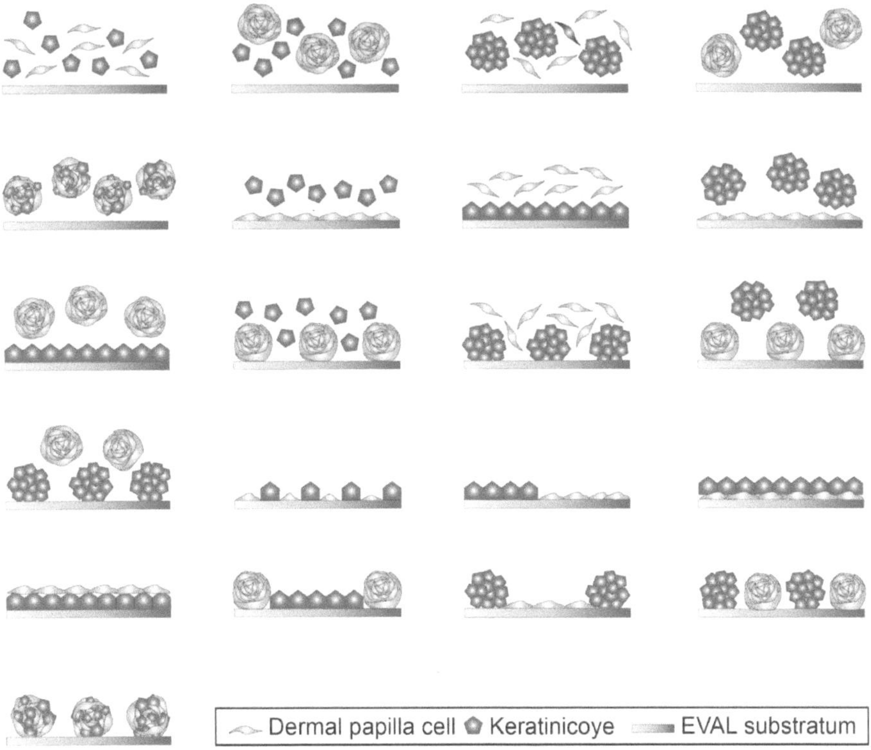

Figure 2-2. Possible patterns of intercellular organization after heterotypic cell seeding on a biomaterial substratum. (Reprinted with permission from [144] Copyright (2010) Springer Nature.)

heterotypic intercellular interaction also dynamically changes cellular function and differentiation. Compared with other systems that apply the extracellular matrix to pattern cells into folliculoid microtissues or HF germs [128, 137], our approach provides a simple and economize strategy for mass production of folliculoid microtissues and other epithelial organ germs (Figure 2-4).

Preparing functional DP aggregates with hair inductivity has been recognized as a gold standard before producing HF germ for such application. However, DP cells isolated from the bald region of hair loss patients exhibit signs of senescent characteristics, such as loss of the hair inductive potential, failure to be expanded *in vitro*, and compromised epithelial-mesenchymal interaction [148, 149]. These issues raise the difficulty and limitation of using auto-derived DP cells from hair loss patients to produce HF organoids for therapy. Therefore, alternative sources of DP cells to support the need for hair loss patients are unmet medical needs.

The technology of induced pluripotent stem cells has brought a new era of regeneration medicine. It has been shown that epidermal precursor cells (EPCs) generated from human-induced pluripotent stem cells (hiPSCs) can cross-talk with DP cells in the co-cultured system. In the hair reconstitution assay, co-injection of these hiPSCs-derived EPCs with DP cells can induce HF neogenesis [150]. In addition, the induced DP-substituting cells (iDPSCs) with trichogenic potential also

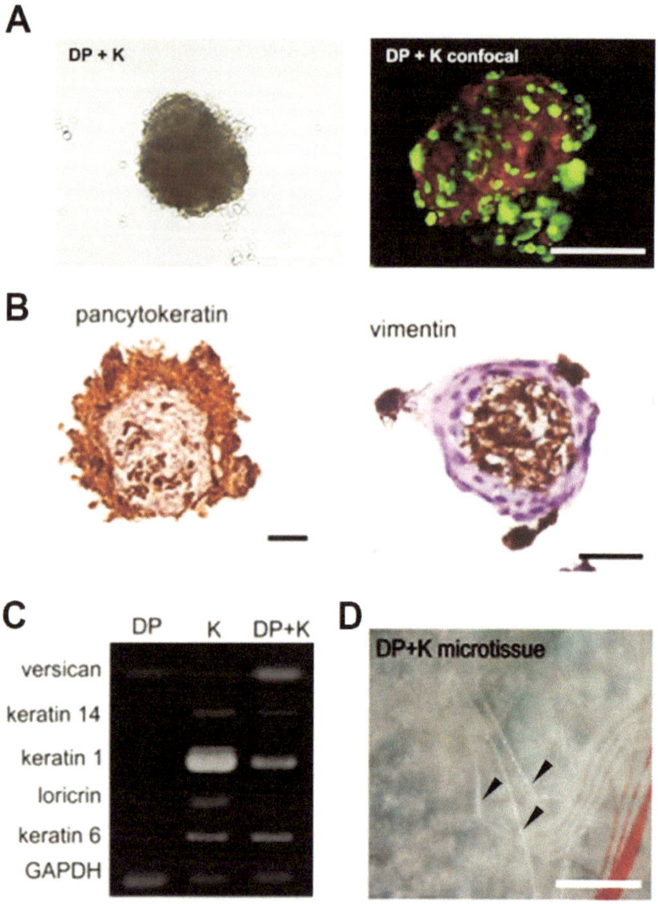

Figure 2-3. Structures, gene expression, and function of microtissues. (A) Cell morphology of microtissue after three days in hanging drop. The confocal image was used for clarifying the distribution of DP cells and keratinocytes within microtissue. red: DP cells, green: keratinocytes. (B) Immunohistochemistry staining of pan-cytokeratin and vimentin. Keratinocytes positive for pan-cytokeratin are sorted to the surface while DP cells positive for vimentin are aggregated in the center. (C) Gene expression of DP cells, keratinocytes, and microtissues. (D) HF neogenesis was observed by the injection of DP-keratinocyte microtissues through patch assay. (Reprinted with permission from [144] Copyright (2010) Springer Nature.)

can be generated from hiPSCs through a two-step differentiation protocol [151]. The HF germs formed by iDPC aggregates and keratinocytes were morphologically analogous to HF germs generated by normal DP aggregates and keratinocytes. The gene expression profiles of DP markers in iDPC aggregates are also comparable to those in normal DP aggregates [152]. These results suggest that hiPSCs have held promise as a substituting cell source for HF organoid-based therapy of hair disorders. These results have also shed light on developing strategies for generating patient-specific iPSCs and iDPSCs that will help to solve the limitation of balding DP cells of hair loss patients.

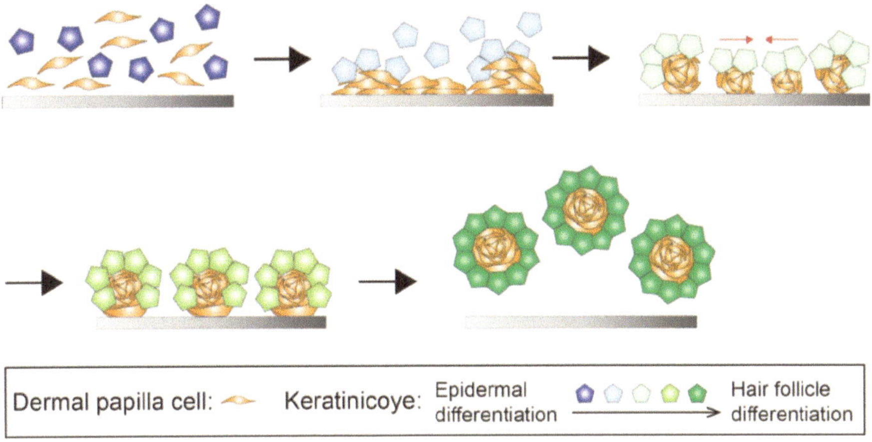

Figure 2-4. Schematic diagram of DP-keratinocyte hybrid spheroidal microtissue formation on EVAL-coated substratum. (Reprinted with permission from [144] Copyright (2010) Elsevier.)

The biomaterial effects on other stem cells

In this chapter, many multipotent adult stem cells have been shown that they can be isolated from various tissues. In this paragraph, we will briefly show some rare cell populations, which are adult progenitor cells with multipotent ability to repair damaged tissues. Different from other somatic stem cells, these types of progenitor cells enable proliferation in the absence of obvious senescence and have differentiation ability. For example, endothelial progenitor cells, also called angioblasts, possess the ability to proliferate, migrate, and differentiate into endothelial lineage cells [153]. In general, endothelial progenitor cells are considered a powerful tool for neovasculogenesis [154] because endothelial progenitor cells exist in a natural reservoir, bone marrow, that allows endothelial progenitor cells to mobilize to damaged vascular tissues [153]. Therefore, endothelial progenitor cells bring the evidence and novel strategy as a source to transplant into the body to supply and replace the injured vascular tissues. Currently, an earlier study utilized decellularized iliac vessels as a scaffold to support the growth of endothelial progenitor cells. Afterwards, compared with direct injection of cells, the cell-laden scaffold could maintain the structure in a sheep model for more than 100 days, and the implanted scaffold can respond to the mediation of nitric oxide and then have a vascular relaxation and express natural artery-like contractility [155]. Furthermore, Royer and co-workers developed biomaterial surfaces with different shapes and sizes of patterns to investigate the effects of surface properties on the orientation, proliferation, replacement, and others of endothelial progenitor cells [156]. They paved photoresistance on a poly(ethylene terephthalate) (PET) substrate and fabricated the micropatterns on the PET substrates by exposing UV light. Afterwards, the active cell molecules, sitagliptin, and GRGDS were immobilized on the micropatterns, and endothelial progenitor cells were cultured on the hybrid GRGDS/sitagliptin-grafted micropatterns, offering a synergistic effect to induce the endothelial progenitor cells presenting a high endothelialization behavior in comparison to PET substrates grafted only on the GRGD peptide. This study provides a strategy to combine the texture of biomaterials and endothelial progenitor cells to fabricate biomimetic biomaterials with the natural vessel-like viscoelastic property for enhancing the survival, growth, and differentiation of endothelial stem cells.

References

[1] Ayala, R. et al. Engineering the cell–material interface for controlling stem cell adhesion, migration, and differentiation. Biomaterials, 2011. 32(15): 3700–3711.
[2] Engler, A.J. et al. Matrix elasticity directs stem cell lineage specification. Cell, 2006. 126(4): 677–89.
[3] Pelham, R.J. and Y.-l. Wang. Cell locomotion and focal adhesions are regulated by substrate flexibility. Proceedings of the National Academy of Sciences, 1997. 94(25): 13661.
[4] Chang, H.H. et al. pH-responsive characteristics of chitosan-based blends for controlling the adhesivity of cells. Journal of the Taiwan Institute of Chemical Engineers, 2020. 111: 34–43.
[5] Chai, C. and K.W. Leong. Biomaterials approach to expand and direct differentiation of stem cells. Mol Ther, 2007. 15(3): 467–80.
[6] Ludwig, T.E. et al. Derivation of human embryonic stem cells in defined conditions. Nat Biotechnol, 2006. 24(2): 185–7.
[7] Yao, S. et al. Long-term self-renewal and directed differentiation of human embryonic stem cells in chemically defined conditions. Proc Natl Acad Sci U S A, 2006. 103(18): 6907–12.
[8] Harrison, J. et al. Colonization and maintenance of murine embryonic stem cells on poly(alpha-hydroxy esters). Biomaterials, 2004. 25(20): 4963–70.
[9] Ahadian, S. et al. Rapid and high-throughput formation of 3D embryoid bodies in hydrogels using the dielectrophoresis technique. Lab Chip, 2014. 14(19): 3690–4.
[10] Irawan, V., A. Higuchi, and T. Ikoma. Physical cues of biomaterials guide stem cell fate of differentiation: The effect of elasticity of cell culture biomaterials. Open Physics, 2018. 16(1): 943–955.
[11] Wobus, A.M., and K.R. Boheler. Embryonic stem cells: prospects for developmental biology and cell therapy. Physiol Rev, 2005. 85(2): 635–78.
[12] Wang, Y.C., J.H. Bahng, and N.A. Kotov. Three-dimensional biomimetic scaffolds for hepatic differentiation of size-controlled embryoid bodies. Journal of Materials Research, 2019. 34(8): 1371–1380.
[13] Bertucci, T.B., and G. Dai. Biomaterial engineering for controlling pluripotent stem cell fate. Stem Cells Int, 2018. p. 9068203.
[14] Tong, Z. et al. Application of biomaterials to advance induced pluripotent stem cell research and therapy. EMBO J, 2015. 34(8): 987–1008.
[15] Lam, M.T., and M.T. Longaker. Comparison of several attachment methods for human iPS, embryonic and adipose-derived stem cells for tissue engineering. J Tissue Eng Regen Med, 2012. 6 Suppl 3: s80–6.
[16] Lyra-Leite, D.M. et al. An updated protocol for the cost-effective and weekend-free culture of human induced pluripotent stem cells. STAR Protoc, 2021. 2(1): 100213.
[17] Aisenbrey, E.A., and W.L. Murphy. Synthetic alternatives to Matrigel. Nat Rev Mater, 2020. 5(7): 539–551.
[18] Zhu, J. Bioactive modification of poly(ethylene glycol) hydrogels for tissue engineering. Biomaterials, 2010. 31(17): 4639–56.
[19] Brafman, D.A. et al. Long-term human pluripotent stem cell self-renewal on synthetic polymer surfaces. Biomaterials, 2010. 31(34): 9135–44.
[20] Mondal, G., S. Barui, and A. Chaudhuri. The relationship between the cyclic-RGDfK ligand and alphavbeta3 integrin receptor. Biomaterials, 2013. 34(26): 6249–60.
[21] Nguyen, E.H. et al. Versatile synthetic alternatives to Matrigel for vascular toxicity screening and stem cell expansion. Nat Biomed Eng, 2017. 1.

[22] Caiazzo, M. et al. Defined three-dimensional microenvironments boost induction of pluripotency. Nat Mater, 2016. 15(3): 344–52.
[23] Gurusamy, N. et al. Adult stem cells for regenerative therapy. Prog Mol Biol Transl Sci, 2018. 160: 1–22.
[24] Pittenger, M.F. et al. Mesenchymal stem cell perspective: cell biology to clinical progress. NPJ Regen Med, 2019. 4: 22.
[25] Suva, D. et al. Non-hematopoietic human bone marrow contains long-lasting, pluripotential mesenchymal stem cells. J Cell Physiol, 2004. 198(1): 110–8.
[26] Pittenger, M.F. et al. Multilineage potential of adult human mesenchymal stem cells. Science, 1999. 284(5411): 143–7.
[27] Bertolo, A. et al. Growth factors cross-linked to collagen microcarriers promote expansion and chondrogenic differentiation of human mesenchymal stem cells. Tissue Eng Part A, 2015. 21(19-20): 2618–28.
[28] Curran, J.M., R. Chen, and J.A. Hunt. The guidance of human mesenchymal stem cell differentiation in vitro by controlled modifications to the cell substrate. Biomaterials, 2006. 27(27): 4783–93.
[29] Estes, B.T., J.M. Gimble, and F. Guilak, Mechanical signals as regulators of stem cell fate. Curr Top Dev Biol, 2004. 60: 91–126.
[30] McBeath, R. et al. Cell shape, cytoskeletal tension, and RhoA regulate stem cell lineage commitment. Dev Cell, 2004. 6(4): 483–95.
[31] Wong, T.Y. et al. Hyaluronan keeps mesenchymal stem cells quiescent and maintains the differentiation potential over time. Aging Cell, 2017. 16(3): 451–460.
[32] Begum, R. et al. Chondroinduction of mesenchymal stem cells on cellulose-silk composite nanofibrous substrates: The role of substrate elasticity. Front Bioeng Biotechnol, 2020. 8: 197.
[33] Voga, M. et al. Silk fibroin induces chondrogenic differentiation of canine adipose-derived multipotent mesenchymal stromal cells/mesenchymal stem cells. J Tissue Eng, 2019. 10: 2041731419835056.
[34] Zhou, L. et al. Combining PLGA scaffold and MSCs for brain tissue engineering: A potential tool for treatment of brain injury. Stem Cells Int, 2018. p. 5024175.
[35] Vashi, A.V. et al. Adipose differentiation of bone marrow-derived mesenchymal stem cells using Pluronic F-127 hydrogel *in vitro*. Biomaterials, 2008. 29(5): 573–9.
[36] Elashry, M.I. et al. Combined macromolecule biomaterials together with fluid shear stress promote the osteogenic differentiation capacity of equine adipose-derived mesenchymal stem cells. Stem Cell Res Ther, 2021. 12(1): 116.
[37] Ramses Ayala, Chao Zhang, Darren Yang, Yongsung Hwang, Aereas Aung, Sumeet S. Shroff, Fernando T. Arce, Ratnesh Lal, Gaurav Arya and Shyni Varghese. Engineering the cell–material interface for controlling stem cell adhesion, migration, and differentiation. Biomaterials, 2011. 32(15): 3700–3711.
[38] Chengjuan Qu, Salla Kaitainen, Heikki Kröger, Reijo Lappalainen and Mikko J. Lammi. Behavior of human bone marrow-derived mesenchymal stem cells on various titanium-based coatings. Materials (Basel), 2016. 9(10): 827.
[39] Gao, L., R. McBeath, and C.S. Chen. Stem cell shape regulates a chondrogenic versus myogenic fate through Rac1 and N-cadherin. Stem Cells, 2010. 28(3): 564–72.
[40] Wang, X. et al. Regulating the stemness of mesenchymal stem cells by tuning micropattern features. Journal of Materials Chemistry B, 2016. 4(1): 37–45.
[41] Guillaume-Gentil, O. et al. pH-controlled recovery of placenta-derived mesenchymal stem cell sheets. Biomaterials, 2011. 32(19): 4376–84.
[42] Huang, C.C. et al. Injectable cell constructs fabricated via culture on a thermoresponsive methylcellulose hydrogel system for the treatment of ischemic diseases. Adv Healthc Mater, 2014. 3(8): 1133–48.
[43] Chen, L.C., H.W. Wang, and C.C. Huang. Modulation of inherent niches in 3D multicellular MSC spheroids reconfigures metabolism and enhances therapeutic potential. Cells, 2021. 10(10).
[44] Mazini, L. et al. Hopes and limits of adipose-derived stem cells (ADSCs) and mesenchymal stem cells (MSCs) in wound healing. International Journal of Molecular Sciences, 2020. 21(4): 1306.
[45] Baer, P.C. et al. Human adipose-derived mesenchymal stem cells in vitro: evaluation of an optimal expansion medium preserving stemness. Cytotherapy, 2010. 12(1): 96–106.

[46] Park, E., and A.N. Patel. Changes in the expression pattern of mesenchymal and pluripotent markers in human adipose-derived stem cells. Cell Biol Int, 2010. 34(10): 979–84.
[47] Pochampally, R.R. et al. Serum deprivation of human marrow stromal cells (hMSCs) selects for a subpopulation of early progenitor cells with enhanced expression of OCT-4 and other embryonic genes. Blood, 2004. 103(5): 1647–52.
[48] Lin, H.R. et al. Purification and differentiation of human adipose-derived stem cells by membrane filtration and membrane migration methods. Sci Rep, 2017. 7: 40069.
[49] Pan, J. et al. Culture and differentiation of purified human adipose-derived stem cells by membrane filtration via nylon mesh filters. J Mater Chem B, 2020. 8(24): 5204–5214.
[50] Cheng, N.C. et al. Sustained release of adipose-derived stem cells by thermosensitive chitosan/gelatin hydrogel for therapeutic angiogenesis. Acta Biomater, 2017. 51: 258–267.
[51] Lin, Y.H. et al. Keratin/chitosan UV-crosslinked composites promote the osteogenic differentiation of human adipose derived stem cells. J Mater Chem B, 2017. 5(24): 4614–4622.
[52] Tsai, C.-Y. et al. Effects of nano-grooved gelatin films on neural induction of human adipose-derived stem cells. RSC Advances, 2017. 7(84): 53537–53544.
[53] Yu, J. et al. Stemness and transdifferentiation of adipose-derived stem cells using L-ascorbic acid 2-phosphate-induced cell sheet formation. Biomaterials, 2014. 35(11): 3516–26.
[54] Lin, H. et al. Influence of decellularized matrix derived from human mesenchymal stem cells on their proliferation, migration and multi-lineage differentiation potential. Biomaterials, 2012. 33(18): 4480–9.
[55] Riis, S. et al. Fabrication and characterization of extracellular matrix scaffolds obtained from adipose-derived stem cells. Methods, 2020. 171: 68–76.
[56] Tang, K.C. et al. Human adipose-derived stem cell secreted extracellular matrix incorporated into electrospun Poly(Lactic-co-Glycolic Acid) nanofibrous dressing for enhancing wound healing. Polymers (Basel), 2019. 11(10).
[57] Cheng, N.C., S. Wang, and T.H. Young. The influence of spheroid formation of human adipose-derived stem cells on chitosan films on stemness and differentiation capabilities. Biomaterials, 2012. 33(6): 1748–58.
[58] Tsai, C.C. et al. Enhancement of human adipose-derived stem cell spheroid differentiation in an in situ enzyme-crosslinked gelatin hydrogel. J Mater Chem B, 2019. 7(7): 1064–1075.
[59] Lu, T.-Y. et al. Enzyme-crosslinked gelatin hydrogel with adipose-derived stem cell spheroid facilitating wound repair in the murine burn model. Polymers, 2020. 12: 2997.
[60] Cheng, N.-C. et al. Chondrogenic differentiation of adipose-derived adult stem cells by a porous scaffold derived from native articular cartilage extracellular matrix. Tissue engineering. Part A, 2008. 15: 231–41.
[61] Tsai, C.-C. et al. Enzyme-crosslinked gelatin hydrogel enriched with articular cartilage extracellular matrix and human adipose-derived stem cells for hyaline cartilage regeneration of rabbits. ACS Biomaterials Science & Engineering, 2020. XXXX.
[62] Yang, B. et al. A conductive PEDOT/alginate porous scaffold as a platform to modulate the biological behaviors of brown adipose-derived stem cells. Biomaterials Science, 2020. 8(11): 3173–3185.
[63] Yamato, M., and T. Okano. Cell sheet engineering. Materials Today, 2004. 7(5): 42–47.
[64] Narita, T. et al. The use of scaffold-free cell sheet technique to refine mesenchymal stromal cell-based therapy for heart failure. Mol Ther, 2013. 21(4): 860–7.
[65] Sakaguchi, K., T. Shimizu, and T. Okano. Construction of three-dimensional vascularized cardiac tissue with cell sheet engineering. Journal of Controlled Release, 2015. 205: 83–88.
[66] Ebihara, G. et al. Cartilage repair in transplanted scaffold-free chondrocyte sheets using a minipig model. Biomaterials, 2012. 33(15): 3846–3851.
[67] Yu, J. et al. Cell sheet composed of adipose-derived stem cells demonstrates enhanced skin wound healing with reduced scar formation. Acta Biomater, 2018. 77: 191–200.
[68] Rani, S. et al. Mesenchymal stem cell-derived extracellular vesicles: Toward cell-free therapeutic applications. Mol Ther, 2015. 23(5): 812–823.
[69] Qiu, H. et al. Prospective application of exosomes derived from adipose-derived stem cells in skin wound healing: A review. Journal of Cosmetic Dermatology, 2020. 19(3): 574–581.

[70] Mo, M. et al. Three-dimensional culture reduces cell size by increasing vesicle excretion. Stem Cells, 2018. 36(2): 286–292.
[71] Riau, A.K. et al. Sustained delivery system for stem cell-derived exosomes. Front Pharmacol, 2019. 10: 1368.
[72] Lin, R.Z., and H.Y. Chang. Recent advances in three-dimensional multicellular spheroid culture for biomedical research. Biotechnology Journal: Healthcare Nutrition Technology, 2008. 3(9-10): 1172–1184.
[73] Cheng, N.C. et al. Short-term spheroid formation enhances the regenerative capacity of adipose-derived stem cells by promoting stemness, angiogenesis, and chemotaxis. Stem Cells Transl Med, 2013. 2(8): 584–94.
[74] Mosaddad, S.A. et al. Stem cells and common biomaterials in dentistry: A review study. J Mater Sci Mater Med, 2022. 33(7): 55.
[75] Magalhaes, F.D. et al. Dental tissue-derived stem cell sheet biotechnology for periodontal tissue regeneration: A systematic review. Arch Oral Biol, 2021. 129: 105182.
[76] Zein, N. et al. Polymer-based instructive scaffolds for endodontic regeneration. Materials (Basel), 2019. 12(15).
[77] Ferrarotti, F. et al. Human intrabony defect regeneration with micrografts containing dental pulp stem cells: A randomized controlled clinical trial. J Clin Periodontol, 2018. 45(7): 841–850.
[78] Apel, C. et al. Differential mineralization of human dental pulp stem cells on diverse polymers. Biomed Tech (Berl), 2018. 63(3): 261–269.
[79] Moshaverinia, A. et al. Bone regeneration potential of stem cells derived from periodontal ligament or gingival tissue sources encapsulated in RGD-modified alginate scaffold. Tissue Eng Part A, 2014. 20(3-4): 611–21.
[80] Hsiao, H.Y. et al. Application of dental stem cells in three-dimensional tissue regeneration. World J Stem Cells, 2021. 13(11): 1610–1624.
[81] Chen, R.S. et al. The behavior of rat tooth germ cells on poly(vinyl alcohol). Acta Biomater, 2009. 5(4): 1064–74.
[82] Chen, R.S., M.H. Chen, and T.H. Young. Induction of differentiation and mineralization in rat tooth germ cells on PVA through inhibition of ERK1/2. Biomaterials, 2009. 30(4): 541–7.
[83] Chang, C.C. et al. Neurogenic differentiation of dental pulp stem cells to neuron-like cells in dopaminergic and motor neuronal inductive media. J Formos Med Assoc, 2014. 113(12): 956–65.
[84] Hsiao, D. et al. Characterization of designed directional polylactic acid 3D scaffolds for neural differentiation of human dental pulp stem cells. J Formos Med Assoc, 2020. 119(1 Pt 2): 268–275.
[85] Carnevale, G. et al. Human dental pulp stem cells expressing STRO-1, c-kit and CD34 markers in peripheral nerve regeneration. J Tissue Eng Regen Med, 2018. 12(2): e774–e785.
[86] van Wachem, P.B. et al. Adhesion of cultured human endothelial cells onto methacrylate polymers with varying surface wettability and charge. Biomaterials, 1987. 8(5): 323–8.
[87] Ge, H. et al. Poly-L-ornithine promotes preferred differentiation of neural stem/progenitor cells via ERK signalling pathway. Sci Rep, 2015. 5: 15535.
[88] Wang, J.H., C.H. Hung, and T.H. Young. Proliferation and differentiation of neural stem cells on lysine-alanine sequential polymer substrates. Biomaterials, 2006. 27(18): 3441–50.
[89] Choi, K.C. et al. Effect of single growth factor and growth factor combinations on differentiation of neural stem cells. J Korean Neurosurg Soc, 2008. 44(6): 375–81.
[90] Leker, R.R., V. Lasri, and D. Chernoguz. Growth factors improve neurogenesis and outcome after focal cerebral ischemia. Journal of Neural Transmission, 2009. 116(11): 1397–1402.
[91] Kaufman, L.M., and J.N. Barrett. Serum factor supporting long-term survival of rat central neurons in culture. Science, 1983. 220(4604): 1394–1396.
[92] Li, Y.C., Y.C. Lin, and T.H. Young. Combination of media, biomaterials and extracellular matrix proteins to enhance the differentiation of neural stem/precursor cells into neurons. Acta Biomater, 2012. 8(8): 3035–48.
[93] Young, T.H. et al. The enhancement of dermal papilla cell aggregation by extracellular matrix proteins through effects on cell-substratum adhesivity and cell motility. Biomaterials, 2009. 30(28): 5031–5040.
[94] Li, Y.C. et al. A neural stem/precursor cell monolayer for neural tissue engineering. Biomaterials, 2014. 35(4): 1192–204.

[95] Li, Y.C. et al. Covalent bonding of GYIGSR to EVAL membrane surface to improve migration and adhesion of cultured neural stem/precursor cells. Colloids Surf B Biointerfaces, 2013. 102: 53–62.
[96] Lakard, B. et al. Effect of ultrasounds on the electrochemical synthesis of polypyrrole, application to the adhesion and growth of biological cells. Bioelectrochemistry, 2009. 75(2): 148–57.
[97] Du, J. et al. Optimal electrical stimulation boosts stem cell therapy in nerve regeneration. Biomaterials, 2018. 181: 347–359.
[98] Park, S., Y.J. Kang, and S. Majd. A review of patterned organic bioelectronic materials and their biomedical applications. Adv Mater, 2015. 27(46): 7583–619.
[99] Lovat, V. et al. Carbon nanotube substrates boost neuronal electrical signaling. Nano Lett, 2005. 5(6): 1107–10.
[100] Cellot, G. et al. Carbon nanotubes might improve neuronal performance by favouring electrical shortcuts. Nat Nanotechnol, 2009. 4(2): 126–33.
[101] Lee, H.J. et al. Amine-modified single-walled carbon nanotubes protect neurons from injury in a rat stroke model. Nat Nanotechnol, 2011. 6(2): 121–5.
[102] Bai, J. et al. Graphene nanomesh. Nat Nanotechnol, 2010. 5(3): 190–4.
[103] Park, S.Y. et al. Enhanced differentiation of human neural stem cells into neurons on graphene. Adv Mater, 2011. 23(36): H263–7.
[104] Schmidt, C.E., and J.B. Leach. Neural tissue engineering: strategies for repair and regeneration. Annu Rev Biomed Eng, 2003. 5: 293–347.
[105] Millesi, H., J. Ganglberger, and A. Berger. Erfahrungen mit der Mikrochirurgie peripherer Nerven, *In*: Axhausen, W., and D.m.D. Buck-Gramcko (Eds.). Chirurgia Plastica et Reconstructiva. 1967, Springer Berlin Heidelberg. pp. 47–55.
[106] Jacobson, S., and L. Guth. An electrophysiological study of the early stages of peripheral nerve regeneration. Exp Neurol, 1965. 11: 48–60.
[107] Jahoda, C.A., A.J. Reynolds, and R.F. Oliver. Induction of hair growth in ear wounds by cultured dermal papilla cells. J Invest Dermatol, 1993. 101(4): 584–90.
[108] Oliver, R.F. The induction of hair follicle formation in the adult hooded rat by vibrissa dermal papillae. J Embryol Exp Morphol, 1970. 23(1): 219–36.
[109] Osada, A. et al. Long-term culture of mouse vibrissal dermal papilla cells and *de novo* hair follicle induction. Tissue Eng, 2007. 13(5): 975–82.
[110] Wobus, A.M., G. Wallukat, and J. Hescheler. Pluripotent mouse embryonic stem cells are able to differentiate into cardiomyocytes expressing chronotropic responses to adrenergic and cholinergic agents and Ca2+ channel blockers. Differentiation, 1991. 48(3): 173–82.
[111] Higgins, C.A. et al. Modelling the hair follicle dermal papilla using spheroid cell cultures. Exp Dermatol, 2010. 19(6): 546–8.
[112] Radisic, M. et al. Cardiac tissue engineering using perfusion bioreactor systems. Nat Protoc, 2008. 3(4): 719–38.
[113] Mahmoudifar, N., and P.M. Doran. Tissue engineering of human cartilage and osteochondral composites using recirculation bioreactors. Biomaterials, 2005. 26(34): 7012–24.
[114] Shvartsman, I. et al. Perfusion cell seeding and cultivation induce the assembly of thick and functional hepatocellular tissue-like construct. Tissue Eng Part A, 2009. 15(4): 751–60.
[115] Khetani, S.R., and S.N. Bhatia. Microscale culture of human liver cells for drug development. Nat Biotechnol, 2008. 26(1): 120–6.
[116] Young, T.H. et al. Self-assembly of dermal papilla cells into inductive spheroidal microtissues on poly(ethylene-co-vinyl alcohol) membranes for hair follicle regeneration. Biomaterials, 2008. 29(26): 3521–30.
[117] Young, T.H. et al. The enhancement of dermal papilla cell aggregation by extracellular matrix proteins through effects on cell-substratum adhesivity and cell motility. Biomaterials, 2009. 30(28): 5031–40.
[118] Hsu, Y.C., L. Li, and E. Fuchs. Transit-amplifying cells orchestrate stem cell activity and tissue regeneration. Cell, 2014. 157(4): 935–49.
[119] Plikus, M.V. et al. Cyclic dermal BMP signalling regulates stem cell activation during hair regeneration. Nature, 2008. 451(7176): 340–4.
[120] Huang, C.F. et al. Assembling composite dermal papilla spheres with adipose-derived stem cells to enhance hair follicle induction. Sci Rep, 2016. 6: 26436.

[121] Enshell-Seijffers, D. et al. beta-catenin activity in the dermal papilla regulates morphogenesis and regeneration of hair. Dev Cell, 2010. 18(4): 633–42.
[122] Chi, W., E. Wu, and B.A. Morgan. Dermal papilla cell number specifies hair size, shape and cycling and its reduction causes follicular decline. Development, 2013. 140(8): 1676–83.
[123] Elliott, K., T.J. Stephenson, and A.G. Messenger. Differences in hair follicle dermal papilla volume are due to extracellular matrix volume and cell number: implications for the control of hair follicle size and androgen responses. J Invest Dermatol, 1999. 113(6): 873–7.
[124] Huang, Y.C. et al. Scalable production of controllable dermal papilla spheroids on PVA surfaces and the effects of spheroid size on hair follicle regeneration. Biomaterials, 2013. 34(2): 442–51.
[125] Ikeda, E. et al. Functional ectodermal organ regeneration as the next generation of organ replacement therapy. Open Biol, 2019. 9(3): 190010.
[126] Stenn, K.S., and G. Cotsarelis. Bioengineering the hair follicle: fringe benefits of stem cell technology. Curr Opin Biotechnol, 2005. 16(5): 493–7.
[127] Asakawa, K. et al. Hair organ regeneration via the bioengineered hair follicular unit transplantation. Sci Rep, 2012. 2: 424.
[128] Nakao, K. et al. The development of a bioengineered organ germ method. Nat Methods, 2007. 4(3): 227–30.
[129] Kang, B.M. et al. Sphere formation increases the ability of cultured human dermal papilla cells to induce hair follicles from mouse epidermal cells in a reconstitution assay. J Invest Dermatol, 2012. 132(1): 237–9.
[130] Takeda, A. et al. Histodifferentiation of hair follicles in grafting of cell aggregates obtained by rotation culture of embryonic rat skin. Scand J Plast Reconstr Surg Hand Surg, 1998. 32(4): 359–64.
[131] Ihara, S. et al. Formation of hair follicles from a single-cell suspension of embryonic rat skin by a two-step procedure *in vitro*. Cell Tissue Res, 1991. 266(1): 65–73.
[132] Qiao, J. et al. Hair morphogenesis *in vitro*: formation of hair structures suitable for implantation. Regen Med, 2008. 3(5): 683–92.
[133] Kageyama, T. et al. Spontaneous hair follicle germ (HFG) formation *in vitro*, enabling the large-scale production of HFGs for regenerative medicine. Biomaterials, 2018. 154: 291–300.
[134] Kageyama, T. et al. Preparation of hair beads and hair follicle germs for regenerative medicine. Biomaterials, 2019. 212: 55–63.
[135] Castro, A.R., and E. Logarinho. Tissue engineering strategies for human hair follicle regeneration: How far from a hairy goal? Stem Cells Transl Med, 2020. 9(3): 342–350.
[136] Abaci, H.E. et al. Tissue engineering of human hair follicles using a biomimetic developmental approach. Nat Commun, 2018. 9(1): 5301.
[137] Havlickova, B. et al. A human folliculoid microsphere assay for exploring epithelial- mesenchymal interactions in the human hair follicle. J Invest Dermatol, 2009. 129(4): 972–83.
[138] Li, Y.C. et al. Programmable laser-assisted surface microfabrication on a Poly(Vinyl Alcohol)-coated glass chip with self-changing cell adhesivity for heterotypic cell patterning. ACS Appl Mater Interfaces, 2015. 7(40): 22322–32.
[139] Honda, M.J. et al. The sequential seeding of epithelial and mesenchymal cells for tissue-engineered tooth regeneration. Biomaterials, 2007. 28(4): 680-–9.
[140] Wei, C. et al. Self-organization and branching morphogenesis of primary salivary epithelial cells. Tissue Eng, 2007. 13(4): 721–35.
[141] Fukuda, J. et al. Micropatterned cell co-cultures using layer-by-layer deposition of extracellular matrix components. Biomaterials, 2006. 27(8): 1479–86.
[142] Rivron, N.C. et al. Tissue assembly and organization: developmental mechanisms in microfabricated tissues. Biomaterials, 2009. 30(28): 4851–8.
[143] Khademhosseini, A. et al. Microscale technologies for tissue engineering and biology. Proc Natl Acad Sci U S A, 2006. 103(8): 2480–7.
[144] Yen, C.M., C.C. Chan, and S.J. Lin. High-throughput reconstitution of epithelial-mesenchymal interaction in folliculoid microtissues by biomaterial-facilitated self-assembly of dissociated heterotypic adult cells. Biomaterials, 2010. 31(15): 4341–52.
[145] Bhatia, S.N. et al. Microfabrication of hepatocyte/fibroblast co-cultures: role of homotypic cell interactions. Biotechnol Prog, 1998. 14(3): 378–87.

[146] McElwee, K.J. et al. Cultured peribulbar dermal sheath cells can induce hair follicle development and contribute to the dermal sheath and dermal papilla. J Invest Dermatol, 2003. 121(6): 1267–75.
[147] Reynolds, A.J., and C.A. Jahoda. Cultured dermal papilla cells induce follicle formation and hair growth by transdifferentiation of an adult epidermis. Development, 1992. 115(2): 587–93.
[148] Huang, W.Y. et al. Stress-induced premature senescence of dermal papilla cells compromises hair follicle epithelial-mesenchymal interaction. J Dermatol Sci, 2017. 86(2): 114–122.
[149] Bahta, A.W. et al. Premature senescence of balding dermal papilla cells *in vitro* is associated with p16(INK4a) expression. J Invest Dermatol, 2008. 128(5): 1088–94.
[150] Veraitch, O. et al. Human induced pluripotent stem cell-derived ectodermal precursor cells contribute to hair follicle morphogenesis *in vivo*. J Invest Dermatol, 2013. 133(6): 1479–88.
[151] Veraitch, O. et al. Induction of hair follicle dermal papilla cell properties in human induced pluripotent stem cell-derived multipotent LNGFR(+)THY-1(+) mesenchymal cells. Sci Rep, 2017. 7: 42777.
[152] Fukuyama, M. et al. Human iPS cell-derived cell aggregates exhibited dermal papilla cell properties in *in vitro* three-dimensional assemblage mimicking hair follicle structures. Front Cell Dev Biol, 2021. 9: 590333.
[153] Takahashi, T. et al. Ischemia- and cytokine-induced mobilization of bone marrow-derived endothelial progenitor cells for neovascularization. Nat Med, 1999. 5(4): 434–8.
[154] Rafii, S., and D. Lyden. Therapeutic stem and progenitor cell transplantation for organ vascularization and regeneration. Nat Med, 2003. 9(6): 702–12.
[155] Asahara, T. et al. Bone marrow origin of endothelial progenitor cells responsible for postnatal vasculogenesis in physiological and pathological neovascularization. Circ Res, 1999. 85(3): 221–8.
[156] Royer, C. et al. Bioactive micropatterning of biomaterials for induction of endothelial progenitor cell differentiation: Acceleration of *in situ* endothelialization. J Biomed Mater Res A, 2020. 108(7): 1479–1492.

3

Scale-Up of Stem Cells

Yi-Chen Ethan Li,[1,*] *Nai-Chen Cheng*,[2] *Min-Huey Chen*,[3] *Wen-Yen Huang*,[4] *Sung-Jan Lin*,[4] *Chia-Ning Shen*[5] and *I-Chi Lee*[6,*]

In the first two chapters, we have introduced stem cells with various properties and their commonly used culture methods. Subsequently, in addition to traditional cultural methods, in the past 20 years, the study of the combination of biology and biotechnology in a multidisciplinary field has made great progress and breakthroughs in tissue engineering and regenerative medicine. In addition, scientists can also use the various properties of biomedical materials to regulate stem cell activity, growth, and differentiation, providing scientists with more diverse ways to create various stem cell models. Subsequently, these models can be applied to the fields of cell therapy, tissue engineering, and regenerative medicine, promoting the development of a variety of cell-based treatments and new and more effective biological agents and improving methods to regenerate damaged tissues.

At the end of the 19th century, the technique of growing cells in a dish has proven invaluable for cell biology experiments. In fact, conventional petri dish cultures are still widely used for anchorage-dependent cells to expand on the surface of substrates, extracellular matrix protein, or feeder cells embryonic fibroblasts and then soaked in a medium containing appropriate nutrients and signaling molecules. The change of cell culture medium is performed in batches, resulting in the composition of the medium changing with time. In a petri dish, cells are grown in

[1] Department of Chemical Engineering, Feng Chia University, Taiwan.
[2] Department of Surgery, National Taiwan University Hospital, Taiwan.
[3] Graduate Institute of Clinical Dentistry, School of Dentistry, National Taiwan University, Taiwan.
[4] Department of Biomedical Engineering, College of Medicine and College of Engineering, National Taiwan University, Taiwan.
[5] Genomics Research Center, Academia Sinica, Taiwan.
[6] Department of Biomedical Engineering and Environmental Sciences, National Tsing Hua University, Taiwan.
* Corresponding authors: iclee@mx.nthu.edu.tw; yicli@fcu.edu.tw

two dimensions. Cells are subcultured for further expansion or differentiation when cell numbers achieve confluence. Although this form of 2D culture has a structure that is similar to the epithelium/endothelial structures of the skin or bladder, it cannot provide most cells with an organ-like environment. In particular, 2D cultures cannot provide biochemical and physical signals and systemic regulation, including cross-talk between different 3D organ systems to cells. Consequently, results obtained in petri dish cultures are not always predictive of whole tissues and organs and are difficult to translate into *in vivo* settings for preclinical animal studies and clinical trials in humans. To overcome the challenges of using 2D petri dishes, Langer and co-workers developed engineered 3D human tissue constructs *in vitro* by culturing cells on a scaffold material to create a biomimetic environment [1]. Based on this idea, it is expected to reproduce relevant developmental and morphological *in vivo* models *in vitro*. However, efficiently generating many cells is necessary to further implement and translate stem cell-related models into clinical treatment and its applications. For example, to produce approximately 110 kg of tissue, around 10^{11}–10^{12} cells are required. However, at a typical laboratory scale, culturing such numbers of cells is extremely challenging because the scalability of adherent cells has never been demonstrated at such. To reduce costs [2], it is necessary to select the correct scale-up process for mass production of cells. Additionally, it is important to choose the proper scale-up process to achieve the large number of cells required and to decrease manufacturing costs [3]. Then, in scaling up the cell numbers, the biggest challenge is that most stem cells need to adhere to substrate surfaces to maintain their viability and proliferate. Therefore, for an effective *in vitro* cell expansion system, it is important to consider alternatives to traditional 2D culture dishes. By combining biomedical materials or using other strategies to culture cells in a 3D space, it is possible to increase the surface-to-volume ratio and to control key growth parameters, resulting in a more efficient system for cell expansion. To efficiently obtain a large number of cells, two strategies are commonly employed: (i) culturing anchorage-dependent cells growing in a suspension, or (ii) using scaffolds to carry cells and culture them in a suspension culture system (Figure 3-1) [3].

In the case of most anchorage-dependent stem cells, growing these cells in a suspension is often laborious, as it can take months to achieve, and because not all cells are fully adapted to the growing conditions of this novel suspension culture, this method has a relatively low mass production efficiency and high failure rate [4]. In addition, if the suspension culture is successful, the newly formed 3D cellular aggregates must also be closely monitored in the system and regularly separated from the aggregates to prevent spontaneous differentiation of the formed cellular aggregates and necrosis of cells close to the cell core in the aggregates because of the nutrient transport issue. On the other hand, compared to suspension culture systems, cell culturing in scaffolds can be used for different biological processes and provide adhesive surfaces. In addition, their mass is small enough to be suspended in the cell culture medium with agitation, allowing the cells to grow entirely and be amplificated on the surface of scaffolds. Although using scaffolds as carriers to expand the number of cells is a common method, the carriers do not have blood vessels to help cells transport nutrients and wastes to induce cell growth

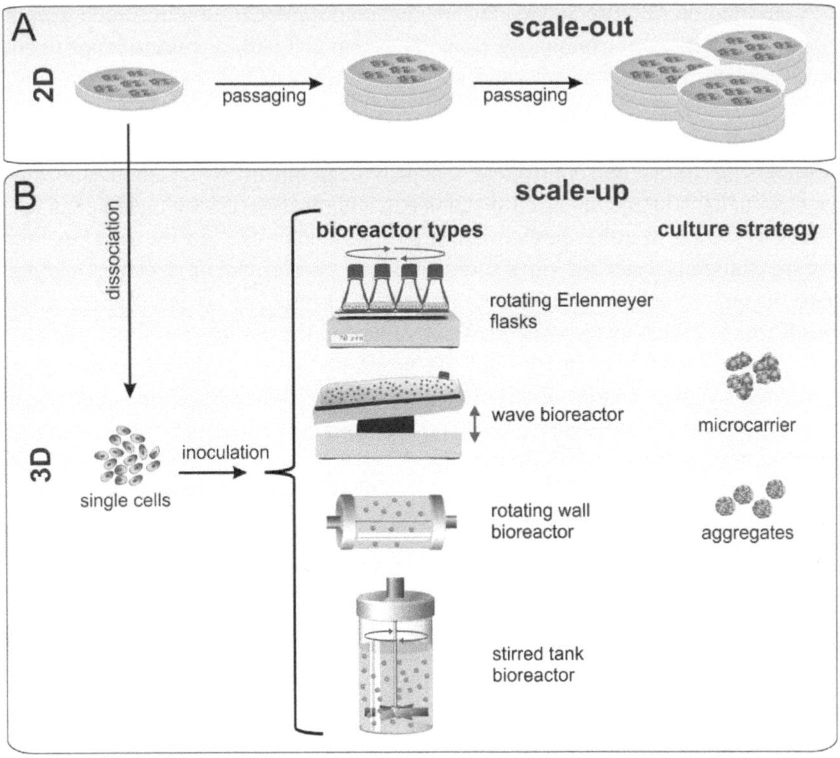

Figure 3-1. The expansion methods for stem cells. (Reprinted with permission from [3] Copyright (2017) Elsevier.)

and support tissue development in three dimensions. The use of microcarriers with a dynamic fluid culture system (or bioreactor) can mimic the internal environment of cells that provide the stem cells located in the carrier with a way for nutrient mass transport. Mass transfer is through diffusion pathways in static cultures compared to dynamic cultures. Due to the decrease of oxygen tension and the increase of toxic concentrations from metabolites (e.g., acidification), these issues limit cellular aggregate size development in three-dimensional culture systems, whose thicknesses may be less than 0.2 mm [5]. In bioreactors, however, culture systems such as agitation, perfusion, and dynamics have been applied to provide convective transport and allow cell aggregates to develop in the millimeter to centimeter range [6, 7]. The direction and rate of media flow and the viscosity of the media can be further tuned to achieve mechanical stimulation through shear stress. In addition, bioreactors can further provide other biophysical stimuli, including compressive and tensile loading, electrical stimulation, and media flow resulting from agitation or perfusion [8–10], individually or in combination. For example, biomechanics have demonstrated that MSCs can differentiate into different cells, so stressing MSCs could help them differentiate into cartilage [11]. If shearing force is applied, it is easy to induce MSCs to become hard bones, and further, if the tensile strength is applied, it will help MSCs differentiate into ligament cells. Further, if the bioreactor is used with online control and monitoring of temperature, pH, oxygen, and medium concentrations of nutrients, the combination of bioreactors, carriers, and cells can be used in research to create a biomimetic tissue environment *in vitro*. The stem cell culture system helps to clarify and establish the mechanisms and knowledge of stem cell behavior regulation in extensive research.

In general, a typical bioreactor consists of three functional parts, which are cell chambers, monitors, and controllers, that provide a controlled and stable environment for stem cells. Bioreactor-based systems can automate necessary media changes and reduce variation in culture batch and microbial contamination by avoiding routine culture maintenance. Cells in a bioreactor can be cultured in a static or dynamic state. As mentioned above, conventional static 2D-based cultures suffer from several disadvantages, such as the change of pH value and the decrease of dissolved oxygen and metabolites concentration, and the proliferation of stem cells is significantly conflicted by their spatial distribution. In contrast, dynamic culture with a carrier with a high surface area can provide sufficient support for stem cells growing in the environment, so bioreactors usually play a key role in establishing stem cell culture systems. Bioreactors offer an *in vivo* biomimetic environment for controlling the quality and quantity of cultured stem cells. According to the requirements of stem cell growth, the bioreactor should have low shear stress, a good transfer effect, and good hydrodynamic properties. Several types of bioreactors have been developed for stem cell expansion. Previous studies have pointed out that most bioreactors can be divided into four operating modes in terms of medium supplementation, such as batch, repeated batch, fed batch, and perfusion (Figure 3-2) [3]. First, the batch mode

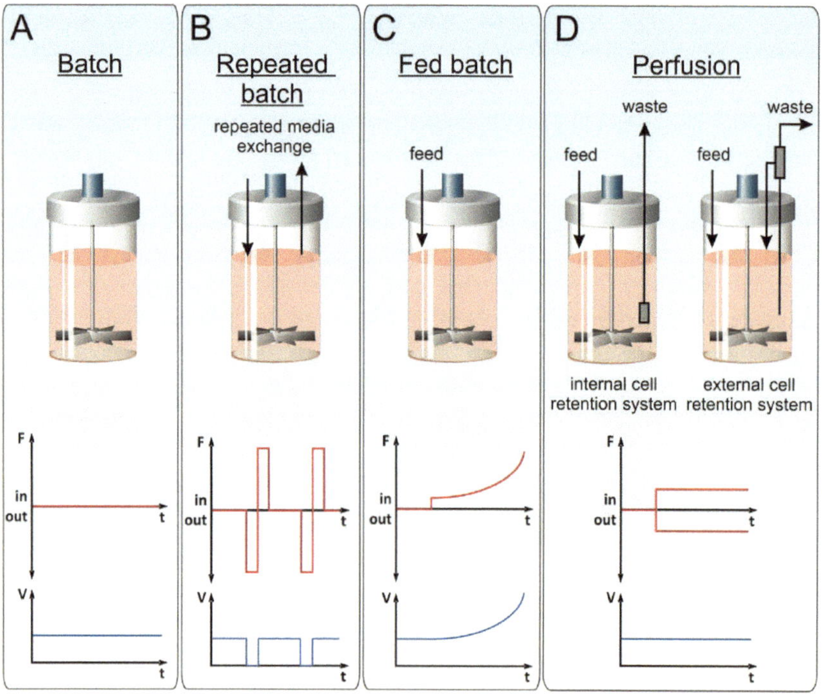

Figure 3-2. The operation models of bioreactors. (Reprinted with permission from [3] Copyright (2017) Elsevier.)

of operation only provides nutrition at the beginning of the process. No other feed is made thereafter, so the liquid working volume V of bioreactors remains unchanged. In batch mode, only add acids or bases, antifoam cleaners, or adjust the gas exchange. Therefore, batch processing is the simplest mode of operation with the lowest risk of contamination. However, moderate cell yields are expected due to the limited nutrient concentrations that most cell types tolerate before causing deleterious effects through hypertonicity [12].

The second repeated batch mode is a semi-discontinuous model. At the initial stage, the part of the medium is recirculated by using a fresh medium, and no cells are harvested during recirculation. The liquid working volume V of the bioreactor is also kept constant, except during medium changes. This model is also a typical system that can be used to culture iPSCs in a traditional petri dish. Several research groups have successfully combined microcarriers and aggregates simultaneously via this type of bioreactor to obtain higher cell yields [13–16]. Therefore, this approach is followed by the application of very short, repeated fed-batch cycles (approximately at 2-hour intervals), followed by partial medium changes to achieve "perfusion-like" conditions called "circulatory perfusion" [13–16].

In a fed-batch model, a concentrated medium or supplement containing one or more nutrient factors such as glucose and amino acids is added to the bioreactor during the cultivation process at the initial stage. The liquid working volume V in the bioreactor is increased over time [17]. Compared with the abovementioned methods, this method is easier to achieve high cell density and has also been successfully used to expand the culture of hematopoietic stem cells. However, the disadvantage of this method is the potential accumulation of toxic metabolites as there is no outlet to remove the fluid.

In a continuous type of bioreactor, the perfusion model, a fresh medium is continuously added to the culture, while the medium is continuously supplied and removed at the inlet and outlet at the same flow rate; thus, the liquid working volume V of the perfusion bioreactor can be kept as a constant volume. The unique feature of perfusion-type bioreactors is that they can continuously replace the medium in the reactor with the fresh medium while the cells are retained in the vessel by a specific system. The perfusion-based bioreactor is a superior mode of operation commonly used in biopharmaceutical production processes to achieve the highest cell yields. However, perfusion feeding also represents the highest level of operational complexity, moderate cost, and risk of contamination [18]. In addition to constantly supplying the perfused cells with fresh nutrients and growth factors, it also flushes out potentially toxic waste, ensuring more uniform conditions in the bioreactor. In addition, the perfusion process supports process automation to improve feedback control of the culture environment, including pH, oxygen, and nutrient concentrations, compared to other methods and does not require process interruption during perfusion, thereby minimizing operator errors. Perfusion model can offer a relatively stable physiological environment for activating the self-regulation of stem cells through their secretion of endogenous factors, ultimately reducing the use of expensive media components in stem cell culture media [19]. In addition to the bioreactor types, several parameters must be considered when using a bioreactor.

For example, oxygen regulation is one of the important factors affecting tissue and cell metabolism in the physiological niche and an important regulator of cell physiology [42]. In the incubator method, stem cells are typically grown at a constant 37°C in a humidified environment with 5% to 10% CO_2. In standard CO_2 incubators, oxygen fraction is a crucial parameter of the gas phase in cell culture. Gases with normal oxygen content are also called normoxic; gases with lower and higher oxygen levels are also called hypoxic and hyperoxic, respectively. Oxygen concentration is an important component for *in vitro* stem cell expansion, and it is also critical for stem cells to maintain stemness and regulated differentiation [42, 43]. Next, bioreactors with dynamic systems are believed to effectively regulate the culture environment, which can stimulate the growth and differentiation of stem cells. However, in dynamic cultures, the fluid in the bioreactor may generate fluid shear forces, which may affect the expansion of stem cells and their physiological functions [23]. Small shear forces can increase cell permeability, nutrient uptake, proliferative activity, and fold expansion. Increased shear stress may inhibit stem cell proliferation, promote their differentiation, or even damage stem cells, resulting in loss of proliferative potential and hematopoietic activity [15], since the stem cells are sensitive to the shear forces generated by agitation. Therefore, in designing a bioreactor (especially the stirring tank type), it is necessary to consider that energy may be transferred from the impeller to the medium, resulting in vigorous movement of the medium, thereby changing the receptors of growth factors on cells and affecting the proliferation efficiency [25]. Therefore, when using this type of bioreactor, selecting a suitable bioreactor for different kinds of stem cells is necessary after considering the maximum shear force that stem cells can withstand [22]. Finally, for the part of the reactor in the nutrient supply and cell metabolism, the survival and growth of cells are highly related to the local environment, and there is a continuous exchange between cells and the surrounding environment so that cells can obtain from the nutrients from the surrounding environment and release metabolites. Nutrients provide the energy necessary for cell growth. Some toxic metabolites (such as lactate, ammonia, and reactive oxygen species) may affect the cell culture environment (such as pH and osmotic pressure), inhibiting cell growth and reducing cell viability. Moreover, in recent years, sensors have been incorporated into bioreactors through different sensors enabling continuous monitoring and adjustment of parameters related to cell growth, creating an optimal environment for cell growth [20].

Here, we have discussed the flow operation models and parameter effect of bioreaction. Next, we will briefly introduce the design of bioreactors commonly used in tissue engineer. The first type is perfusion-based bioreactors. The perfusion-based bioreactor is shown in Figure 3-3 [21, 22]. This perfusion-based bioreactor uses a pump to drive the culture medium continuously passing through the scaffold to transfer nutrients to cells and remove metabolism waste. Moreover, using perfusion-based bioreactors can simulate an *in vivo* biomimetic environment by adjusting different pressure, pulse, etc., so that the cells can be uniformly distributed on the scaffolds. However, in this type of perfusion bioreactor, if the pore distribution of scaffolds is not uniform, it is easy to cause a channeling effect. At this time, the culture medium will easily flow in the direction of the large pores in scaffolds,

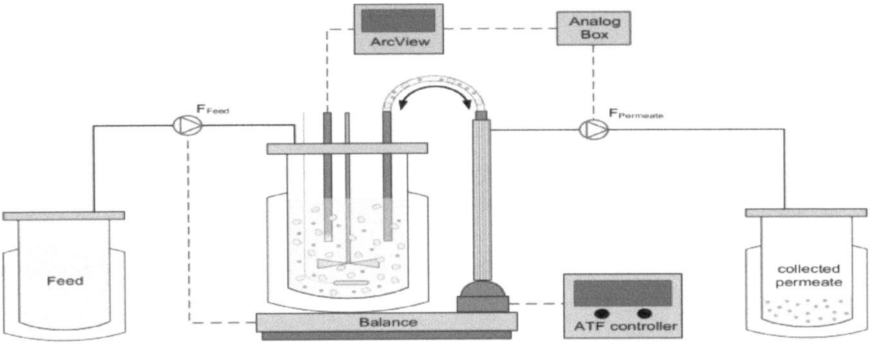

Figure 3-3. The schematic diagram of perfusion-based bioreactor [21, 22].

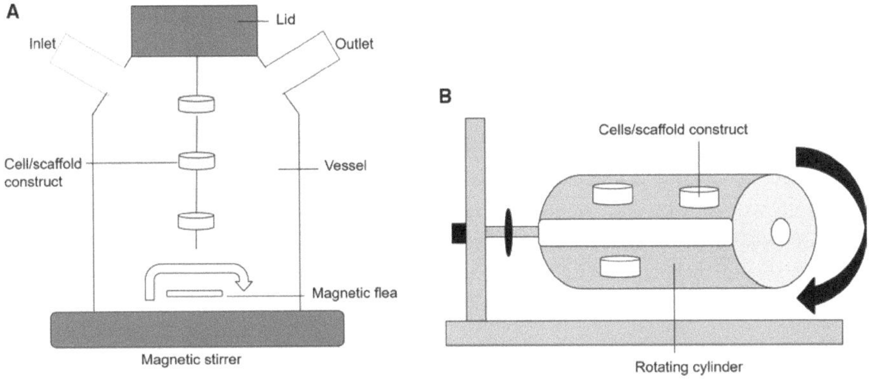

Figure 3-4. The schematic diagram of (A) spinner flask bioreactor and (B) rotating-wall vessels. Reprinted with permission from [24] Copyright (2018) Springer Nature.)

resulting in an insufficient nutrient supply in local areas. Currently, perfusion-based bioreactors are most commonly used and suitable for the culture of bone cells [23]. Spinner flask bioreactors, also called stirred bioreactors, use stirring to promote the mixing behavior of culture medium to increase the effect of mass transfer (Figure 3-4(A)) [24]. When using spinner flask bioreactors, the liquid has a turbulent flow, so the liquid can generate high shear forces and cause a negative effect on cells. Furthermore, the scaffolds are usually fixed in the stirring bottle by hanging wires in this system. Next, the cell implantation method engrafts the cells on the scaffolds by a convection flow generated by the fluid. This method has a high cell engraftment rate, but most cells are distributed on the surface of scaffolds rather than in the scaffolds. Therefore, Spinner flask bioreactors are more commonly used to culture chondrocytes.

A rotating-wall vessel is a bioreactor using concentric circle rotation to drive the flow of culture medium, thereby forming different shear forces (Figure 3-4(B)) [24]. The rotating-wall vessel is the most commonly used reactor among the various commercial bioreactors. Rotating-wall vessels provide lower shear forces to cells than other types of reactors. A rotating microgravity bioreactor developed by NASA is based on the principle of a rotating-wall vessel [25]. The shear force is produced by rotating the culture medium, which can balance with gravity to generate a microgravity state and then suspend the cell-laden scaffolds in the culture medium to prevent falling to the bottle wall. The rotating-wall vessels will also be equipped with a membrane exchanger to exchange oxygen and carbon dioxide. In addition, a challenge of using rotating-wall vessels that needs to be overcome is that the density of the scaffold increases due to cell expansion when the incubation time increases. At this point, the cell-laden scaffolds may fall on the bottle wall because they cannot resist gravity.

Pulsatile flow bioreactors are shown in Figure 3-5 [26]. This type of bioreactor uses a pulsed pump to drive the flow of the medium so that the medium can pass through the tubular scaffolds at a fixed frequency [27]. In general, pulsatile flow bioreactors are suitable for the cultivation of tubular vascular scaffolds because the pulsed flow of the medium can simulate the flow of human blood and stimulate the cells in the lumen for maturation.

Dynamic compression bioreactor is a kind of bioreactor that enables applying a compressive force to the cell-laden scaffolds [20]. By using force to change the shape of the scaffolds, the force is enabled to transmit to the cells laden on the scaffolds to achieve the effect of mechanical stimulation on cells (Figure 3-6). The advantage of a dynamic compression bioreactor is that it can simulate an environment similar to that of force working on the human joints in the body. In addition to utilizing the force of compression, another similar bioreactor is designed for tendons or ligaments, which periodically applies stress or tension to the artificial tissues to stimulate the maturation of the artificial tissues [28].

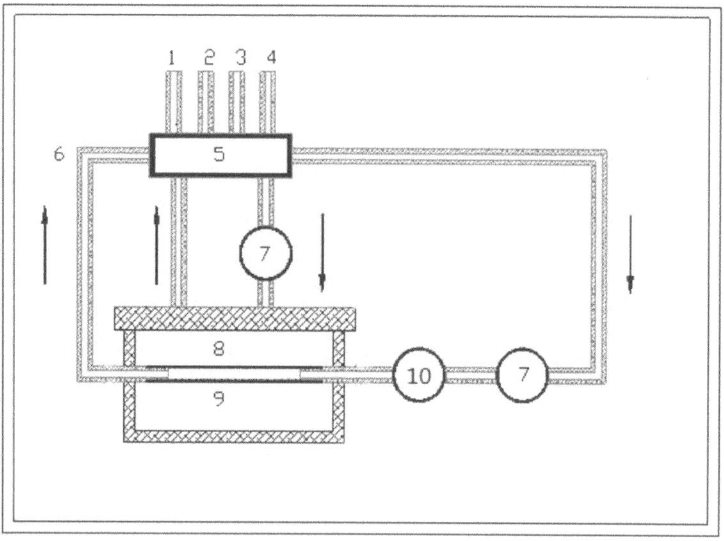

Figure 3-5. The schematic diagram of the pulsatile flow bioreactor. The unit (1) and (3) are the air inlet and liquid inlet; the unit (2) and (4) are the air outlet and fluid outlet; the unit (5) is the gas/liquid and liquid/liquid exchanger; the unit (6) is a silicone tube; the unit (7) is a stepper motor-driven pump; the unit (8) is a cell culture chamber; the unit (9) is the tissue-engineered vascular graft and the arrow is liquid flow; and the unit (10) is a pressure sensor. (Reprinted with permission from [26] Copyright (2012) PLOS.)

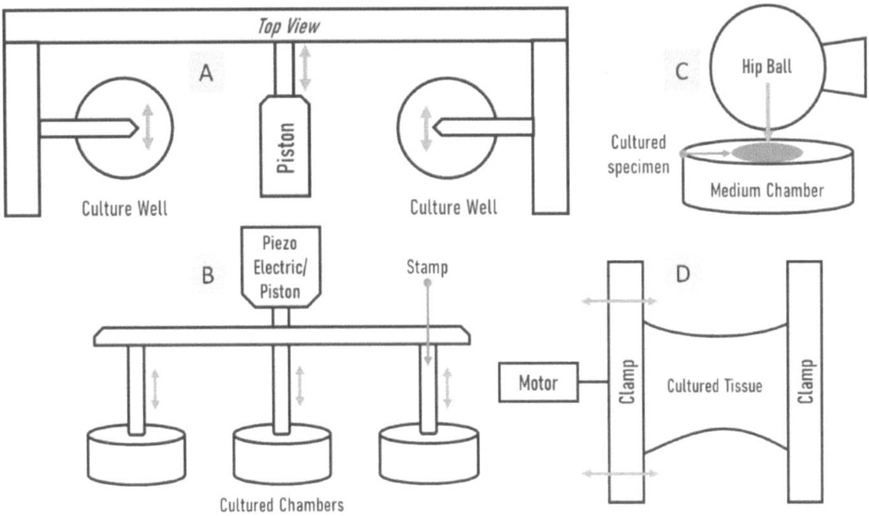

Figure 3-6. The schematic diagram of different types of dynamic compression bioreactor and bioreactors with other dynamical mechanical stimulation. (a–d) The types of mechanical stimulation mechanisms are parallel piston, perpendicular piston, hip ball type, and stretch clamp. (Reprinted with permission from [28] Copyright (2020) Faculty of Engineering.)

Based on the introduction of bioreactors, the cells from various body tissues have their own suitable reactors for cultivation because cells face different microenvironments in each body tissue. Therefore, Table 3-1 provides examples of shear forces generated from each type of bioreactor related to the stress distribution of the corresponding cultured bone-related tissues [29]; and Table 3-2 is the example of the force types and stimulation parameters for cells [30].

Table 3-1. The shear fore distribution and applications of different bioreactor types for culturing bone-related tissues [29].

Bioreactor types	Flow rate and stirring speed	Shear force ranage (dyn/cm^2)	Applications
Perfusion-based bioreactors	0.5 mL/min	0 ~ 0.8	Bone
Spinner flask bioreactors	50 rpm	0 ~ 1.2	Cartilage and Bone
Rotating-wall vessels	13 and 37 rpm	0 ~ 0.8	Cartilage and Cardiovascular tissues

Table 3-2. The parameters of different mechanical forces for stimulation of cells [30].

Cell types	Forces for stimulation	Parameters
Smooth muscle cells	Fluid and Pressure	120/60 mmHg, 0.1 L/min 10 mmHg, 0.6/1.2 L/min
Cartilage cells	Hydrostatic pressure	0.55–5.03 MPa/1Hz
Endothelial cells	Fluid shear forces	15 dyne/cm^2

From this chapter, we can understand that the design of the bioreactor is developed to simulate the *in vivo* environment of various tissue parts in the human body and their applications in scaling up the number of stem cells. The bioreactors can generate different fluid flow methods to drive the transmission and supply of nutrients, ensure uniform dissolved oxygen concentration in the medium, and remove cellular metabolites that can ensure the viability of cells growing inside the scaffold. Furthermore, bioreactors can incorporate physical stimuli, such as changing fluid flow behaviors to create different shear stresses and flow fields or using intermittent compression or tension as reciprocating stimuli, to induce ECM secretion, proliferation, differentiation, and other functions of stem cells for producing functionalized artificial cell-based tissues. In the following chapters, we will introduce the interactions between the properties of biomaterials and stem cells for the design of stem cell culture systems.

References

[1] Langer, R., and J.P. Vacanti. Tissue engineering, Science, 1993. 260(5110): 920–6.

[2] Ovics, P., D. Regev, P. Baskin, M. Davidor, Y. Shemer, S. Neeman, Y. Ben-Haim, and O. Binah. Drug development and the use of induced pluripotent stem cell-derived cardiomyocytes for disease modeling and drug toxicity screening. Int J Mol Sci, 2020. 21(19).

[3] Kropp, C., D. Massai, and R. Zweigerdt. Progress and challenges in large-scale expansion of human pluripotent stem cells. Process Biochemistry, 2017. 59: 244–254.

[4] Dos Santos, F., A. Campbell, A. Fernandes-Platzgummer, P.Z. Andrade, J.M. Gimble, Y. Wen, S. Boucher, M.C. Vemuri, C.L. da Silva, and J.M. Cabral. A xenogeneic-free bioreactor system for the clinical-scale expansion of human mesenchymal stem/stromal cells. Biotechnol Bioeng, 2014. 111(6): 1116–27.

[5] Tandon, N., D. Marolt, E. Cimetta, and G. Vunjak-Novakovic. Bioreactor engineering of stem cell environments, Biotechnol Adv, 2013. 31(7): 1020–31.

[6] Grayson, W.L., D. Marolt, S. Bhumiratana, M. Frohlich, X.E. Guo, and G. Vunjak-Novakovic. Optimizing the medium perfusion rate in bone tissue engineering bioreactors. Biotechnol Bioeng, 2011. 108(5): 1159–70.

[7] Grayson, W.L., M. Frohlich, K. Yeager, S. Bhumiratana, M.E. Chan, C. Cannizzaro, L.Q. Wan, X.S. Liu, X.E. Guo, and G. Vunjak-Novakovic. Engineering anatomically shaped human bone grafts, Proc Natl Acad Sci U S A, 2010. 107(8): 3299–304.

[8] Baker, B.M., R.P. Shah, A.H. Huang, and R.L. Mauck. Dynamic tensile loading improves the functional properties of mesenchymal stem cell-laden nanofiber-based fibrocartilage. Tissue Eng Part A, 2011. 17(9-10): 1445–55.

[9] Bian, L., D.Y. Zhai, E.C. Zhang, R.L. Mauck, and J.A. Burdick. Dynamic compressive loading enhances cartilage matrix synthesis and distribution and suppresses hypertrophy in hMSC-laden hyaluronic acid hydrogels. Tissue Eng Part A, 2012. 18(7-8): 715–24.

[10] Maidhof, R., N. Tandon, E.J. Lee, J. Luo, Y. Duan, K. Yeager, E. Konofagou, and G. Vunjak-Novakovic. Biomimetic perfusion and electrical stimulation applied in concert improved the assembly of engineered cardiac tissue. J Tissue Eng Regen Med, 2012. 6(10): e12–23.

[11] Tsai, A.C., and T. Ma. Expansion of human mesenchymal stem cells in a microcarrier bioreactor. Methods Mol Biol, 2016. 1502: 77–86.

[12] Wang, D., W. Liu, B. Han, and R. Xu. The bioreactor: a powerful tool for large-scale culture of animal cells. Curr Pharm Biotechnol, 2005. 6(5): 397–403.

[13] Serra, M., C. Brito, M.F. Sousa, J. Jensen, R. Tostoes, J. Clemente, R. Strehl, J. Hyllner, M.J. Carrondo, and P.M. Alves. Improving expansion of pluripotent human embryonic stem cells in perfused bioreactors through oxygen control. J Biotechnol, 2010. 148(4): 208–15.

[14] Krawetz, R., J.T. Taiani, S. Liu, G. Meng, X. Li, M.S. Kallos, and D.E. Rancourt. Large-scale expansion of pluripotent human embryonic stem cells in stirred-suspension bioreactors. Tissue Eng Part C Methods, 2010. 16(4): 573–82.
[15] Fridley, K.M., M.A. Kinney, and T.C. McDevitt. Hydrodynamic modulation of pluripotent stem cells. Stem Cell Res Ther, 2012. 3(6): 45.
[16] Abbasalizadeh, S., M.R. Larijani, A. Samadian, and H. Baharvand. Bioprocess development for mass production of size-controlled human pluripotent stem cell aggregates in stirred suspension bioreactor. Tissue Eng Part C Methods, 2012. 18(11): 831–51.
[17] Csaszar, E., D.C. Kirouac, M. Yu, W. Wang, W. Qiao, M.P. Cooke, A.E. Boitano, C. Ito, and P.W. Zandstra. Rapid expansion of human hematopoietic stem cells by automated control of inhibitory feedback signaling. Cell Stem Cell, 2012. 10(2): 218–29.
[18] Bellani, C.F., J. Ajeian, L. Duffy, M. Miotto, L. Groenewegen, and C.J. Connon. Scale-up technologies for the manufacture of adherent cells. Front Nutr, 2020. 7: 575146.
[19] Kumar, A., and B. Starly. Large scale industrialized cell expansion: producing the critical raw material for biofabrication processes. Biofabrication, 2015. 7(4): 044103.
[20] Martin, I., D. Wendt, and M. Heberer. The role of bioreactors in tissue engineering. Trends Biotechnol, 2004. 22(2): 80–6.
[21] Hein, M.D., A. Chawla, M. Cattaneo, S.Y. Kupke, Y. Genzel, and U. Reichl. Cell culture-based production of defective interfering influenza A virus particles in perfusion mode using an alternating tangential flow filtration system. Appl Microbiol Biotechnol, 2021. 105(19): 7251–7264.
[22] Nikolay, A., A. Leon, K. Schwamborn, Y. Genzel, and U. Reichl. Process intensification of EB66(R) cell cultivations leads to high-yield yellow fever and Zika virus production. Appl Microbiol Biotechnol, 2018. 102(20): 8725–8737.
[23] Chen, H.C., and Y.C. Hu. Bioreactors for tissue engineering. Biotechnol Lett, 2006. 28(18): 1415–23.
[24] Ahmed, S., V.M. Chauhan, A.M. Ghaemmaghami, and J.W. Aylott. New generation of bioreactors that advance extracellular matrix modelling and tissue engineering. Biotechnol Lett, 2019. 41(1): 1–25.
[25] Begley, C.M., and S.J. Kleis. The fluid dynamic and shear environment in the NASA/JSC rotating-wall perfused-vessel bioreactor. Biotechnol Bioeng, 2000. 70(1): 32–40.
[26] Song, L., Q. Zhou, P. Duan, P. Guo, D. Li, Y. Xu, S. Li, F. Luo, and Z. Zhang. Successful development of small diameter tissue-engineering vascular vessels by our novel integrally designed pulsatile perfusion-based bioreactor. PLoS One, 2012. 7(8): e42569.
[27] Plunkett, N., and F.J. O'Brien. Bioreactors in tissue engineering. Technol Health Care, 2011. 19(1): 55–69.
[28] Nadhif, M.H., H. Assyarify, A.K. Waafi, and Y. Whulanza. Reflecting on mechanical functionalities in bioreactors for tissue engineering purposes. International Journal of Technology, 2020. 11(5): 1066–1075.
[29] Bilgen, B., and G.A. Barabino. Location of scaffolds in bioreactors modulates the hydrodynamic environment experienced by engineered tissues. Biotechnol Bioeng, 2007. 98(1): 282–94.
[30] Engelmayr, G.C., Jr., V.L. Sales, J.E. Mayer, Jr., and M.S. Sacks. Cyclic flexure and laminar flow synergistically accelerate mesenchymal stem cell-mediated engineered tissue formation: Implications for engineered heart valve tissues. Biomaterials, 2006. 27(36): 6083–95.

4

The Effects of Biomaterial Properties on the Behaviors of Stem Cells

I-Chi Lee,[1,*] *Nai-Chen Cheng,*[2] *Chia-Ning Shen,*[3] *Wen-Yen Huang,*[4] *Sung-Jan Lin,*[4] *Min-Huey Chen*[5] and *Yi-Chen Ethan Li*[6,*]

During the past decade, stem cell biology and biomaterial technology have promoted a synergistic effect on stem cells-based tissue engineering and regenerative therapies. Stem cells with the self-renewal capacity and multipotential differentiation properties provide a natural and outstanding source for cell therapy. Creating reserves of undifferentiated stem cells and controlling their differentiation to a specific lineage in an efficient and scalable manner is critical for their clinical application in regenerative medicine. Therefore, understanding the interaction of stem cell and biomaterials will promote the creation of new biomaterials that can be applied in medicine. Stem cells are unspecialized cells of the human body, having the potential to differentiate into any cell of an organism, and have the ability of self-renewal. Stem cells has gradually become valuable cell resource in tissue engineering and disease treatment. Stem cell therapy has also become a very promising and advanced scientific research topic.

[1] Department of Biomedical Engineering and Environmental Sciences, National Tsing Hua University, Taiwan.
[2] Department of Surgery, National Taiwan University Hospital, Taiwan.
[3] Genomics Research Center, Academia Sinica, Taiwan.
[4] Department of Biomedical Engineering, College of Medicine and College of Engineering, National Taiwan University, Taiwan.
[5] Graduate Institute of Clinical Dentistry, School of Dentistry, National Taiwan University, Taiwan.
[6] Department of Chemical Engineering, Feng Chia University, Taiwan.
* Corresponding authors: iclee@mx.nthu.edu.tw; yicli@fcu.edu.tw

Stem cells can be divided into three specific categories depending on their developmental potency: embryonic stem cells (ESCs), induced pluripotent stem cells (iPSCs), and adult stem cells. Several therapeutic issues have been discussed vis-à-vis stem cell regulation and their application, such as differentiation protocols, teratoma or tumor issues, and immune compatibility, especially with ESCs and iPSCs. Adipose-derived stem cells (ASCs) are a kind of adult stem cells that are generally used and can be found abundantly in fat tissue, thus representing a rich source of mesenchymal stem cells (MSCs), which can be easily harvested with minimally invasive procedures, like liposuction [1–2]. ASCs exhibit multipotency to differentiate into several lineages, including osteogenic, adipogenic, and chondrogenic lineages [3]. They are even capable of trans-differentiation into lineages of distinct origins, such as hepatocyte and neuron [4–5]. In addition, it has been found that ASCs can secrete proteins involved in angiogenesis, wound healing, tissue regeneration, and immunomodulation [6].

In a living organism, cells are surrounded by other cells and embedded in an extracellular matrix (ECM) that modulates cell activity and function by the construction, signaling, and biomechanics of the cellular microenvironment. Therefore, the chemical and biophysical cues of the surrounding matrix and the interaction with surrounded cells affect stem cell adhesion, proliferation, and differentiation. Biomaterials open up a new avenue on the regulation of stem cell behaviors via surface properties modulation for mimicking the *in vivo* microenvironment. Biomaterials are usually designed with a combination of suitable biological and mechanical properties such as biocompatibility, high porosity, and suitable mechanical strength to mimic the natural ECM and nature microenvironment [7]. Different kinds of proteins and peptides in the ECM regulate cellular behaviors and receptor binding. Besides, mechanical properties and the construction of ECM play a key role in stem cell differentiation. Advances in material synthesis, processing, and modification have opened up a range of synthetic and natural materials for use in controlled microenvironments.

The cellular microenvironment is known to play a significant role in determining cell fates. For example, topography, hydrophilicity, protein adsorption ability, and stiffness of the substrates are critical parameters for directing cell fate and have been reported to modulate stem cells attachment, proliferation, and differentiation [8–9]. Another example is the association between matrix elasticity and lineage specification of stem cells [10]. With advances in technology, biomaterials can be designed as active and response biomaterials or synthesized and processed to create a microenvironment with desired characteristics that can support stem cell growth, maintain the stemness, or induce specific lineage commitment. In this chapter, we focus on the influence of biomaterial hydrophilicity/hydrophobicity, stiffness, and protein adsorption ability on the behaviors of stem cells. In order to mimic the natural microenvironment, researchers have fabricated complex artificial extracellular matrices and the design of biomaterials to control stem cell behavior, such as cell-responsive ligands, mechanical signals, and delivery of soluble factors, which are also highlighted in this chapter.

Biomaterials used for cells and stem cell culture includes polymers, ceramics, metals and composites. These materials derived from synthetic, natural, or a combination of sources must respond to biological signals and interact with the stem cells to influence stem cell behaviors such as cell adhesion, proliferation, migration, differentiation, and signaling. These biomaterials can interact with the stem cells through their biophysical and biochemical properties, which can alter local microenvironments and stem cell fate by modulating the cell-materials interaction and protein adsorption. Stem cells sense and react to the surface properties of biomaterials as they anchor and pull on their surrounding ECM. Cells can respond to microenvironment and regulate ECM expression resulting in a rapid remodeling of the matrix in both the nature and quantity of constituent molecules. With the goal of mimicking 3D ECM to regulate stem cell behavior, the following natural ECMs and their derivates have been reported to support stem cell proliferation and regulate cell differentiation: collagen, laminin, gelatin, hyaluronan, fibrin, glycosaminoglycans (GAGs), alginate, Matrigel, silk fibroin, hydroxyapatite (HA), etc. Although natural ECMs and their derivates demonstrate specific advantages such as specific ligands and similar mechanical properties as the natural tissue, batch variability, and short degradation period also make up the main disadvantages in comparison with synthetic biomaterials.

Collagens, ubiquitous proteins found throughout the body, are the most abundant ECM macromolecule. Among all collogen types, collagen IV is a network-forming collagen which has been largely implicated in mesodermal differentiation, such as the hematopoietic, endothelial, and smooth muscle lineages [11–12]. In addition, collagen I has also showed evidence in playing a role in guiding cardiomyocyte differentiation in mouse ES cells [13]. Fibronectin is an ECM molecule expressed during the early stages of development, especially during the development of the mesoderm and neural tubes. An earlier study showed that binding $\alpha5\beta1$ integrin to specific fibronectin domains demonstrated differentiation into mesodermal and ectodermal lineages [14]. Many of the adult stem cell niches are also rich in laminin. The basement membrane of the epidermis consists of laminin 332 and 511 with skin stem cells shown to express the integrins, $\alpha2\beta1$, $\alpha3\beta1$, and $\alpha6\beta4$ [15]. In addition, collagen XVII deposition by hair follicle stem cells provides a niche for the self-renewal and maintenance of both hair follicle stem cells and melanocyte stem cells [16]. Also, neural stem cells (NSCs) and their progeny are anchored to the ECM emanating from blood vessels and the parenchyma of the ventricular cavity [17]. It is revealed that the neurogenic niche is a complex cellular and extracellular microenvironment. A study found that NSCs and precursors are embedded in a laminin-rich ECM in comparison with the surrounding non-neurogenic areas [18]. Furthermore, Matrigel was isolated from the natural basement membrane, which contained several kinds of ECM components including laminin, collagen IV, enactin, and growth factors. Matrigel is generally used to retain the stem cells in an undifferentiated state and to maintain self-renewal and pluripotency, especially for the growth of human ESCs [19]. Fibrinogen and fibrin are another source of tissue-derived natural materials and have been applied in 3D scaffold and 3D bioprining construction [20]. Fibrin hydrogels are a popular choice for use as a 3D engineered construct based on their good biocompatibility. Gandhi et al. compared the behaviors of iPSC culture on 2D fibrinogen and 3D fibrin hydrogels. They revealed that iPSCs successfully proliferated and maintained on fibrin hydrogels but tend to differentiate into endothelial cells on 2D fibrinogencoated plates [21]. In another study, a fibrin-based bioink was designed for bioprinting with the human dental pulp stem cells and they demonstrated that odontogenic differentiation was also regulated according to the fibrinogen concentration [22].

Alginate is a hydrophilic polysaccharide with good biocompatibility, derived from seaweed. Alginate hydrogel is generally used for biomaterial scaffolds, cell encapsulation, and *in vitro* differentiation. It is revealed that the properties of alginate hydrogel enhanced neural differentiation of encapsulated hADSCs in comparison with monolayer culture [23]. Also, RGD-alginate hydrogel scaffolds showed enhanced retinal ontogenesis from hiPSCs [24].

Furthermore, ECM are important in many cellular phenomena, including cell shape, function, and stem cell differentiation. Integrins are the main mediators of cell-ECM interaction. Design and modulation of ECM analogs to ligate specific integrins is a promising approach to control cellular processes. Integrins are transmembrane proteins consisting of α- and β-subunits, and bind to the ECM and cellular cytoskeleton to provide biophysical and biochemical signaling. An earlier study controlled the integrin-dependent cellular interactions with the ECM through production of FN III9-10 variants with variable stabilities to regulate stem cell behavior [25]. Schwab E.H. et al. have also assayed MSCs for their osteogenesis in the presence of the ECM glycoproteins, which contain an integrin-binding sequence that includes fibronectin, vitronectin, and osteopontin. Taken together, this study revealed that β1-integrins appear to be the predominant receptor utilized by MSCs to adhere and proliferate and β3-integrins seem to suppress MSCs to the osteoblast lineage [26].

Although natural ECMs showed good biocompatibility and self-existing biosignals, the weakness of mechanical strength still limits their applications. In contrast, synthetic polymers have some advantages over the natural materials, such as their polymer composition can be accurately controlled, mechanical strength is tunable, easy modification is possible, and the materials are more uniform when there is a sufficient source of raw materials. However, synthetic biomaterials lack cell adhesion receptors and biological signals that may not direct cell fate on their own. Notably, biodegradable and bioresorbable polymers are commonly used polymers for scaffold preparation and stem cell culture, such as polylactic acid (PLA), poly (lactic-co-glycolic acid) (PLGA), polycaprolactone (PCL), polyethylene glycol (PEG), polyhydroxyl ethyl methacrylate (PHEMA), and polyvinyl alcohol (PVA) [27]. Polyester-based scaffolds have been thoroughly investigated for tissue engineering, as they are biodegradable, biocompatible, and yield degradation byproducts that can be metabolized by the body. Previous literature have developed co-electrospun fibers of PLA and gelatin with a degradation rate and mechanical properties to mimic peripheral nerve tissue and demonstrated the differentiation of motor neuronal lineages and neurite outgrowth promotion [28]. PLA has also been fabricated as a nanofibrous scaffold to induce MSCs differentiated into chondrocyte-like cells and induce odontogenic differentiation of human dental pulp stem cells (DPSCs), respectively [29–30]. PCL is one of the biocompatible and biodegradable polymers, has a slower degradation rate, and is widely applied on the fibers fabricated on the stem cells scaffold. Aligned PCL fibers are fabricated in different sizes using a microfluidic platform to facilitate cell attachment and regulate NSCs differentiation [31]. A combination of PCL and natural materials such as a polycaprolactone/hydroxyapatite (PCL/HAp) scaffold has also been applied on osteogenic differentiation from three types of mesenchymal stem cells, including bone marrow-derived mesenchymal stem cells (BMSCs), DPSCs, and ADSCs [32]. Synthetic hydrogel polymer are generally prepared with different mechanical properties on the determination of stem cell differentiation. A synthetic PEG-based hydrogel was constructed to replace feeder layer culture system and they demonstrated the hydrogel have the potential to support ESCs self-renewal [33]. They revealed that modulation of 3D properties can create various models for stem cell transformation and differentiation. PEG hydrogel contains a series of Young's modulus gradients and to examine changes in the Young's modulus of the culture substrate on NSC neurite extension and neural differentiation [34]. PVA is a water soluble polymer that has been well studied for use in tissue engineering scaffolds due to its relatively high strength, creep resistance, water retention, and porous structure [35]. PVA comprises a large amount of hydroxyl groups, which interact with the water molecules through hydrogen bonds. A PVA/chitosan porous hydrogel was designed and loaded with BMSCs, then applied to the treatment of osteochondral defects [36]. Another PVA/sulfated alginate nanofibers scaffold has been fabricated and the substrate enhance MSCs growth and is capable of inducing neuronal differentiation [37].

A material's interfacial properties strongly affect stem cell behavior, including adhesion, proliferation, migration, and differentiation. Cells interact with the materials through the cell-materials interface, the "interfacial" properties of the interfacial energy of the substrate, hydrophobicity, and surface topography, all of which strongly regulate stem cells' functions. Hydrophilic/hydrophobic characteristics of a biomaterial are commonly evaluated by the water contact angle. A lower contact angle indicates higher hydrophilicity. It is known that cell adhesion is maximized on surfaces with intermediate wettability [38]. For example, bone marrow-derived MSCs have been shown to attach better on moderately wettable polyethylene (PE) surfaces with a contact angle of about 65°[39]. ASCs have similar characteristics. In a study, a sodium hydroxide (NaOH) treatment was used to enhance the hydrophilicity of poly(ε-caprolactone) (PCL)/pristine graphene scaffolds, thereby further enhancing ASC attachment [40]. In another study, surface-modified PE was applied to demonstrate a parabolic relationship between cell attachment and the water contact angle and establish a correlation between surface wettability and biological response in ASCs [41]. Mei et al. have developed a high-throughput analysis to determine structure–function relationships between material properties and biological performance. They revealed that matrix hydrophobicity exhibits a strong correlation with the stem cell behaviors. They found that optimal human ESC substrates with a high acrylate content have a moderate wettability and employ integrin αvβ3 and αvβ5 engagement with adsorbed vitronectin to promote colony formation [42]. In addition, a tunable, synthetic matrix in which the interfacial hydrophobicity is controlled in a systematic manner by varying the alkyl chain length of pendant side chains has been developed as a platform to determine the hydrophobicity of the extracellular matrix on the bioeffect of stem cells [43]. Ahn et al. have designed a wettability gradient surface (water contact angle, 90° to ~50°) to study the adhesion and proliferation of ASCs cultured on polyethylene (PE) surfaces. They found that hASCs adhered better and showed a higher proliferation rate on hydrophilic and rough PE surfaces in comparison with hydrophobic and smooth surfaces [41]. In addition, a previous study tried to define the relation of hydrophobicity and EB development; it revealed that the hydrophobicity of polydimethylsiloxane (PDMS) and self-assembled monolayers of various lengths of alkanethiolates may help deliver uniform EB populations and may significantly improve the efficiency of ES cell differentiation. However, the cell attachment was decreased [44]. During the process of altering surface hydrophilicity, other biomaterial properties are often modified simultaneously; thus the observed cell behaviors might result from a mixed effect of different characteristics. To specifically observe the influence of surface hydrophilicity on stem cell behaviors, a tunable, synthetic matrix whose hydrophilicity can be controlled without altering the chemical and mechanical properties was developed, and a striking effect of matrix interfacial hydrophobicity on stem cell adhesion, motility, cytoskeletal organization, and differentiation was noted [43].

Furthermore, researchers considered that surface topography also impacts stem cell differentiation and it is necessary to assess the relevance of specific topographic parameters with systematic defined structures. Cells can respond to topographical features as small as 5 nm, which affect the cells' orientation and alignment on the patterns at a nanoscale resolution. In general, there are two factors affecting substrate surface topography, roughness, and patterns on the surface [45]. Different types of patterns have been fabricated by current nano- and microfabrication methods, including grooves, aligned fibers, pillars, pits, and tubes [45]. Numbers of literatures have reported that surface roughness regulate stem cell behaviors such as adhesion, migration, proliferation, and differentiation, especially on MSCs. Faia-Torres et al. designed a PCL gradient substrate with average roughness (Ra) varying from the sub-micron to the micrometer range (~ 0.5–4.7 μm), and mean distance (from 214 μm–33 μm) to study the roughness's effect on MSCs differentiation. Their results show that the higher ALP, COL1, and mineralization expression was found on a region of substrate gradient at a position 5 mm that corresponded to $R_a \approx 0.93$ μm, which revealed that the osteogenesis of MSCs are related to the surface's roughness [46]. Moreover, a pattern issue is generally based on the orientation of topography.

Cell orientation and migration along the anisotropic direction of ridges and grooves have long been observed on a microscale. For example, MSCs on a microgrooved polyimide chip featuring a combination of 25 different structures with a systematic variation of the width of grooves and ridges was used to observe cell spreading and orientation and to determine the MSCs differentiation lineages. The result is interesting in that 15 μm ridges increased adipogenic differentiation whereas 2 μm ridges enhanced osteogenic differentiation [47]. In another study, a series of submicron-grooved polystyrene substrates with equal groove-to-ridge ratio but different width and depth were fabricated to determine the osteogenesis, adipogenesis, and myogenesis of rat MSCs. Their results demonstrated that MSCs committed to adipogenic and myogenic tendency on microscale grooves, especially when the groove scale was less than 500 nm. In contrast, osteogenesis was not significantly modulated by the grooved substrates in this study [48]. In addition, researchers have also tried to analyze the signal pathway of the topography effect and it is revealed that topography influences MSC wnt signaling through the regulation of the primary cilia structure and function, which affects intracellular actin-myosin tension and the stem cell fate [49]. Ruiz et al. have fabricated PLL-micropatterned surfaces with a plasma polymerized PEO on petri dishes using the microcontact printing method and to investigate NSCs differentiation. The results revealed that NSCs cultured on the PLL-micropatterned surface expressed highly neural markers and more axon-like outgrowth [50]. Fiber alignment is proof of neural differentiation. Poly(l-ornithine) (PLO) and laminin coating on aligned nanofibers are prepared for neuron differentiation, and the direction of fiber alignment may induce NSC polarization and cellular elongation [51]. Besides NSCs, human MSCs cultured on aligned PCL/gelatin nanofibers with and without RA encapsulation also demonstrated upregulation of neural markers in comparison with that on randomly orientated nanofibers [52]. In another study, Beduer et al. determined the bioeffect of groove and terrace width of micropatterned surfaces with striped groove morphologies on the differentiation of NSCs, the differentiation efficiency of NSCs into neurons, and the intercommunication between neurons[53]. For isotropic patterns, a previous review has revealed that cell response to an isotropic pattern is often inconsistent and difficult for in-depth analysis due to the variation factors such as stem cell culture conditions, materials, and pattern types [45].

Similarly, the surface charge of the substrate is another important factor with a high impact on stem cell regulation. Many substrates with differently charged surfaces and prepared by different methods have been reported. Bodhak et al. investigated the influence of surface charge and polarity on *in vitro* bone cell adhesion, proliferation, and differentiation on electrically polarized hydroxyapatite-coated Ti. Their results showed that negatively charged surfaces enhanced rapid cell attachment and faster tissue ingrowth while a positive charge on HAp surfaces restricted apatite nucleation with a limited cellular response [54]. In addition, a ferroelectric crystal platform has been developed to provide oppositely charged surfaces and the 2D culture substrate was applied to promote the osteogenesis of MSCs [55]. Their results demonstrate that positively charged surfaces may enhance the enlargement of the area of MSCs adhesion which promote the osteogenic differentiation of MSCs that have been cultured in the osteogenesis medium. In contrast, most of the experiments showed evidence that NSCs demonstrated high differentiation into functional neurons on surfaces with positive charge. There are also a few reports, which focus on investigating the cellular response to charged functional groups such as –COOH, $-CH_2NH_2$, $-CH_2OH$, etc., Lee et al. have published several literatures to use polyelectrolyte multilayer (PEM) films to regulate NSCs differentiation [56–57]. A biomimetic material comprising a supported lipid bilayer (SLB) or on ITO surface with adsorbed sequential PEM films has been fabricated to induce NSPCs to form functional neurons. The results demonstrated that the process outgrowth length, the percentage of differentiated neurons, and the synaptic function were regulated by the number of layers and the surface charge of the outermost layer.

Mechanical properties and stiffness

The natural environment consists of ECM proteins, polysaccharides, and water. The structure of tissue microenvironment is dependent on the location and function of the tissue. For instance, cartilage has a high proteoglycan content within a strong collagen network, a matrix that gives this tissue its important mechanical properties to sustain natural loading. In contrast, a tendon is highly anisotropic with respect to ECM orientation, making mechanical properties directionally dependent. Various proteins and peptides in the ECM control cellular interactions and receptor binding. Also, factors such as ECM mechanics and constriction of cell shape can play a role in the differentiation of stem cells and can be engineered into synthetic matrices for directed differentiation.

Besides surface properties of the matrix, the importance of substrate mechanical properties on stem cell differentiation has been demonstrated [10]. The mechanical properties of matrix affect the stem cell attachment, migration, proliferation, and differentiation. Early *in vitro* work demonstrated the role of stiffness in controlling adhesion, spreading, and the differentiation of stem cells in 2D culture conditions [58]. Biomaterials can be fabricated into an engineered construct that exhibits a wide range of mechanical properties with moduli from the Pa through the GPa range. Hydrogels are primary tools used to study the stem cells response to stiffness *in vitro*. The hydrogel stiffness can be prepared and regulated from very soft at less than 0.1 kPa to very hard at about 500 kPa. Mechanical stimuli represent major regulators of the development and function of many tissues, such as bone, cartilage, ligament, and smooth muscle and cardiovascular tissues. Human MSCs were the first stem cells used to present the effect of matrix stiffness on stem cell differentiation. Polyacrylamide hydrogels modified with collagen have been fabricated to exhibit a wide range of stiffness that correlate to the mechanical properties of native tissues. On soft matrices that mimic brain elasticity, bone marrow-derived MSCs tend to differentiate into the neurogenic lineage; on stiffer matrices that mimic muscle elasticity, MSCs tend to differentiate into the myogenic lineage; on the stiffest matrices that mimic bone elasticity, MSCs tend to differentiate into the osteogenic lineage without adding any soluble induction factor [58]. Taken together, MSC differentiation commits to the lineage specified by matrix elasticity in the absence of any known inducing soluble factors. Similarly, researchers have investigated the influence of ECM components and their mechanical properties on the stimulation of adipogenesis of ASCs [59]. On gels that mimic the stiffness of adipose tissue, ASCs showed an elevated expression of adipogenic transcription factors, while the cells failed to express adipogenic markers with increased substrate stiffness. In addition, a previous study has designed a novel hydrogel array with differential wettability surfaces to determine the effect of stiffness on MSCs. The results demonstrated enhancement of MSCs adhesion, migration, and proliferation as hydrogel stiffness has increased [60]. In addition, the effect of the matrix stiffness on the phenotype and differentiation pathway of MSCs has been evaluated in thixotropic gels of varying rheological properties (7, 25, and 75 Pa). The results revealed that the stiffer gels immobilized with RGD showed evidence to promote both proliferation and the differentiation potential of MSCs [61]. A patterned PEG hydrogel nanocomposite with the stiffness of liver tissues was prepared and the MSCs' mechanoresponse to the material was investigated. It is showed that hMSCs' locomotion is influenced by the nature of the hydrogel layer [62].

Also, 3D scaffolds with different degrees of stiffness but the same 3D microstructure have been fabricated by using mixtures of collagen and hydroxyapatite coating on decellularized cancellous bone to determine the stiffness effect in 2D culture conditions [63]. Their results demonstrated that these 3D microstructure with different levels of stiffness can sustain the adhesion and growth of rat MSCs and promote osteogenic differentiation. Furthermore, the subcutaneous implantation of these scaffolds showed that they attract endogenous stem cells. Overall, it is showed that scaffold with higher stiffness increased the production of osteo-related proteins, which demonstrated that the matrix stiffness could be sensed by stem cells and facilitated deposition of ECM.

Matrix stiffening occurs in many biological events, including tissue development, tissue fibrosis, and disease progression. In a study, a controllable stiffening hydrogel system had been applied to investigate the response of ASCs toward a dynamic change of microenvironment stiffness [64]. It was revealed that the change in mechanical properties only regulate lineage specification in ASCs that are undifferentiated, whereas already differentiated cells are not responsive to the change in stiffness. The results here indicated the importance of timing to exert the substrate stiffening effect on ASC behaviors.

Besides MSCs, human ESCs cultured on fibronectin-coated PDMS with stiffness variation ranging from 0.078 to 1.167 MPa has also provided evidence that substrate stiffness promote ESCs adhesion and proliferation [65]. Furthermore, cardiac muscle has a more limited adult stem-cell progenitor pool that cardiomyocytes derived from iPSCs is an alternative on regenerative medicine application. Many researchers try to culture iPSCs on rigid plastic to generate functional cardiomyocytes, however, the efficiency and functionality are limited [66]. In addition, it is revealed that rigid postinfarct regions limit pumping by the adult heart. Moreover, iPSC-derived cardiomyocytes have been proved to be mechanosensitive to the matrix and thus generalize the main observation that myosin II organization and contractile function are optimally matched to the load contributed by matrix stiffness [67]. An *in vivo* model also demonstrated that transplanting human iPSC-derived cardiomyocytes into a neonatal rat heart with a soft microenvironment facilitates nearly full maturation of the cardiomyocytes in comparison to when the cardiomyocytes are transplanted into a stiffer adult rat heart [68]. In addition, embryonic cardiomyocytes also tune their beating to the stiffness of their substrate and it is shown that too stiff gel may inhibit the beating [67].

Furthermore, a combination effect of surface properties on stem cell regulation was investigated in the past decade. A micro-fabricated polyacrylamide hydrogel substrate with two elasticities, two topographies, and three dimensions have been developed to systematically explore MSCs behaviors includes proliferation, spreading, differentiation, and cytoskeletal re-organization [69]. It is demonstrated that topography is a key factor for manipulating cell morphology and spreading and substrate stiffness or dimension is predominant in regulating cell proliferation, respectively. In another study, Annulus fibrosus (AF)-derived stem cells were cultured on four types of electrospinning poly (ether carbonate urethane) urea (PECUU) scaffolds of various stiffness and fiber size to determine and mimic the mechanical and topographical features of native AF tissue. The results revealed that fiber size of scaffolds affect the topography and result in the variations in cell shape, spreading area, and extracellular matrix expression; the matrix stiffness regulates the differentiation [70].

Conductive biomaterials represent a huge revolution on biomedical engineering due to their excellent electrically conducting, elasticity, and tensile strength [71]. Conductive materials can be classified based on their structure and nanoscale dimensions such as zero-dimensional structures such as fullerenes and diamond clusters, one-dimensional materials such as carbon nanotubes and carbon nanofibers, two-dimensional such as nanosheets and nanoplatelets, and three-dimensional structures such as nanocrystalline diamond films. Electroactive nanomaterials are gradually developed and applied on stem cells differentiation, such as conducting polymers, carbon nanotubes, graphene, silicon nanowires, and MXenes. Electrical stimulation of conducting polymer scaffolds greatly affects the cell functionality and enhanced cell growth.

Conductive polymers display several advantages vis-à-vis stem cell regulation, such as encapsulating with biomolecules and release, functionalization with bioactive molecules, and altering the electrical properties of the surface to regulate stem cell behavior and enhance cell functionality [71]. Among these polymers, poly(pyrrole) (Ppy) and poly(3,4-ethylenedioxythiophene) (PEDOT)-based materials and some of their derivatives have received more attention for biomedical applications, especially on MSCs differentiation, neurogenesis, and cardiomyogenic differentiation. Hardy et al. fabricated biomineralized conducting polymer-based Ppy/PCL scaffolds seeded with ADSCs and facilitate the electrical stimulation effect on ASCs. They found that the electrical stimulation of ADSCs with 200 μA of direct current enhanced the cell migration into deeper regions of the scaffold and improved their osteogenic differentiation [72].

Furthermore, the incorporation of biological components into the Ppy film such as brain-derived nerve growth factor (BDNF) and nerve growth factor (NGF) have significantly showed enhancement of neurite extension on PC12 cells [73]. A PEDOT/alginate porous scaffold has been prepared and the PEDOT exists in the alginate matrix as particles to enhance the conductivity of scaffold up to 6×10^{-2} S cm^{-1} [74]. The results demonstrated that adding PEDOT supports the attachment and proliferation of BADSCs and promotes the cardiomyogenic differentiation of BADSCs under electrical stimulation. In another study, PEDOT:PSS showed the benefit of differentiation of mouse ESCs and this platform showed the evidence to produce higher electrical current with the pulsed-direct-current (DC) electrostimulation mode and also enhance cardiomyogenesis [75]. Besides, neural cells are electro-active ones; NSCs and CNS neural cells are exposed to electrical signaling during development which influences their organization and maturation into adult tissue. Many researchers have showed evidence that a conductive scaffold can provide appropriate synergistic cell guidance cues on NSCs. Conductive conjugated polymers such as PEDOT:PSS have been showed as non-cytotoxic materials on NSCs differentiation and they revealed that pulsed electrical current applied to PEDOT:PSS promotes differentiation of NSC and neurites elongation of neurons [76]. In order to improve the biocompatibility of conductive polymers, an electrically conductive, porous, and biodegradable scaffold prepared by PEDOT/chitosan/gelatin with a nanostructured layer of PEDOT assembling on the channel surface of porous Cs/Gel scaffold was fabricated and the effect on NSCs differentiation were determined. It was found that PEDOT layer not only greatly promoted NSCs adhesion and proliferation, but also enhanced the NSCs differentiation towards neurons and astrocytes under the differentiation condition [77]. Also, the entrapment of NGF into PEDOT doped with peptide have been proven to enhance electrical and mechanical stability and promote neurite outgrowth [78] .

Carbon nanotubes (CNT) have remarkable optical, mechanical, and electrical properties and have also been tested on biomedical application. CNT have been refined including single wall nanotubes (SWNT), multiwalled nanotubes (MWNT), and functionalized nanotubes. In an earlier study, different types of CNT including SWNT, MWNT, COOH-functionalized SWNT, and OH-functionalized MWNT, to identify the optimum type of CNT for use with hMSC for electrically stimulating hMSC to promote differentiation toward a cardiomyocyte lineage [79]. In addition, the arrangement of individual CNTs in the CNT networks enhanced proliferation and osteogenic differentiation of hMSCs on aligned CNT networks and allowed the controlling of the growth direction [80]. CNT-based substrates on neural differentiation have also been determined and the results demonstrated that differentiation, neurite outgrowth, and electrophysiological maturation of ESC-derived and NSC-derived neurons were observed [81–82].

Atomically thin graphene (G) and graphene oxides (GO) sheets are soft membranes with high in-plane stiffness and can potentially serve as a biocompatible materials and have been shown to have the potential to mediate stem cell fate on the application of tissue regeneration. As a monatomic layer of carbon atoms in a honeycomb lattice, G has been intensively studied with excellent electrical conduction ability (carrier mobility: 10000 cm2·V−1·s−1) and superior mechanical strength (Young's modulus: 1100 GPa) [83]. G is characterized by a purely carbon, aromatic network that presents an open surface for noncovalent interaction with biomolecules. In contrast, GO can be derived by the chemical oxidation and exfoliation of graphite. Hydrophilic functional groups such as hydroxyl groups present on the basal plane of GO enable greater interactions with proteins through covalent bonding or electrostatic properties [84]. A number of studies have observed the use of G-based materials in stem cell application, with a particular focus on promoting stem cell growth, guiding stem cell differentiation into specific lineages, stem cell delivery, and monitoring of stem cell differentiation. Herein, we focused on materials properties on the regulation of stem cell behaviors. The regulation of differentiation lineages of G and GO on the MSCs have been investigated early. The results revealed that G accelerate MSCs growing and toward the osteogenic lineage. Besides, the molecular origin of accelerated differentiation is investigated by studying the binding abilities of G and GO toward different growth agents. The results are interesting in that the differentiation to adipocytes is greatly suppressed on G but not on GO since different binding interactions such as π–π stacking and electrostatic and hydrogen bonding mediated by G and GO influence on the stem cell growth and differentiation [84]. An earlier study revealed that NSCs differentiated on the G films exhibited not only an improvement of neuronal differentiation, but also enhanced functional maturation into electrically active neurons [85]. G and GO have been proofed to regulate the mouse iPSCs proliferation and allow for spontaneous differentiation [86]. iPSCs cultured on the G surface exhibited similar degrees of cell adhesion and proliferation. In contrast, iPSCs on the GO surface adhered and proliferated at a faster rate. They also revealed that G favorably maintained the iPSCs in the undifferentiated state while GO expedited the differentiation, especially the endodermal differentiation [86]. Additionally, GO was also shown to effectively promote dopamine neuron differentiation derived from mouse ESCs [87] and it is revealed that among many different carbon-based materials including CNTs, GO, and graphene; only GO was found to be effective in guiding the generation of dopaminergic neurons from mESCs [88].

A three-dimensional structure composed of a graphene and nickel template has been defined as a graphene foam and found to induce elongated morphologies of MSCs and enhance osteogenesis [89]. Besides, graphene foams have been demonstrated to allow electrical stimulation based on the excellent electrical properties of graphene on differentiated neuronal cells and result in changes in intracellular calcium ion concentrations [90]. In another study, a graphene-based nanomaterial was designed as hybrid nanofibrous scaffolds to guide NSC differentiation into oligodendrocytes [91]. They demonstrated that the overexpression of some key integrin-related intracellular signaling molecules that are well known to promote oligodendrocyte differentiation in normal development were investigated on GO-coating scaffolds. Furthermore, graphene polymer scaffolds have also showed effects on stem cells proliferation and differentiation. The enhancement of oligodendrogenesis of GO-modified PCL nanofibers on the differentiation of NSCs [92] and chondrogenic differentiation of MSCs was also investigated on polymer-GO hybrid scaffolds [93], and the cell proliferations showed a 30%–50% increase in the rGO/ PEG-poly(l-alanine) diblock copolymer (PEG-l-PA) hybrid system than in the GO/PEG-l-PA hybrid system [93].

GO/PEDOT nanocomposite films have also been prepared as a NSCs scaffold to selectively drive NSC differentiation toward either neuronal or oligodendrocyte lineage with different factors [94]. The results demonstrated that the surfaces support a larger population of neurons when modified with IFNγ and a larger population of oligodendrocytes when modified by PDGF. A self-powered electrical stimulation-assisted neural differentiation system with reduced graphene oxide (rGO)-PEDOT hybrid microfiber as a scaffold for MSCs differentiation was fabricated by Guo et al.; they demonstrated that MSCs cultured on the highly conductive rGO–PEDOT hybrid microfiber showed a significantly enhanced proliferation ability and neural differentiation tendency [95].

Gogotsi's group was the pioneering team to fabricate 2D Titanium Carbide MXene (Ti3C2) nanosheets in 2011 [96]. After that, MXene-based nanomaterials gained a lot of attention and became a popular two-dimensional material henceforward. MXenes possesses some attractive properties that makes them suitable for biomedical applications such as a large specific surface area, rich surface functional groups, high conductivity, hydrophilicity, and photothermal conversion efficiency [97]. These outstanding properties enabled MXenes to be applied in numerous research fields, also includes biomedical study. Most of the biomedical application of MXenes are related with biosensing, photothermal therapy, theranostic nanoplatform, and antibacterial activity [98], the application of MXenes on tissue engineering and stem cells regulation have barely been studied. The cytotoxicity of MXenes on NSCs has been tested and they revealed that 12.5 μg/mL MXenes had no observable adverse effect but 25 μg/mL MXenes induced significant cytotoxicity [99]. In addition, NSCs cultured on MXenes nanosheets with laminin coating has been shown to form more active and synchronous network activity than those cultured on TCPS substrates and they also demonstrated electrical stimulation for 10 min a day for three consecutive days coupled with MXenes film significantly enhance the proliferation of NSCs [100]. Recently, smart biomaterials comprising MXene nanofibers were fabricated for tissue engineering and cell culture. MXene composite nanofibers were fabricated by electrospinning for culture of BMSCs and it wasdemonstrated that the obtained MXene composite nanofibers showed good biocompatibility, improved cellular activity, and enhanced BMSCs' differentiation to osteoblasts [101]. Also, cation-induced assembly of conductive MXene fibers was prepared by the fast extrusion of alginate fibers followed by electrostatic assembly of MXene nanosheets. The fibers with laminin coating through physical adsorption interactions also showed guidance of neural stem/progenitor cells differentiation and the promotion of neurite outgrowth [102]. In addition, cell-laden bioink encompassing MXene nanosheets was fabricated in HA/alginate hydrogel and the MXene nanocomposite ink showed remarkable rheological characteristics and biocompatibility for 3D bioprinting, and introduces electrical conductivity to the ink [103]. In a recent study, hydrogels with inter-connected porous architecture assembling from 2D materials was fabricated by inducing chemical inter-sheet crosslinks via an ethylenediamine mediated reaction between MXenes and rGO to obtain a rGO-MXene hydrogel [104]. It is suggested that 3D hydrogels based on 2D materials may provide an alternative to biological application and tissue engineering.

Protein adsorption ability

Protein adsorption on the biomaterial surface is believed to be the initial event when a material comes into contact with a biological environment. After adsorbed to the surface, some proteins, like fibronectin, laminin, collagen, and growth factors can interact with cells. These proteins bind with receptors on the cell membrane (mainly integrin), and induce downstream signaling. Thus, the protein adsorption ability of biomaterials can have a great influence on cell fate, such as adhesion, proliferation, and differentiation [105–107].

In native ECM, the sulfated polysaccharides (chondroitin sulfate or dermatan sulfate) are expected to bind and enrich growth factors, which are essential for cell differentiation. Carrageenan, a chondroitin sulfate-like material, was added to chitosan-based scaffolds in order to provide an environment similar to natural ECM [8]. The carrageenan/chitosan/gelatin scaffolds contained multiple functional groups of ECM, such as $-NH_2$, –OH, –COOH, and –SO3H, to adsorb more proteins, so they showed better support for ASC attachment, proliferation, and osteogenic differentiation.

Interaction between cells and the ECM in the natural environment occurs on a nanometer scale, so nanoscaled biomaterials should have positive effects on the cell functions. The ability of multi-walled carbon nanotubes (MWNTs) and graphite (GP) to induce osteogenic differentiation of ASCs was evaluated [108]. While MWNTs have nanostructures and GP does not, MWNTs compacts display better protein adsorption ability relative to GP compacts, and ASCs on MWNTs attached and proliferated better than those on GP compacts. In addition, ASCs cultured on MWNTs compacts had higher concentration of ALP and total protein than GP compacts, indicating a better osteogenic potential.

In a study examining the effects of surface chemical functionalities on the behavior of ASCs, hydroxyapatite (HAp) substrates were modified with amine ($–NH_2$), carboxyl (–COOH), and methyl ($–CH_3$) function groups [109]. Substrates with hydrophilic functional groups (–COOH and $–NH_2$) were found to adsorb more proteins than those with relatively hydrophobic functional groups ($–CH_3$). The results showed that –NH2 modified surfaces encouraged osteogenic differentiation, which may result from the conformational change of adsorbed proteins and their ability to bind lineage-special integrins. This study demonstrated that the surface chemical status can affect protein adsorption on the biomaterials, with further alteration in ASC behaviors.

In contrast, some researchers have designed antifouling surfaces applied on the maintenance of stem cells phenotype. Because the correlation of a given protein with cells cannot be evaluated when the protein is adsorbed on a surface in an uncontrolled manner. Antifouling surface can be used to immobilize specific recognition ligands or ligand-free materials inert to nonspecific adsorption and provide a protein resistant interface. The stiffness of PEG hydrogels can be tuned with the PEG length and their antifouling property will prevent unspecific protein adsorption [110]. In addition, poly-l-glutamic acid/poly-l-lysine alternating films were prepared for fetal liver stem/progenitor cells maintenance and enrichment based on the protein adsorption regulation [111]. Lee et al. have further designed a series of layer-by-layer polypeptide adsorbed supported lipid bilayer (SLB) films as a novel and label-free platform for the isolation and maintenance of rare populated stem cells [112].

The regulation of stem cell differentiation into specific lineages is complex and includes a variety of cellular signaling moieties. It is also well known that the cell's microenvironment plays an important role in determining the progenitor cell's fate and function. The cellular microenvironment has a profound influence on the stem cells within it. Physical, biochemical, and biomechanical cues of cell culture materials can lead to distinct cell reactions. Hence, it is of importance for biomaterials to possess favorable chemical and physical characteristics in order to control downstream signals within the attached cells. The development of biomaterials requires a multidisciplinary approach, includes appropriate biomaterial choosing and ordered scaffold structures, adequate stiffness regulation, and suitable biochemical signal modulation to ultimately mimic the stem cell niche which will open the door to the guided differentiation of stem cells into specific lineages.

Generally, surface hydrophilicity/hydrophobicity can mediate cell attachment; matrix stiffness can direct cell differentiation; electroconductivity can enhance specific cells functionality, and the protein adsorption ability of the substrate can regulate multiple cell response. In addition to these important factors, a variety of biomaterial properties also can affect cell behaviors, which gives us wide-ranging targets too, and create many more possibilities. Understanding the influence of different biomaterial properties on stem cell behaviors and the possible underlying mechanisms is essential in engineering biomaterials for further clinical applications. Although it is still challenging to regulate stem cell differentiation fate, with the development of new materials high throughput arrays may provide the alternative to understand the material properties, distinguish biomaterial functionality, explore cell-materials interaction, and mimic stem cell microenvironments in short periods of time, wasting few materials and ultimately allowing for a more rapid screening.

References

[1] Gimble, J.M., and F. Guilak. Differentiation potential of adipose derived adult stem (ADAS) cells. Current Topics in Developmental Biology, 2003. 58(225): 137–60.

[2] Aust, L., B. Devlin, S. Foster, Y. Halvorsen, K.d. Hicok, T. Du Laney, A. Sen, G. Willingmyre, and J. Gimble. Yield of human adipose-derived adult stem cells from liposuction aspirates. Cytotherapy, 2004. 6(1): 7–14.

[3] Guilak, F., K.E. Lott, H.A. Awad, Q. Cao, K.C. Hicok, B. Fermor, J.M. Gimble. Clonal analysis of the differentiation potential of human adipose-derived adult stem cells. Journal of Cellular Physiology, 2006. 206(1): 229–237.

[4] Anghileri, E., S. Marconi, A. Pignatelli, P. Cifelli, M. Galié, A. Sbarbati, M. Krampera, O. Belluzzi, and B. Bonetti. Neuronal differentiation potential of human adipose-derived mesenchymal stem cells. Stem Cells and Development, 2008. 17(5): 909–916.

[5] Banas, A., T. Teratani, Y. Yamamoto, M. Tokuhara, F. Takeshita, G. Quinn, H. Okochi, and T. Ochiya. Adipose tissue-derived mesenchymal stem cells as a source of human hepatocytes. Hepatology, 2007. 46(1): 219–228.

[6] Kapur, S.K., and A.J. Katz. Review of the adipose derived stem cell secretome. Biochimie, 2013. 95(12): 2222–2228.

[7] Engineered Microenvironments for Controlled Stem Cell Differentiation. Tissue Engineering Part A, 2009. 15(2): 205–219, DOI: 10.1089/ten.tea.2008.0131.

[8] Li, J., B. Yang, Y. Qian, Q. Wang, R. Han, T. Hao, Y. Shu, Y. Zhang, F. Yao, and C. Wang. Iota-carrageenan/chitosan/gelatin scaffold for the osteogenic differentiation of adipose-derived MSCs *in vitro*. Journal of Biomedical Materials Research Part B: Applied Biomaterials, 2015. 103(7): 1498–1510.

[9] Martino, S., F. D'Angelo, I. Armentano, J.M. Kenny, and A. Orlacchio. Stem cell-biomaterial interactions for regenerative medicine. Biotechnology Advances, 2012. 30(1): 338–351, DOI: https://doi.org/10.1016/j.biotechadv.2011.06.015.

[10] Engler, A.J., S. Sen, H.L. Sweeney, and D.E. Discher. Matrix elasticity directs stem cell lineage specification. Cell, 2006. 126(4): 677–689.

[11] Vo, E., D. Hanjaya-Putra, Y. Zha, S. Kusuma, and S. Gerecht. Smooth-muscle-like cells derived from human embryonic stem cells support and augment cord-like structures *in vitro*. Stem Cell Rev Rep, 2010. 6(2): 237–47, DOI: 10.1007/s12015-010-9144-3.

[12] Yamashita, J., H. Itoh, M. Hirashima, M. Ogawa, S. Nishikawa, T. Yurugi, M. Naito, K. Nakao, and S. Nishikawa. Flk1-positive cells derived from embryonic stem cells serve as vascular progenitors. Nature, 2000. 408(6808): 92–6, DOI: 10.1038/35040568.

[13] Sato, H., M. Takahashi, H. Ise, A. Yamada, S.-i. Hirose, Y.-i. Tagawa, H. Morimoto, A. Izawa, and U. Ikeda. Collagen synthesis is required for ascorbic acid-enhanced differentiation of mouse embryonic stem cells into cardiomyocytes. Biochemical and Biophysical Research Communications, 2006. 342(1): 107–112, DOI: https://doi.org/10.1016/j.bbrc.2006.01.116.

[14] Singh, M.D., M. Kreiner, C.S. McKimmie, S. Holt, C.F. van der Walle, and G.J. Graham. Dimeric integrin alpha5beta1 ligands confer morphological and differentiation responses to murine

embryonic stem cells. Biochem Biophys Res Commun, 2009. 390(3): 716–21, DOI: 10.1016/j.bbrc.2009.10.035.

[15] Watt, F.M., and H. Fujiwara. Cell-extracellular matrix interactions in normal and diseased skin. Cold Spring Harb Perspect Biol, 2011. 3(4), DOI: 10.1101/cshperspect.a005124.

[16] Tanimura, S., Y. Tadokoro, K. Inomata, N.T. Binh, W. Nishie, S. Yamazaki, H. Nakauchi, Y. Tanaka, J.R. McMillan, D. Sawamura, K. Yancey, H. Shimizu, and E. K. Nishimura. Hair follicle stem cells provide a functional niche for melanocyte stem cells. Cell Stem Cell, 2011. 8(2): 177–87, DOI: 10.1016/j.stem.2010.11.029.

[17] Silva-Vargas, V., E.E. Crouch, and F. Doetsch. Adult neural stem cells and their niche: a dynamic duo during homeostasis, regeneration, and aging. Curr Opin Neurobiol, 2013. 23(6): 935–42, DOI: 10.1016/j.conb.2013.09.004.

[18] Kazanis, I., J.D. Lathia, T.J. Vadakkan, E. Raborn, R. Wan, M.R. Mughal, D.M. Eckley, T. Sasaki, B. Patton, M.P. Mattson, K.K. Hirschi, M.E. Dickinson, and C. ffrench-Constant. Quiescence and activation of stem and precursor cell populations in the subependymal zone of the mammalian brain are associated with distinct cellular and extracellular matrix signals. The Journal of Neuroscience, 2010. 30(29): 9771–9781, DOI: 10.1523/jneurosci.0700-10.2010.

[19] Hughes, C.S., L.M. Postovit, and G.A. Lajoie. Matrigel: A complex protein mixture required for optimal growth of cell culture. PROTEOMICS, 2010. 10(9): 1886–1890, DOI: https://doi.org/10.1002/pmic.200900758.

[20] de Melo, B.A.G., Y.A. Jodat, E.M. Cruz, J.C. Benincasa, S.R. Shin, and M.A. Porcionatto. Strategies to use fibrinogen as bioink for 3D bioprinting fibrin-based soft and hard tissues. Acta, 2020. 117: 60–76, DOI: https://doi.org/10.1016/j.actbio.2020.09.024.

[21] Jarel K. Gandhi, Travis Knudsen, Matthew Hill, Bhaskar Roy, Lori Bachman, Cynthia Pfannkoch-Andrews, Karina N. Schmidt, Muriel M. Metko, Michael J. Ackerman, Zachary Resch, Jose S. Pulido and Alan D. Marmorstein. Human fibrinogen for maintenance and differentiation of induced pluripotent stem cells in two dimensions and three dimensions. Stem Cells Translational Medicine, 2019. 8(6): 512–521, DOI: 10.1002/sctm.18-0189.

[22] Han, J., D.S. Kim, H. Jang, H.-R. Kim, and H.-W. Kang. Bioprinting of three-dimensional dentin–pulp complex with local differentiation of human dental pulp stem cells. Journal of Tissue Engineering, 2019. 10, 2041731419845849, DOI: 10.1177/2041731419845849.

[23] Khosravizadeh, Z., S. Razavi, H. Bahramian, and M. Kazemi. The beneficial effect of encapsulated human adipose-derived stem cells in alginate hydrogel on neural differentiation. Journal of Biomedical Materials Research Part B: Applied Biomaterials, 2014. 102(4): 749–755, DOI: https://doi.org/10.1002/jbm.b.33055.

[24] Hunt, N.C., D. Hallam, A. Karimi, C.B. Mellough, J. Chen, D.H.W. Steel, and M. Lako. 3D culture of human pluripotent stem cells in RGD-alginate hydrogel improves retinal tissue development. Acta Biomaterialia, 2017. 49: 329–343, DOI: https://doi.org/10.1016/j.actbio.2016.11.016.

[25] Martino, M.M., M. Mochizuki, D.A. Rothenfluh, S.A. Rempel, J.A. Hubbell, and T.H. Barker. Controlling integrin specificity and stem cell differentiation in 2D and 3D environments through regulation of fibronectin domain stability. Biomaterials, 2009. 30(6): 1089-97, DOI: 10.1016/j.biomaterials.2008.10.047.

[26] Schwab, E.H., M. Halbig, K. Glenske, A.-S. Wagner, S. Wenisch, and E.A. Cavalcanti-Adam. Distinct effects of RGD-glycoproteins on Integrin-mediated adhesion and osteogenic differentiation of human mesenchymal stem cells. Int J Med Sci, 2013. 10(13): 1846–1859, DOI: 10.7150/ijms.6908.

[27] Xu, Y., C. Chen, P.B. Hellwarth, and X. Bao. Biomaterials for stem cell engineering and biomanufacturing. Bioact Mater, 2019. 4: 366–379, DOI: 10.1016/j.bioactmat.2019.11.002.

[28] Binan, L., C. Tendey, G. De Crescenzo, R. El Ayoubi, A. Ajji, and M. Jolicoeur. Differentiation of neuronal stem cells into motor neurons using electrospun poly-l-lactic acid/gelatin scaffold. Biomaterials, 2014. 35(2): 664–674, DOI: https://doi.org/10.1016/j.biomaterials.2013.09.097.

[29] Richardson, S.M., J.M. Curran, R. Chen, A. Vaughan-Thomas, J.A. Hunt, A.J. Freemont, and J. Hoyland. A. The differentiation of bone marrow mesenchymal stem cells into chondrocyte-like cells on poly-l-lactic acid (PLLA) scaffolds. Biomaterials, 2006. 27(22): 4069–4078, DOI: https://doi.org/10.1016/j.biomaterials.2006.03.017.

[30] Wang, J., X. Liu, X. Jin, H. Ma, J. Hu, L. Ni, and P.X. Ma. The odontogenic differentiation of human dental pulp stem cells on nanofibrous poly(l-lactic acid) scaffolds *in vitro* and *in vivo*. Acta Biomaterialia, 2010. 6(10): 3856–3863, DOI: https://doi.org/10.1016/j.actbio.2010.04.009.
[31] Sharifi, F., B.B. Patel, A.K. Dzuilko, R. Montazami, D.S. Sakaguchi, and N. Hashemi. Polycaprolactone microfibrous scaffolds to navigate neural stem cells. Biomacromolecules, 2016. 17(10): 3287–3297, DOI: 10.1021/acs.biomac.6b01028.
[32] Chuenjitkuntaworn, B., T. Osathanon, N. Nowwarote, P. Supaphol, and P. Pavasant. The efficacy of polycaprolactone/hydroxyapatite scaffold in combination with mesenchymal stem cells for bone tissue engineering. Journal of Biomedical Materials Research Part A, 2016. 104(1): 264–271, DOI: https://doi.org/10.1002/jbm.a.35558.
[33] Jang, M., S.T. Lee, J.W. Kim, J.H. Yang, J.K. Yoon, J.-C. Park, H.-M. Ryoo, A.J. van der Vlies, J.Y. Ahn, J.A. Hubbell, Y.S. Song, G. Lee, and J.M. Lim. A feeder-free, defined three-dimensional polyethylene glycol-based extracellular matrix niche for culture of human embryonic stem cells. Biomaterials, 2013. 34(14): 3571–3580, DOI: https://doi.org/10.1016/j.biomaterials.2013.01.073.
[34] Mosley, M.C., H.J. Lim, J. Chen, Y.-H. Yang, S. Li, Y. Liu, and L.A. Smith Callahan. Neurite extension and neuronal differentiation of human induced pluripotent stem cell derived neural stem cells on polyethylene glycol hydrogels containing a continuous Young's Modulus gradient. Journal of Biomedical Materials Research Part A, 2017. 105(3): 824–833, DOI: https://doi.org/10.1002/jbm.a.35955.
[35] Harpaz, D., T. Axelrod, A.L. Yitian, E. Eltzov, R.S. Marks, and A.I.Y. Tok. Dissolvable polyvinyl-alcohol film, a time-barrier to modulate sample flow in a 3D-printed holder for capillary flow paper diagnostics. Materials, 2019. 12(3): 343.
[36] Peng, L., Y. Zhou, W. Lu, W. Zhu, Y. Li, K. Chen, G. Zhang, J. Xu, Z. Deng, and D. Wang. Characterization of a novel polyvinyl alcohol/chitosan porous hydrogel combined with bone marrow mesenchymal stem cells and its application in articular cartilage repair. BMC Musculoskeletal Disorders, 2019. 20(1): 257, DOI: 10.1186/s12891-019-2644-7.
[37] Hazeri, Y., S. Irani, M. Zandi, and M. Pezeshki-Modaress. Polyvinyl alcohol/sulfated alginate nanofibers induced the neuronal differentiation of human bone marrow stem cells. International Journal of Biological Macromolecules, 2020. 147: 946–953, DOI: https://doi.org/10.1016/j.ijbiomac.2019.10.061.
[38] Somasundaran, P. Encyclopedia of Surface And Colloid Science. CRC press: 2006; Vol. 5.
[39] Shin, Y.N., B.S. Kim, H.H. Ahn, J.H. Lee, K.S. Kim, J.Y. Lee, M.S. Kim, G. Khang, and H.B. Lee. Adhesion comparison of human bone marrow stem cells on a gradient wettable surface prepared by corona treatment. Applied Surface Science, 2008. 255(2): 293–296.
[40] Wang, W., G. Caetano, W.S. Ambler, J.J. Blaker, M.A. Frade, P. Mandal, C. Diver, and P. Bártolo. Enhancing the hydrophilicity and cell attachment of 3D printed PCL/graphene scaffolds for bone tissue engineering. Materials, 2016. 9(12): 992.
[41] Ahn, H.H., I.W. Lee, H.B. Lee, and M.S. Kim. Cellular behavior of human adipose-derived stem cells on wettable gradient polyethylene surfaces. Int J Mol Sci, 2014. 15(2): 2075–2086, DOI: 10.3390/ijms15022075.
[42] Mei, Y., K. Saha, S.R. Bogatyrev, J. Yang, A.L. Hook, Z.I. Kalcioglu, S.-W. Cho, M. Mitalipova, N. Pyzocha, F. Rojas, K.J. Van Vliet, M.C. Davies, M.R. Alexander, R. Langer, R. Jaenisch, and D.G. Anderson. Combinatorial development of biomaterials for clonal growth of human pluripotent stem cells. Nature Materials, 2010. 9(9): 768–778, DOI: 10.1038/nmat2812.
[43] Ayala, R., C. Zhang, D. Yang, Y. Hwang, A. Aung, S.S. Shroff, F.T. Arce, R. Lal, G. Arya, and S. Varghese. Engineering the cell–material interface for controlling stem cell adhesion, migration, and differentiation. Biomaterials, 2011. 32(15): 3700–3711, DOI: https://doi.org/10.1016/j.biomaterials.2011.02.004.
[44] Valamehr, B., S.J. Jonas, J. Polleux, R. Qiao, S. Guo, E.H. Gschweng, B. Stiles, K. Kam, T.J. Luo, O.N. Witte, X. Liu, B. Dunn and H. Wu. Hydrophobic surfaces for enhanced differentiation of embryonic stem cell-derived embryoid bodies. Proc Natl Acad Sci U S A, 2008. 105(38): 14459–64, DOI: 10.1073/pnas.0807235105.
[45] Metavarayuth, K., P. Sitasuwan, X. Zhao, Y. Lin, and Q. Wang. Influence of surface topographical cues on the differentiation of mesenchymal stem cells *in vitro*. ACS Biomaterials Science & Engineering, 2016. 2(2): 142–151, DOI: 10.1021/acsbiomaterials.5b00377.

[46] Faia-Torres, A.B., M. Charnley, T. Goren, S. Guimond-Lischer, M. Rottmar, K. Maniura-Weber, N.D. Spencer, R.L. Reis, M. Textor, and N.M.Neves. Osteogenic differentiation of human mesenchymal stem cells in the absence of osteogenic supplements: A surface-roughness gradient study. Acta Biomater, 2015. 28: 64–75, DOI: 10.1016/j.actbio.2015.09.028.

[47] Abagnale, G., M. Steger, V.H. Nguyen, N. Hersch, A. Sechi, S. Joussen, B. Denecke, R. Merkel, B. Hoffmann, A. Dreser, U. Schnakenberg, A. Gillner, and W. Wagner. Surface topography enhances differentiation of mesenchymal stem cells towards osteogenic and adipogenic lineages. Biomaterials, 2015. 61: 316–326, DOI: https://doi.org/10.1016/j.biomaterials.2015.05.030.

[48] Wang, P.Y., W.T. Li, J. Yu, and W.B. Tsai. Modulation of osteogenic, adipogenic and myogenic differentiation of mesenchymal stem cells by submicron grooved topography. J Mater Sci Mater Med, 2012. 23(12): 3015–28, DOI: 10.1007/s10856-012-4748-6.

[49] McMurray, R.J., A.K.T. Wann, C.L. Thompson, J.T. Connelly, and M.M. Knight. Surface topography regulates wnt signaling through control of primary cilia structure in mesenchymal stem cells. Scientific Reports, 2013. 3(1): 3545, DOI: 10.1038/srep03545.

[50] Solanki, A., S. Shah, K.A. Memoli, S.Y. Park, S. Hong, and K.-B. Lee. Controlling differentiation of neural stem cells using extracellular matrix protein patterns. Small, 2010. 6(22): 2509–2513, DOI: 10.1002/smll.201001341.

[51] Bakhru, S., A.S. Nain, C. Highley, J. Wang, P. Campbell, C. Amon, and S. Zappe. Direct and cell signaling-based, geometry-induced neuronal differentiation of neural stem cells. Integr Biol (Camb), 2011. 3(12): 1207–14, DOI: 10.1039/c1ib00098e.

[52] Jiang, X., H.Q. Cao, L.Y. Shi, S.Y. Ng, L.W. Stanton, and S.Y. Chew. Nanofiber topography and sustained biochemical signaling enhance human mesenchymal stem cell neural commitment. Acta Biomater, 2012. 8(3): 1290–302, DOI: 10.1016/j.actbio.2011.11.019.

[53] Béduer, A., C. Vieu, F. Arnauduc, J.C. Sol, I. Loubinoux, and L. Vaysse. Engineering of adult human neural stem cells differentiation through surface micropatterning. Biomaterials, 2012. 33(2): 504–14, DOI: 10.1016/j.biomaterials.2011.09.073.

[54] Bodhak, S., S. Bose, and A. Bandyopadhyay. Electrically polarized HAp-coated Ti: *in vitro* bone cell-material interactions. Acta Biomater, 2010. 6(2): 641–51, DOI: 10.1016/j.actbio.2009.08.008.

[55] Li, J., X. Mou, J. Qiu, S. Wang, D. Wang, D. Sun, W. Guo, D. Li, A. Kumar, X. Yang, A. Li, and H. Liu. Surface charge regulation of osteogenic differentiation of mesenchymal stem cell on polarized ferroelectric crystal substrate. Advanced Healthcare Materials, 2015. 4(7): 998–1003, DOI: https://doi.org/10.1002/adhm.201500032.

[56] Lee, I.C., and Y.C. Wu. Assembly of polyelectrolyte multilayer films on supported lipid bilayers to induce neural stem/progenitor cell differentiation into functional neurons. ACS Appl Mater Interfaces, 2014. 6(16): 14439–50, DOI: 10.1021/am503750w.

[57] Lei, K.F., I.C. Lee, Y.C. Liu, and Y.C. Wu. Successful differentiation of neural stem/progenitor cells cultured on electrically adjustable indium tin oxide (ITO) surface. Langmuir, 2014. 30(47): 14241–9, DOI: 10.1021/la5039238.

[58] Engler, A.J., S. Sen, H.L. Sweeney, and D.E. Discher. Matrix elasticity directs stem cell lineage specification. Cell, 2006. 126(4): 677–89, DOI: 10.1016/j.cell.2006.06.044.

[59] Young, D.A., Y.S. Choi, A.J. Engler, and K.L. Christman. Stimulation of adipogenesis of adult adipose-derived stem cells using substrates that mimic the stiffness of adipose tissue. Biomaterials, 2013. 34(34): 8581–8588.

[60] Le, N.N.T., S. Zorn, S.K. Schmitt, P. Gopalan, and W.L. Murphy. Hydrogel arrays formed via differential wettability patterning enable combinatorial screening of stem cell behavior. Acta Biomaterialia, 2016. 34: 93–103, DOI: https://doi.org/10.1016/j.actbio.2015.09.019.

[61] Pek, Y.S., A.C. Wan, and J.Y. Ying. The effect of matrix stiffness on mesenchymal stem cell differentiation in a 3D thixotropic gel. Biomaterials, 2010. 31(3): 385–91, DOI: 10.1016/j.biomaterials.2009.09.057.

[62] Randriantsilefisoa, R., Y. Hou, Y. Pan, J.L.C. Camacho, M.W. Kulka, J. Zhang, and R. Haag. Interaction of human mesenchymal stem cells with soft nanocomposite hydrogels based on polyethylene glycol and dendritic polyglycerol. Advanced Functional Materials, 2020. 30(1): 1905200, DOI: https://doi.org/10.1002/adfm.201905200.

[63] Chen, G.,C. Dong, L. Yang, and Y. Lv. 3D Scaffolds with different stiffness but the same microstructure for bone tissue engineering. ACS Applied Materials & Interfaces, 2015. 7(29): 15790–15802, DOI: 10.1021/acsami.5b02662.

[64] Guvendiren, M., and J.A. Burdick. Stiffening hydrogels to probe short-and long-term cellular responses to dynamic mechanics. Nature Communications, 2012. 3(1): 1–9.

[65] Eroshenko, N., R. Ramachandran, V.K. Yadavalli, and R.R. Rao. Effect of substrate stiffness on early human embryonic stem cell differentiation. J Biol Eng, 2013. 7(1): 7, DOI: 10.1186/1754-1611-7-7.

[66] Robertson, C., D.D. Tran, and S.C. George. Concise review: maturation phases of human pluripotent stem cell-derived cardiomyocytes. Stem Cells, 2013. 31(5): 829–37, DOI: 10.1002/stem.1331.

[67] Majkut, S., T. Idema, J. Swift, C. Krieger, A. Liu, and D.E. Discher. Heart-specific stiffening in early embryos parallels matrix and myosin expression to optimize beating. Curr Biol, 2013. 23(23): 2434–9, DOI: 10.1016/j.cub.2013.10.057.

[68] Cho, G.S., D.I. Lee, E. Tampakakis, S. Murphy, P. Andersen, H. Uosaki, S. Chelko, K. Chakir, I. Hong, K. Seo, H.V. Chen, X. Chen, C. Basso, S.R. Houser, G.F. Tomaselli, B. O'Rourke, D.P. Judge, D.A. Kass, and C. Kwon. Neonatal transplantation confers maturation of PSC-derived cardiomyocytes conducive to modeling cardiomyopathy. Cell Rep, 2017. 18(2): 571–582, DOI: 10.1016/j.celrep.2016.12.040.

[69] Li, Z., Y. Gong, S. Sun, Y. Du, D. Lü, X. Liu and M. Long. Differential regulation of stiffness, topography, and dimension of substrates in rat mesenchymal stem cells. Biomaterials, 2013. 34(31): 7616–7625, DOI: https://doi.org/10.1016/j.biomaterials.2013.06.059.

[70] Chu, G., Z. Yuan, C. Zhu, P. Zhou, H. Wang, W. Zhang, Y. Cai, X. Zhu, H. Yang, and B. Li. Substrate stiffness- and topography-dependent differentiation of annulus fibrosus-derived stem cells is regulated by Yes-associated protein. Acta Biomaterialia, 2019. 92: 254–264, DOI: https://doi.org/10.1016/j.actbio.2019.05.013.

[71] Fattahi, P., G. Yang, G. Kim, and M.R. Abidian. A review of organic and inorganic biomaterials for neural interfaces. Advanced Materials, 2014. 26(12): 1846–1885, DOI: https://doi.org/10.1002/adma.201304496.

[72] Hardy, J.G., R.C. Sukhavasi, D. Aguilar, M.K. Villancio-Wolter, D.J. Mouser; S.A. Geissler, L. Nguy, J.K. Chow, D.L. Kaplan and C.E. Schmidt. Electrical stimulation of human mesenchymal stem cells on biomineralized conducting polymers enhances their differentiation towards osteogenic outcomes. Journal of Materials Chemistry B, 2015. 3(41): 8059–8064, DOI: 10.1039/C5TB00714C.

[73] Kim, D.-H., S.M. Richardson-Burns, J.L. Hendricks, C. Sequera, and D.C. Martin. Effect of immobilized nerve growth factor on conductive polymers: Electrical properties and cellular response. Advanced Functional Materials, 2007. 17(1): 79–86, DOI: https://doi.org/10.1002/adfm.200500594.

[74] Yang, B., F. Yao, L. Ye, T. Hao, Y. Zhang, L. Zhang, D. Dong, W. Fang, Y. Wang, X. Zhang, C. Wang, and J. Li. A conductive PEDOT/alginate porous scaffold as a platform to modulate the biological behaviors of brown adipose-derived stem cells. Biomaterials Science, 2020. 8(11): 3173–3185, DOI: 10.1039/C9BM02012H.

[75] Šafaříková, E., J. Ehlich, S. Stříteský, M. Vala, M. Weiter, J. Pacherník, L. Kubala, and J. Víteček. Conductive polymer PEDOT:PSS-based platform for embryonic stem-cell differentiation. Int J Mol Sci, 2022. 23(3): 1107.

[76] Pires, F., Q. Ferreira, C.A.V. Rodrigues, J. Morgado, and F.C. Ferreira. Neural stem cell differentiation by electrical stimulation using a cross-linked PEDOT substrate: Expanding the use of biocompatible conjugated conductive polymers for neural tissue engineering. Biochimica et Biophysica Acta (BBA) - General Subjects, 2015. 1850(6): 1158–1168, DOI: https://doi.org/10.1016/j.bbagen.2015.01.020.

[77] Wang, S., S. Guan,W. Li, D. Ge, J. Xu, C. Sun, T. Liu, and X. Ma. 3D culture of neural stem cells within conductive PEDOT layer-assembled chitosan/gelatin scaffolds for neural tissue engineering. Materials Science and Engineering: C, 2018. 93: 890–901, DOI: https://doi.org/10.1016/j.msec.2018.08.054.

[78] Green, R.A., N.H. Lovell, and L.A. Poole-Warren. Cell attachment functionality of bioactive conducting polymers for neural interfaces. Biomaterials, 2009. 30(22): 3637–44, DOI: 10.1016/j.biomaterials.2009.03.043.

[79] Mooney, E., P. Dockery, U. Greiser, M. Murphy, and V. Barron. Carbon nanotubes and mesenchymal stem cells: Biocompatibility, proliferation and differentiation. Nano Letters, 2008. 8(8): 2137–2143, DOI: 10.1021/nl073300o.

[80] Namgung, S., K.Y. Baik, J. Park, and S. Hong. Controlling the growth and differentiation of human mesenchymal stem cells by the arrangement of individual carbon nanotubes. ACS Nano, 2011. 5(9): 7383–7390, DOI: 10.1021/nn2023057.

[81] Shao, H., T. Li, R. Zhu, X. Xu, J. Yu, S. Chen, L. Song, S. Ramakrishna, Z. Lei, Y. Ruan, and L. He. Carbon nanotube multilayered nanocomposites as multifunctional substrates for actuating neuronal differentiation and functions of neural stem cells. Biomaterials, 2018. 175: 93–109, DOI: https://doi.org/10.1016/j.biomaterials.2018.05.028.

[82] Chao, T.-I., S. Xiang, C.-S. Chen, W.-C. Chin, A.J. Nelson, C. Wang, and J. Lu. Carbon nanotubes promote neuron differentiation from human embryonic stem cells. Biochemical and Biophysical Research Communications, 2009. 384(4): 426–430, DOI: https://doi.org/10.1016/j.bbrc.2009.04.157.

[83] Peng, Q., A.K. Dearden, J. Crean, L. Han, S. Liu, X. Wen, and S. De. New materials graphyne, graphdiyne, graphone, and graphane: review of properties, synthesis, and application in nanotechnology. Nanotechnol Sci Appl, 2014. 7: 1–29, DOI: 10.2147/NSA.S40324.

[84] Lee, W.C., C.H.Y.X. Lim, H. Shi, L.A.L. Tang, Y. Wang, C.T. Lim, and K.P. Loh. Origin of enhanced stem cell growth and differentiation on graphene and graphene oxide. ACS Nano, 2011. 5(9): 7334–7341, DOI: 10.1021/nn202190c.

[85] Qiaojun Fang, Yuhua Zhang, Xiangbo Chen, He Li, Liya Cheng, Wenjuan Zhu, Zhong Zhang, Mingliang Tang, Wei Liu, Hui Wang, Tian Wang, Tie Shen and Renjie Chai. Three-dimensional graphene enhances neural stem cell proliferation through metabolic regulation. Front. Bioeng. Biotechnol. 2020; 7: 436.

[86] Lin Xia, Wenjuan Zhu, Yunfeng Wang, Shuangba He, Renjie Chai. Regulation of neural stem cell proliferation and differentiation by graphene-based biomaterials. Neural Plast. 2019; 2019: 3608386.

[87] Yang, D., T. Li, M. Xu, F. Gao, J. Yang, Z. Yang, and W. Le. Graphene oxide promotes the differentiation of mouse embryonic stem cells to dopamine neurons. Nanomedicine (Lond), 2014. 9(16): 2445–55, DOI: 10.2217/nnm.13.197.

[88] Kim, T.-H., T. Lee, W.A. El-Said, and J.-W. Choi. Graphene-based materials for stem cell applications. Materials, 2015. 8(12): 8674–8690.

[89] Crowder, S.W., D. Prasai, R. Rath, D.A. Balikov, H. Bae, K.I. Bolotin, and H.J. Sung. Three-dimensional graphene foams promote osteogenic differentiation of human mesenchymal stem cells. Nanoscale, 2013. 5(10): 4171–6, DOI: 10.1039/c3nr00803g.

[90] Li, N., Q. Zhang, S. Gao, Q. Song, R. Huang, L. Wang, L. Liu, J. Dai, M. Tang, and G. Cheng. Three-dimensional graphene foam as a biocompatible and conductive scaffold for neural stem cells. Sci Rep, 2013. 3: 1604, DOI: 10.1038/srep01604.

[91] Shah, S., P.T. Yin, T.M. Uehara, S.-T.D. Chueng, L. Yang, and K.-B. Lee. Guiding stem cell differentiation into oligodendrocytes using graphene-nanofiber hybrid scaffolds. Advanced Materials, 2014. 26(22): 3673–3680, DOI: https://doi.org/10.1002/adma.201400523.

[92] Shah, S., P.T. Yin, T.M. Uehara, S.T. Chueng, L. Yang, and K.B. Lee. Guiding stem cell differentiation into oligodendrocytes using graphene-nanofiber hybrid scaffolds. Adv Mater, 2014. 26(22): 3673–80, DOI: 10.1002/adma.201400523.

[93] Park, J., I.Y. Kim, M. Patel, H.J. Moon, S.-J. Hwang, and B. Jeong. 2D and 3D hybrid systems for enhancement of chondrogenic differentiation of tonsil-derived mesenchymal stem cells. Advanced Functional Materials, 2015. 25(17): 2573–2582, DOI: https://doi.org/10.1002/adfm.201500299.

[94] Weaver, C.L. and X.T. Cui. Directed neural stem cell differentiation with a functionalized graphene oxide nanocomposite. Advanced Healthcare Materials, 2015. 4(9): 1408–1416, DOI: https://doi.org/10.1002/adhm.201500056.

[95] Guo, W., X. Zhang, X. Yu, S. Wang, J. Qiu, W. Tang, L. Li, H. Liu, and Z.L. Wang. Self-powered electrical stimulation for enhancing neural differentiation of mesenchymal stem cells on graphene–Poly(3,4-ethylenedioxythiophene) Hybrid Microfibers. ACS Nano, 2016. 10(5): 5086–5095, DOI: 10.1021/acsnano.6b00200.

[96] Naguib, M., M. Kurtoglu, V. Presser, J. Lu, J. Niu, M. Heon, L. Hultman, Y. Gogotsi, and M.W. Barsoum. Two-dimensional nanocrystals produced by exfoliation of Ti3AlC2. Advanced Materials, 2011. 23(37): 4248–4253, DOI: https://doi.org/10.1002/adma.201102306.

[97] Lim, G.P., C.F. Soon, N.L. Ma, M. Morsin, N. Nayan, M.K. Ahmad, and K.S. Tee. Cytotoxicity of MXene-based nanomaterials for biomedical applications: A mini review. Environmental Research, 2021. 201: 111592, DOI: https://doi.org/10.1016/j.envres.2021.111592.

[98] Huang, K., Z. Li, J. Lin, G. Han, and P. Huang. Two-dimensional transition metal carbides and nitrides (MXenes) for biomedical applications. Chemical Society Reviews, 2018. 47(14): 5109–5124, DOI: 10.1039/C7CS00838D.

[99] Wu, W., H. Ge, L. Zhang, X. Lei, Y. Yang, Y. Fu, and H. Feng. Evaluating the cytotoxicity of Ti3C2 MXene to neural stem cells. Chemical Research in Toxicology, 2020. 33(12): 2953–2962, DOI: 10.1021/acs.chemrestox.0c00232.

[100] Guo, R., M. Xiao, W. Zhao, S. Zhou, Y. Hu, M. Liao, S. Wang, X. Yang, R. Chai, and M. Tang. 2D Ti3C2TxMXene couples electrical stimulation to promote proliferation and neural differentiation of neural stem cells. Acta Biomaterialia, 2022. 139: 105–117, DOI: https://doi.org/10.1016/j.actbio.2020.12.035.

[101] Huang, R., X. Chen, Y. Dong, X. Zhang, Y. Wei, Z. Yang, W. Li, Y. Guo, J. Liu, Z. Yang, H. Wang, and L. Jin. MXene composite nanofibers for cell culture and tissue engineering. ACS Applied Bio Materials, 2020. 3(4): 2125–2131, DOI: 10.1021/acsabm.0c00007.

[102] Fu, X., H. Yang, Z. Li, N.-C. Liu, P.-S. Lee, K. Li, S. Li, M. Ding, J.S. Ho, Y.-C.E. Li, I.C. Lee, and P.-Y. Chen. Cation-induced assembly of conductive MXene fibers for wearable heater, wireless communication, and stem cell differentiation. ACS Biomaterials Science & Engineering, 2021. DOI: 10.1021/acsbiomaterials.1c00591.

[103] Rastin, H., B. Zhang, A. Mazinani, K. Hassan, J. Bi, T.T. Tung, and D. Losic. 3D bioprinting of cell-laden electroconductive MXene nanocomposite bioinks. Nanoscale, 2020. 12(30): 16069–16080, DOI: 10.1039/D0NR02581J.

[104] Wychowaniec, J.K., J. Litowczenko, K. Tadyszak, V. Natu, C. Aparicio, B. Peplińska, M.W. Barsoum, M. Otyepka, and B. Scheibe. Unique cellular network formation guided by heterostructures based on reduced graphene oxide - Ti3C2Tx MXene hydrogels. Acta Biomaterialia, 2020. 115: 104–115, DOI: https://doi.org/10.1016/j.actbio.2020.08.010.

[105] Hynes, R.O. Integrins: versatility, modulation, and signaling in cell adhesion. Cell, 1992. 69(1): 11–25.

[106] Albelda, S.M., and C.A. Buck. Integrins and other cell adhesion molecules. The FASEB Journal, 1990. 4(11): 2868–2880.

[107] Berrier, A.L., and K.M. Yamada. Cell–matrix adhesion. Journal of Cellular Physiology, 2007. 213(3): 565–573.

[108] Li, X., H. Liu, X. Niu, B. Yu, Y. Fan, Q. Feng, F.-z. Cui, and F. Watari. The use of carbon nanotubes to induce osteogenic differentiation of human adipose-derived MSCs *in vitro* and ectopic bone formation *in vivo*. Biomaterials, 2012. 33(19): 4818–4827.

[109] Liu, X., Q. Feng, A. Bachhuka, and K. Vasilev. Surface chemical functionalities affect the behavior of human adipose-derived stem cells *in vitro*. Applied Surface Science, 2013. 270: 473–479.

[110] Randriantsilefisoa, R., J.L. Cuellar-Camacho, M.S. Chowdhury, P. Dey, U. Schedler, and R. Haag. Highly sensitive detection of antibodies in a soft bioactive three-dimensional bioorthogonal hydrogel. Journal of Materials Chemistry B, 2019. 7(20): 3220–3231, DOI: 10.1039/C9TB00234K.

[111] Tsai, H.-A., R.-R. Wu, I.C. Lee, H.-Y. Chang, C.-N. Shen, and Y.-C. Chang. Selection, enrichment, and maintenance of self-renewal liver stem/progenitor cells utilizing polypeptide polyelectrolyte multilayer films. Biomacromolecules, 2010. 11(4): 994–1001, DOI: 10.1021/bm901461e.

[112] Lee, I.C. Y.C. Liu, H.A. Tsai, C.N. Shen, and Y.C. Chang. Promoting the selection and maintenance of fetal liver stem/progenitor cell colonies by layer-by-layer polypeptide tethered supported lipid bilayer. ACS Appl Mater Interfaces, 2014. 6(23): 20654–63, DOI: 10.1021/am503928u.

5

The 2D Membrane Technology for Fabrication of 2D Stem Cell Culture System

Nai-Chen Cheng,[1] *Wen-Yen Huang,*[2] *Sung-Jan Lin,*[2]
Min-Huey Chen,[3] *I-Chi Lee,*[4,*] *Chia-Ning Shen*[5]
and *Yi-Chen Ethan Li*[1,*]

Since the past century, 2D culture systems have been considered the gold standard for cell culture. The simplest 2D culture system is directly sending cells on flat tissue-culture polystyrene (TCPS) or cell culture disks. Afterwards, the seeded cell can adhere, spread, and grow. 2D culture systems offer many advantages. For example, the materials for TCPS or culture disk are usually polymer-based materials, which can be made through rapid batch and mass production through production lines in manufacturing factories. Therefore, traditional culture plates/dishes are inexpensive and economical for general laboratories. Second, the developmental history of 2D culture systems started and dated back to the 1900s. Therefore, many related references about the protocols for 2D cell culture systems have been well established and developed, meaning that these 2D cell culture systems have gained widespread acceptance. In addition, due to the maturation of research in this field, abundant works of literature provide various databases so users can easily obtain comparative

[1] Department of Surgery, National Taiwan University Hospital, Taiwan.
[2] Department of Biomedical Engineering, College of Medicine and College of Engineering, National Taiwan University, Taiwan.
[3] Graduate Institute of Clinical Dentistry, School of Dentistry, National Taiwan University, Taiwan.
[4] Department of Biomedical Engineering and Environmental Sciences, National Tsing Hua University, Taiwan.
[5] Genomics Research Center, Academia Sinica, Taiwan.
[6] Department of Chemical Engineering, Feng Chia University, Taiwan.
* Corresponding authors: iclee@mx.nthu.edu.tw; yicli@fcu.edu.tw

results for their systems. Furthermore, the fabrication process of 2D culture systems usually has no complex manufacturing procedure and offers user-friendly and easily operated protocols for everyone working with cell culture- or biology-related research in laboratories. Afterwards, the cells in 2D culture systems can be directly and easily observed through a microscope. Therefore, till now, 2D culture systems are still popular and widely accepted by most researchers and scientists.

Despite 2D culture systems being the universal method, some inherent flaws still require combining this method with other techniques or external devices as alternative strategies. For example, typical plastic 2D culture systems are flat dishes or multi-wells that enable stem cells to grow on the surface of dishes or wells in the presence or absence of special coating or treatments. However, 2D culture systems find it difficult to recapture the natural body-like environments for stem cells, so the growth, differentiation, and functions of cells in a human body are not good strategies to investigate and decipher the growth and functionalization mechanisms of stem cells in tissues/organs, where stem cells are surrounded by other various types of cells in 3D environments. Furthermore, the results from 2D culture systems provide only a preliminary understanding and have less predictable models, which may raise the failure risk and cost when scientists discover new drugs, screen their effect, and test them in clinical trials, leading to pharmaceutical manufacturers investing millions of dollars into the failed drug development per year. In addition, 2D culture systems are usually designed as a close static system. Therefore, the accumulation of exuded waste and consumption of nutrients in the culture medium may restrict the expansion of stem cells in limited 2D flat culture dishes or wells. Fortunately, these shortcomings provide opportunities to improve the current commonly used 2D culture systems in the future. In terms of 2D culture systems, although the issues mentioned above exist in the use of 2D cell culture, it provides a fast way to obtain preliminary results to understand the cell-matrix interactions at least, and the results contribute confidence experience to scientists to design 2.5D, 3D, even more, complex culture systems for creating biomimetic culture environments [1]. Therefore, the requirement of designing 2.5D and 3D culture systems sets a foundation for the indispensability of 2D culture systems. According to the design of 2D cell culture systems, surface modification of plastic culture dish/well surfaces endow 2D culture systems with versatile functions, which can satisfy the different requirements for culturing stem cells. For example, to improve the mechanical and biological properties of culture dish/well surfaces, different polymers has been widely investigated. Chitosan polymers are widely used for biomedical applications, and chitosan-based 2D substrates have been used as various culture systems or tissue-like models [2–5]. Primary chondrocytes cultured on chondroitin 4-sulfate-augmented chitosan maintained the capability to synthesize cartilage-specific collagens [6]. Chitosan can also be used as the modification agent to improve cell attachment to PLA and alginate, increase cell adhesion, proliferation, and biosynthetic activity [7, 8]. Moreover, Lin, Y.H. et al. also fabricated keratin/chitosan-azide composite films via UV crosslinking, showing the substrates with long-term stability, elasticity, and hydrophilicity in PBS that motivated intercellular interaction, improved cellular viability, promoted proliferation, and facilitated osteogenic differentiation into mature osteoblasts from ASCs [9]. Additionally, while chitosan is relatively hydrophobic in nature, gelatin, a denatured collagen composed of high content of the amino acids such as glycine, proline, and hydroxyproline, is a hydrophilic material. The mixture of gelatin and chitosan forms polyelectrolytic complexes in different gelation states [10]. With a similar structure to GAGs and collagen, chitosan/gelatin (C/G) composites mimic the natural components of ECM. Moreover, blending chitosan with gelatin can increase the surface wettability compared with pure chitosan films, resulting in better ASC attachment and improved cell proliferation. Plus, gradual gelatin release from the C/G blend films increased the ratio of chitosan in the blends and encouraged ASC detachment [11]. The consequently decreased adhesiveness led to spheroid formation after day 4. After comparing different C/G proportions, a C/G film composed of 75% chitosan was found to have significantly more cells transferred into the overlying collagen substrates, which could be beneficial as a culture surface in ASC-based cell therapy.

The principle of surface modification of culture dishes or wells is paving thin films or thick membranes on the tissue culture surface of the flat plane through physical or chemical methods. In general, most thin-layer membranes can be easily produced through a coating process in each laboratory. Briefly, a coating process is based on adding the material solution to cover the surface of culture plates, and then the material-based thin films could form on the surface through physical or chemical interactions such as Vann der Wall forces and electrostatic interactions, and chemical reactions. Therefore, the standard techniques used for fabricating thin film are coating/dip-coating, layer-by-layer assembly technique, and chemical modification [12]. Compared with the thin films, thick membranes used for culturing cells are usually fabricated through manufacturing technology to obtain the membrane-type substrates, which are cut and placed in culture dishes/wells as the surface for cell adhesion and growth.

Next, we want to introduce the current techniques for preparing membranes that are widely used to culture cells. In terms of creating thin films, the coating/dip-coating technique is a friendly method for everyone in the laboratory, which can be widely applied on different biomaterial solutions, such as protein, ECM, synthetic and natural polymers, and others. The solutions are directly added to culture dishes/wells; then, the solvents are dried and removed by using airflow or an oven. Afterward, the biomaterials can be deposited, and they then form a thin film on the culture well surfaces. For example, a commercial ECM-based matrigel is a typical biomaterial that can fabricate an ECM layer on the culture dishes. Zhang and co-workers coated the culture dishes with a Matrigel solution (1:500). After the Matrigel film formed, reprogrammed iPSCs obtained from the human synovial fluid MSC (SFMSC) transfected with Oct-3/4, c-Myc., Sox2, and klf4l were cultured on the Matrigel surface, and the results indicated that the Matrigel-coated surface allowed for the MSC differentiation of SFMSC-iPSCs. Moreover, the Matrigel-treated SFMSC-iPSC-MSCs showed a higher proliferation ability than SFMSCs and enabled osteogenic and chondrogenic differentiation [13]. In addition to Matrigel-based thin films, other proteins from ECM also could be fabricated into a thin film to modify the surface properties of substrates to regulate the behaviors of stem cells and investigate the effect and mechanism of ECM on stem cells. For example, fibronectin is an ECM containing a cell-anchoring segment, arginine-glycine-aspartic (RGD). Therefore, Kalaskar and co-workers investigated the effect of culture surfaces adsorbing different concentrations of fibronectin on the regulation of hESCs [14]. In their study, the fibronectin-functionalized culture surfaces were prepared by coating a series of fibronectin concentrations from (0 to 500 mg/ml) to obtain fibronectin-coated surfaces with different surface saturation. Then, 2×10^4 cells/cm^2 hESC were seeded on the fibronectin-coated surfaces in a serum-free environment to prevent other proteins in the serum from influencing the effect of the fibronectin-coated surface. In addition, Kalaskar and co-workers also coated the peptides with the sequences derived from fibronectin on the culture surface to culture hESCs. Interestingly, the results indicate that the RGD sequence enabled only attachment of the hESCs but was no effects on the regulation of stem cell stemness. In the comparison to the short RGD sequence, both the central fibronectin and the 120 kDa fragment central binding domain from fibronectin were able to support the adhesion and maintain the undifferentiation phenotype of hESCs via the $\alpha5\beta1$ integrin receptor. The results confirmed that hESCs required a different functional domain from fibronectin to mediate their growth, spread, and stemness. Moreover, the study also demonstrated that the culture surface containing only 25 percent of fibronectin saturation was enough to allow hESCs to grow and maintain the undifferentiated phenotype. Therefore, the results showed that the coating technique provides a convenient way to functionalize the cell culture surface.

In addition to fibronectin, an exogenous additive, the fetal bovine serum contains many ECM and growth factors which are widely added to the medium to provide nutrients for cells. Therefore, another case also presents the application of the coating technique for the investigation of the effect of serum in the substrate-coated and soluble form on neural stem cells (NSCs) [15]. As mentioned in Chapter 2, NSCs are a suspensive type of stem cells that can differentiate into neurons, astrocytes, and oligodendrocytes once NSCs anchor on the culture dishes/wells or substrates. Previous studies have demonstrated that the surface properties of substrates could influence the behaviors of NSCs [16, 17] under a serum-free culture environment. Notably, compared to NSCs, the culture medium recipe usually contains the serum components for supplying nutrients to most cells, but the culture medium for NSCs is usually a serum-free condition. Therefore, Hung and co-workers are eager to understand the effect of the serum on NSCs [15]. In the study, they designed four groups for culturing NSCs: a cell-anchored substrate and a low-attachment substrate at the medium in the (1) absence or (2) presence of 10% fetal bovine serum (FBS); a serum-precoated cell-anchored substrate and a low-attachment substrate at the medium in the (3) absence or (4) presence of 10% fetal bovine serum. The results showed that the behaviors and fate of NSCs were determined mainly by the substrates when the medium was in the absence of serum; the medium containing 10% FBS can induce NSCs into the protoplasmic cell before adhesion so that the effect of substrates on NSCs was blocked and then the differentiation of NSCs were no significant difference. Interestingly, the FBS-precoated substrates offered an ECM-based architecture thin film to alter the low attachment substrate effect, promote the NSC migration, and differentiate. Therefore, these results indicated that the coating technique brings a versatile strategy to provide ECM or proteins with more applications to regulate the behaviors of stem cells. Next, in addition to ECM or protein components, synthetic polymers are also widely used to design the desired culture surfaces for stem cells through coating techniques. For example, PVA is a well-known hydrophilic polymer that can be used to create a low-attachment environment. Based on the properties of PVA polymer, Chen and co-workers designed a low-attachment thin film by using PVA polymer to obtain tooth germ cell organoids [17]. The human body consists of cells, tissues, and organs which are highly organized with various non-cellular and cellular materials. The highly organized structure contributes to endogenous cell-cell and cell-matrix interactions and exogenous cell-environment interactions to regulate multi-level physiological behaviors from cell to organs. To recapitulate these intrinsic interactions *in vitro* stem cell-based systems are widely fabricated as various *in vitro* models for use in tissue engineering and regenerative medicine applications. Recently, stem cell-based organoid systems are viewed as a new strategy to achieve a great number of biological features and functions such as heterogeneous cell-based spatial architecture, cell-cell, and cell-matrix interactions. To achieve organoid structures, Chen and co-workers dissolved a 5% (w/v) solution of PVA in distilled water at 95°C. Then, 140 μL PVA solution was added to a 24-well culture plate and dried at 60°C o overnight to form a thin film. After seeding tooth germ cells on PVA film for 7 days of incubation, tooth germ cells could form the organoid structure at both low and high cell density (5×10^3 cells/ml and 1×10^5 cells/ml). Furthermore, the tooth germ cell-formed organoids expressed higher cell viability and better alkaline phosphatase activity and mineralization compared to cells cultured on TCPS. Therefore, the coating technique also provides a way to fabricate synthetic polymer film for culturing stem cells and creating a variety of *in vitro* models.

Besides the monolayer-coating technique, the polyelectrolyte multilayers (PEMs) technique was first introduced in 1966 and has been widely used since the 1990s [18]. The principle of the polyelectrolyte multilayer technique is that polycations and polyanions are alternatively deposited on a surface through a layer-by-layer assembly because the positive charge of polycations can spontaneously form a static bond with the negative charge of polyanions. Due to layer-by-layer deposition with a key advantage of simplicity and versatility, the layer-by-layer technique has gained increasing attention and has become a popular strategy for preparing functionalized surfaces. Until now, the layer-by-layer technique has been in a widely used for different applications. PEM thin films could be formed on almost any surface if the surface and the matters had complementary charges [19, 20]. In biomedical engineering applications, many researchers reported that PEM films enabled the creation of multilayer capsules or were used as thin layers on surfaces to create drug delivery systems, which could be triggered to release growth factors, biomolecules, or drugs by dissolving PEMs or changing the temperature, pH value, and ionic strength of ambient conditions [21, 22]. In addition, the layer-by-layer technique can be further used to modulate biomaterial surface properties via the chemical compositions, number of depositing layers, and the layer sequences [23].

Since 2008, the previous study have demonstrated that poly (sodium-4-styrenesulfonate) (PSS)- and poly (allylamine hydrochloride) (PAH)-based multilayer films provided an anchorable environment to solve the low adhesion issue of endothelial progenitor cells and enhanced the differentiation of endothelial progenitor cells [24]. Furthermore, poly (sodium-4-styrenesulfonate)/poly (allylamine hydrochloride) thin films also confirmed their biocompatibility, which can activate the survival and proliferation of MSCs without any cytotoxicity. In addition, the combination of poly(D-lysine) and poly(glutamic acid) multilayers can not only promote the endothelial cells attaching to the substrate but can also be used to maintain the stemness expression, such as β-catenin and CD133, of fetal live stem cells [25]. Moreover, the surface features of biomaterials enable the determination of the cell-matrix interactions to regulate the behaviors of stem cells. To predict the cell-matrix interactions between the PEM multilayer films and non-anchored NSPCs and to control the differentiation of NSCs, previous studies fabricated polylysine surfaces with different positive charge densities by coating polylysine with varying concentrations on the substrates to observe the differentiation of NSCs [26]. The results indicated that NSPCs could sense the net surface charge density and showed different neurite outgrowth behaviors. Based on this information, Tsai and co-workers tried using PEM thin films to develop a differentiation-induction environment for culture NSCs and to obtain an *in vitro* neuron-rich model [27]. In their study, the PAH polymer contained cationic functional groups, and PSS and polyacrylic acid (PAA) with anionic functional groups were prepared as the solutions. Then, the PEM thin films were fabricated by alternatively immersing the coverslips in each polymeric solution (300 μL, 5 mg/ml) for 10 mins. Afterwards, NSCs were cultured on the PAH/PSS or PAH/PAA PEM thin films in a serum-free environment. Interestingly, the results showed that NSCs favor adhering to the PEM thin films when cationic PAH was the ending layer, possessed a low adhesion on the anionic PAA ending films, and activated neuronal differentiation on the anionic PSS ending films. These results provides insight into the possible mechanisms between PEM films and NSCs: (1) The cationic PAH out-layer with positive surface density can attract the negative ligand on the cell membrane of NSCs and dominate the adhesion of NSCs; (2) When anionic polymers were the outer layer, the negative charge density was higher than the positive charge density, and it may dominate the surface characterization and decrease the adhesion of NSCs; (3) In addition, compared with PAA polymers, the sulfonate functional group on PSS polymers can support neuronal differentiation and neurite outgrowth. They also indicated that stem cells could respond to the effect of chemical functional groups and surface charge density to express various regulable cellular behaviors, showing that PEM thin films can be culture systems to provide varied microenvironments for different needs of stem cells.

In 2D culture systems, most cells are anchorage-dependent and require substrates allowing for adhesion and survival. The adhesion of cells plays an indispensable role in cell communication and cell activities involving proliferation, differentiation, and migration [28]. In the human body, ECM provides cells with appropriate substrates to adhere to and regulate the functions of cells. However, the culture dishes or culture wells have no ECM on the surface, so the standard *in vitro* 2D culture system usually requires further surface modification to functionalize the culture surface to regulate cell behaviors in tissue engineering and regenerative medicine. In the above paragraphs, we introduced two basic surface modification methods using the coating processes via physical interactions. However, coating materials on the culture surfaces via physical interactions may easily resuspend in the culture medium and not support long-term culture. Therefore, another method, chemical surface modification, provides an alternative strategy to fix growth factors, ECM, or other biomolecules on the culture surfaces through covalent bonding interactions.

The major advantage of chemical surface modification is that the method provides thin films for long-term culture because of the target active biomolecules with a strong and permanent chemical bonding on the substrates. Therefore, chemical surface modification is widely used to create a specific functional thin layer on any culture surface. In addition, many different biomaterials, such as metals, polymers, ceramics, or composite materials are widely used to design the desired culture surfaces. Compared with metal- and ceramic-based substrates, polymer-based substrates have a long history of being a substrate for culturing cells and can be machined to matter with different porosity, size, and shape. Therefore, polymeric substrates have a long history of being chosen as the fundament substrates to perform the modifications and investigate the effect of various physiochemical properties of materials on cells. For example, Senguta and co-workers modified the surface of polyetherimide films to create an ECM-like surface to enhance cell adhesion [29]. The polyetherimide solution was prepared by dissolving polyetherimide polymers in chloroform, and the solution was poured into a glass culture dish and left overnight at room temperature to form a thin film. Then, an N2/H2 mixed plasma (the flow ratio of N2 : H2 = 2 : 3) was used to activate the polyetherimide surface at 60 W for 20 mins. Afterward, gold nanoparticles were synthesized by reducing 0.1 mM chloroauric acid through 10 mg/mL citric acid at boiling water, and the plasma-treated polyetherimide film was immediately immersed in a citrate-stabilized gold nanoparticle sol at room temperature overnight. Then, a uniform gold layer can be obtained on the polyetherimide film (PEI-gold film), and the PEI-gold film was further dipped into 0.1% (w/v) arginine solution to form a PEI-gold-Arg film. Compared with the fibronectin-coated polyetherimide surface, hydrophilic arginine-modified PEI surfaces changed the hydrophobic surface property of pure PEI films and significantly improved the adhesion of murine fibroblasts. Moreover, the results also indicated that grating small molecules on polymeric surfaces as ECM-like surfaces could offer the same purpose of coating ECM on the substrate surfaces of the cells. In addition, in tissue engineering field, previous studies has reported that MSCs can sense the stiffness of culture matrix and the substrates with scar-like stiffness (35–70 kPa) or soft substrate (1 kPa) may induce native effect onf MSCs, such as myocardical calcification of MSC, low beating frequency, or stopping beating [30–32]. Therefore, substrates with a optimal modulus could provide a friendy environment to cardiac cells and facilitate cardiac cells with best cardiac behaviors [32]. Recently, poly(dimethylsiloxane) (PDMS) polymers are popular candidates for caridiac tissue engineering field because PDMS polymers possess a biocompatible and flexible properties widely used as a different type of cardiac culture substrates. Notably, one of the features of PDMS polymers is that

PDMS polymers have tunable mechanical properties by controlling the crosslinkers concentration of PDMS polymers, which the mechanical properties can be arranged from 50 kPa to 1 MPa. Therefore, the feature plays an important role in various application requirements from different tissues [33]. Although PDMS polymers have tunable mechanical properties, PDMS-based substrates have super hydrophobic properties leading to substrates with a few adhesive cues for cells attaching to the surface. Therefore, surface modification is necessary to improve cell adhesion and extend the applications of PDMS polymers. Recently, Öztürk-Öncel and co-workers developed the PDMS substrates with amino acid-conjugated surface modification for inducing the cardiac differentiation of iPSCs [34]. To prepare an amino acid-conjugated PDMS surface, two amino acids, leucine (Leu) and histidine (His), were firstly functionalized with carboxy groups to obtain Cbz-protected amino acids, Cbz-Leu-OH and Cbz-His-OH. Then, benzotriazole Cbz-Leu-OH and Cbz-His-OH reacted with thionyl chloride to form intermediate products, and the intermediate products were further performed a Cbz-deprotection procedure to create the final products, Leu-SAM and His-SAM. Subsequently, the PDMS substrates were treated with oxygen plasma under 200 mTorr pressure and 50 sccm oxygen flow for 1 min to activate the surface. The active PDMS substrates were immersed in the 1% Leu-SAM, His-SAM, octadecyltrimethoxysilane (OTS), and 3-aminopropyl) triethoxysilane (APTES) in absolute ethanol solutions overnight at 25°C, respectively. After the modification, the modified PDMS surfaces were washed with ethanol and deionized water to remove unreacted SAM solutions. Furthermore, Öztürk-Öncel and co-workers seeded murine iPSCs on the functionalized PDMS surfaces with a cell density of 2×10^5 cells/each 12 well [35]. Interestingly, the results show that the PDMS substrate formed by using 1:50 crosslinker-prepolymer concentration offers a stiffness similar to a normal cardiac environment. Moreover, compared with the PDMS substrate without modification, all modified PDMS substrates significantly promoted the adhesion of iPSCs and enhanced cardiac differentiation. Especially, the Leu-modified PDMS substrate highly serves the cardiac differentiation behavior of iPSCs, which raises the expression of cardiac cells. The results indicated that regulating substrate stiffness and surface chemical properties enabled the creation of an *in vitro* tissue-like environment and offered synergistic effects on modulating the cellular behaviors of iPSCs.

In addition to plasma modification, several conventional modification methods through Cu-catalyzed azide-alkyne cycloaddition reactions, 1-Ethyl-3-(3-dimethyl aminopropyl)carbodiimide/N-hydroxy succinimide (EDC/NHS), thiol-Michael addition, and 4-(4,6-dimethoxy-1,3,5-triazine-2-yl)-4-methylmorpholinium chloride (DMTMM) were widely used to create a functionalized culture surfaces for stem cells [36, 37]. Therefore, Golunova and co-workers grafted a peptide sequence on alginate substrate surfaces by using the abovementioned methods to evaluate the efficiency of grafted processes and improve the adhesion of stem cells [38]. In their study, alginate polymers were selected as a base substrate because alginate polymers are polysaccharide-based polymers but without cell-adhesive sites on alginate backbones. Therefore, a peptide, TYRAY, from bone sialoprotein [39] is promising to support cell adhesion that is conjugated on the surface of alginate substrates for culturing stem cells. The four methods are briefly described below. (1) The first group: one of the reactions through click chemistry is the Cu-catalyzed azide-alkyne cycloaddition reaction. CuSO4 and Tris(3-hydroxypropyl triazolyl methyl)amine (THPTA) were mixed together in the water to prepare a Cu-complex stock solution. Then, alginate polymers were first modified with propargylamine in water overnight. Afterwards, the peptide, TYRAY, was mixed with an alginate-propargylamine solution, and an ascorbate solution was immediately added into the alginate-proargylamine-TYRAY solution with a 15-mins degas process. Then, the solution was reacted under an ambient condition for 24 hours and further lyophilized for the generation of alginate-TRAY (Cu-complex). (2) The second group, alginate-TYRAY (EDC): EDC and NHS were added to the alginate solution at pH 6.5 to react for 3 hours to activate alginate polymers. Subsequently, the TYRAY peptide was added to the active alginate solution within 24 hours of reaction. Then, the alginate-TYRAY solution was purified by dialyzing against deionized water with NaCl, ethanol, and distilled water, and subsequently dried using lyophilization. (3) The third modified alginate group: another reaction in click chemistry is the thiol Michael addition that was used to prepare the modified alginate polymer. First, alginate polymers were reacted with N-(2-Aminoethyl)maleimide trifluoroacetate salt (AEMI) in 2-Morpholinoethanesulfonic acid monohydrate (MES) buffer overnight. Then a cell-adhesion peptide CRGD was mixed with an alginate-AEMI solution, and the mixture was placed at ambient conditions for 24 hours. Subsequently, the modified alginate-RGD (click) was lyophilized for use. (4) The alginate-TYRAY (DMTMM) group: After dissolving sodium alginate polymers, the DMTMM and target peptide, TYRAY, were mixed with the alginate solution to react for 24 hours and followed by a lyophilized process to obtain the alginate-TYRAY (DMTMM). After the preparation of modified alginate polymers with the four different processes, undifferentiated human ESCs were cultured on these substrates to identify the effects of four different peptide-conjugated alginate substrates on stem cells. Compared with the other three peptide-conjugated alginate substrates, the alginate-TYRAY (Cu-complex) obtained from a two-step functionalized process including alginate amidation and Cu-catalyzed azide-alkyne cycloaddition reaction contributed a high efficient peptide conjugation yield. Furthermore, using the alginate-TYRAY (Cu-complex) substrate successfully endowed the native alginate polymers with the active cues and provided a cell-anchored environment for culturing human ESCs. The results indicated that the chemical surface reactions offer a promising and efficient strategy to perform the surface modification of culture substrates to fabricate *in vitro* stem cell culture systems.

In the above paragraphs, we introduced the thin films as the *in vitro* 2D culture stem cell culture systems. Compared with thin films, thick membrane-based systems could support a broad range of applications for culturing stem cells. It is well known that cell-signal and cell-substrate interactions play an important role in the regulation of stem cellular behaviors. In terms of both interactions, the human body usually enables sensing specific biomolecules and physiochemical, structural, and chemical features of substrates, and these interactions usually can transit the bio-signals to cells and further regulate the following related cellular behaviors. Therefore, efficiently designing functional substrates with the ability to generate synergistic effects with biomolecules on cells is fundamental to creating a cell-friendly environment.

Compared with thin films, thick membranes provide more versatile properties such as shape, microstructure, porosity, or mechanical strength to combine with biomolecules for use in the fabrication of 2D culture systems.

As we know, compared with glass, metal, and ceramic-based biocompatible materials, polymeric-based materials are economical for general laboratories and are used in rapid batch and mass production to fabricate thick membranes in manufacturing factories. Here, we will focus on polymeric materials, introduce the source of the common materials for fabricating thick membranes, and then discuss the type of microstructure formation in thick membranes and the techniques for producing thick membranes. In general, the source of materials for fabricating thick membranes could be natural or synthetic materials. According to the physiochemical properties of these materials, the properties of thick membranes could be categorized as hydrophilic, hydrophobic, positively charged, and negatively charged membranes. Moreover, the polymeric-based thick membranes used for cell culture applications could be divided into organic-, inorganic-, and organic/inorganic complex-based membranes. First, the common organic polymers used in biomedical applications are gelatin, alginate, cellulose, chitosan, polyolefine, polyester, polyamide, vinyl polymers, polytetrafluoroethylene, polydimethylsiloxane, and others. Compared with gelatin and alginate, cellulose is a natural polymer with high mechanical properties that are widely used to develop thick membranes in biomedical applications. However, pure cellulose is a low solubility in polar solvents, so pure cellulose is difficult to manufacture to form the thick membrane. Therefore, many cellulose-derivates such as cellulose, ethyl cellulose, cellulose triacetate, and others are developed via the modification of cellulose, as shown in Figure 5-1. Besides cellulose, chitosan and its derivatives usually exist in the shell of the arthropod. Similarly, the uniform and well-arranged polymeric structures endow the chitosan-based membrane with high mechanical properties.

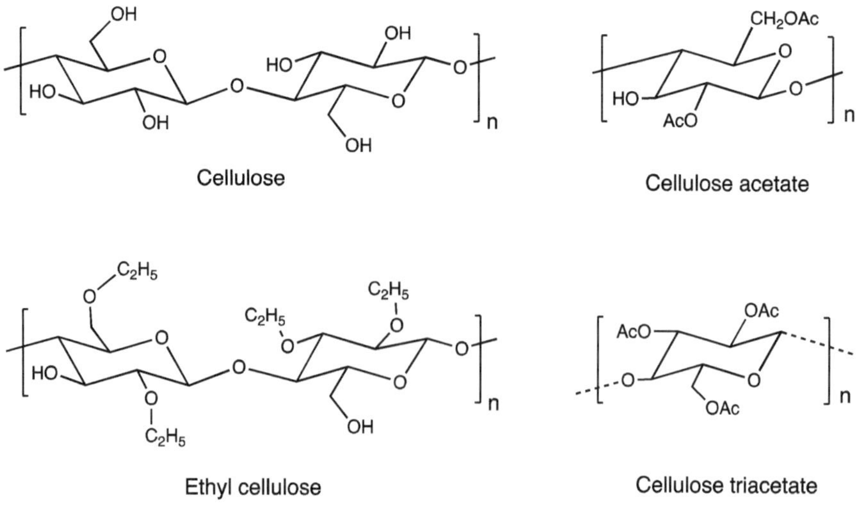

Figure 5-1. The structures of cellulose derivates.

In addition to natural polymers, synthetic polymers can have versatile physiochemical properties by designing their molecular structures. Currently, many different types of synthetic polymers have been used in designing cell culture systems or biomedical applications. For example, polyolefine, polyester, polysulfone, polyamide, vinyl polymers, and polydimethylsiloxane (PDMS) are widely used to form membranes. Polyolefine-based polymers such as polyethylene can be divided into low-density, high-density, and ultra-high molecule polyethylene-based polymers products through the synthetic processes under specific polymeric conditions. Ultra-high molecule polyethylene (UHMWPE) has high strength and inert surface properties and is usually used as an implant in knee or hip replacement. Furthermore, UHMWPE also enables fabricating porous membranes, which can provide a 2.5D structure for the growth of neuron cells and MSCs [40, 41]. Polyester-based polymers possess high mechanical properties, stability, and well solvent and thermal resistance. Therefore, one of the polyesters, polyethylene terephthalate (PET), has been developed as the cell culture wells such as Transwell®. Moreover, polysulfone polymers can generate porous membranes and enable mass production. Therefore, the polycarbonate polymer has been used to culture MSCs and shown to induce a high proliferation rate and prevent cell apoptosis [42]. Next, typical polyamide or polyimide, nylon-6, nylon-66, or nylon-11, provide the semi-crystalline structure, which contributes excellent fatigue resistance, high toughness, and stiffness to be additives for enhancing the mechanical properties of other tissue culture materials. For example, Keirouz and co-workers combined nylon-6 and chitosan to generate a core-shell antimicrobial nanofiber-based membrane for use on the surgical site to prevent bacterial infection [43]. Furthermore, the combination of nanohydroxyapatite and nylon-66 can produce a thick membrane with tunable porosity and bone-like modulus for culturing MSCs and can further enhance the osteogenesis of MSCs [44]. In terms of vinyl polymers, PVA is a typical hydrophilic vinyl polymer that provides a high water-absorption ability to form a barrier on the surface to prevent the adhesion of stem cells and enhance the spheroid formation of stem cells for organoid research [45]. In addition, another popular vinyl polymer, polyvinylidene fluoride (PVDF), contains piezoelectric properties, which can make the membrane with patterns to align the ESC, induce cardiogenesis, and the ESC-derived cardiomyocytes enable contraction spontaneously, indicating that the functional vinyl polymers can be a candidate to culture stem cells [46]. Next, an organosilicon polymer, PDMS, has excellent biocompatibility, high stability, and low manufacturing cost. Importantly, PDMS has been shown to have a wide range of stiffness, making it suitable for use in flexible systems [47]. Therefore, Etezadi and co-workers showed that a PDMS-based membrane could induce the cardiomyocyte differentiation of iPSCs under a xeno-free condition [48]. Based on the abovementioned information, the polymeric-based thick membrane shows its potential applications for making versatile 2D stem cell culture systems.

Next, compared with organic polymer membranes, thick inorganic membranes provide higher chemo- and thermo-stability. The advantages of thick inorganic membranes as cell culture systems are that (1) the membrane can prevent acid erosion in the physiological environment; (2) the high mechanical properties of thick inorganic membranes contribute to fatigue resistance. However, the high production cost and low machinable properties are the challenges of thick inorganic membranes used in cell culture systems. In general, thick inorganic membranes could be categorized into ceramic, metallic, and glass membranes. Ceramic membranes usually consist of metal (Al, Zr, or Ti) and non-metal molecules (carbide, nitride, or oxide), which form aluminum oxide or zirconium oxide membranes through the sol-gel method for culturing stem cells and activating the osteogenic differentiation of BMSCs [49, 50]. Metallic or alloy-based membrane usually possesses high stiffness, so these membranes are widely designed as scaffolds in bone tissue engineering. For example, Arpornmaeklong and co-workers investigated the micromorphological effect of titanium surfaces on the differentiation of human MSCs. Their study shows that the morphology of titanium membrane sandblasted with large grits and acid-etched (SLA)-the surface is similar to the roughness and porous structure of bone that can promote the osteogenic differentiation of MSCs in comparison to smooth titanium surface [51]. In terms of glass membranes, silicon oxide-based membranes could fabricate the membrane with nano-pores, which can provide versatile and tunable properties to create a variety of planar shapes with pore sizes ranging from 3 nm to 80 nm through the modulation of silicon crystal growth and other manufacturing processes [52]. Therefore, ultrathin silicon oxide membranes with porosity exceeding 20% and high optical transparency were used to co-culture systems to separate two different cells and control the transition of cytoplasmic cargo or cytokine. Compared with the traditional transwell system, the highly porous silicon oxide membrane provide a visual alternative as a cell culture barrier for co-culture systems [53]. Moreover, although organic polymer or inorganic membranes offer various features for different stem cell culture systems, to endow the current thick membranes with new characteristics or improve the membrane properties, organic/inorganic complex membranes are developed to complement the weakness of single component-based membranes. For example, regulating the dispersion of inorganic additives in the organic polymeric membrane keeps the features of original organic and inorganic materials, respectively, and enhances the physical and chemical properties of native organic polymeric membranes [54].

After introducing the common materials for making thick membranes, we will briefly introduce the micromorphology of thick membranes and their possible applications for cell culture. The micromorphology of thick membranes enables the design of various spatial microstructures for mimicking different partial tissue structures. According to the geometric shapes, the microstructure of thick membranes could be divided into dense, porous, and composite membranes. The porous structures can especially be further categorized into symmetric and asymmetric membranes (Figure 5-2) [55]. Furthermore, the porous morphology of asymmetric structures can be classified into finger-like, sponge-like, nodular-like, teardrop-like, and lace-like structures. These structures have potential applications such as mimicing tissue morphology for culturing stem cells. In terms of symmetric membranes, the microstructures, such as porous structures, are uniform from the top to bottom of the membrane. The average thickness of symmetric membranes range from 10 to 200 mm, and the symmetric membranes are widely used for applications in reverse osmosis or blood-contact devices [56]. Compared with symmetric membranes, asymmetric membranes offer broader applications in tissue engineering. Generally,

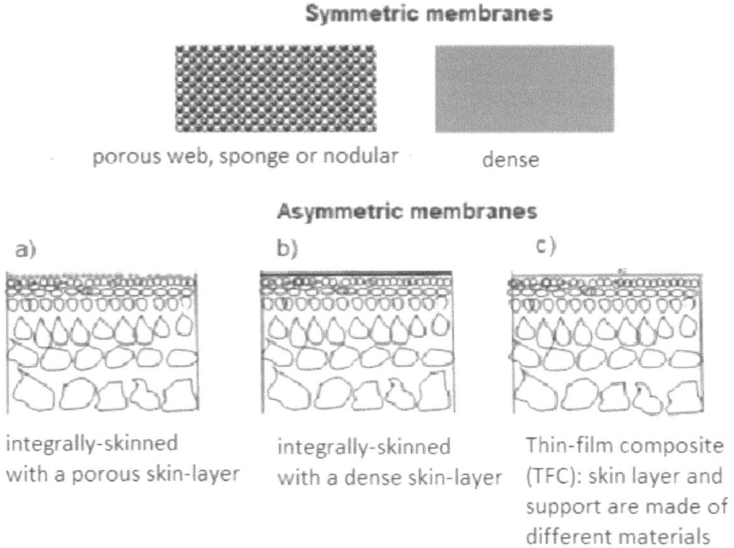

Figure 5-2. The symmetric and asymmetric structures of porous membranes. (Reprinted with permission from [55] Copyright (2025) MDPI.)

a typical sponge-like structure with non-uniform pore-size distribution in which the pore sizes are nano-scale to macro-scale [57], enables the creation of a bone-like environment for culturing stem cells and reconstructing the bone defect [58]. Subsequently, other typical asymmetric membranes consist of a 0.1–0.5 mm skin layer and a 50–150 mm porous supporting layer. This type of skin-porous asymmetric membrane provides more applications in cell culture systems for different purposes. In addition, the skin-porous asymmetric membrane offers a strategy to fabricate composite membranes with two or more different materials. In terms of asymmetric composite membranes, the skin layers of materials with selectivity can determine the functions of membranes, and the control of the supporting layer porosity allows the regulation of the mechanical properties and mass transfer efficiency. Therefore, asymmetric composite membranes have received increased attention for their use in cell culture systems. For example, periodontitis usually causes the defect of cementum, alveolar bone, periodontal ligament, and gingiva [59]. The general treatment for periodontitis is to completely induce the supporting bone regeneration and eliminate the interference of non-osteogenic tissues on tooth bone regeneration. Currently, guide bone regeneration provides a strategy to create a barrier to prevent non-osteogenic cells from soft tissues to migrate into and grow in the defect site of tooth bone in comparison to the conventional approach [60]. Therefore, Niu and co-workers developed electrospun dense-porous composite membranes for the periodontitis bone defect [61]. Due to the low porosity of the dense site, the skin layer of the membrane could exclude the non-osteogenic cells migrating in the membrane, and the other porous site in the composite membrane allows it to serve as a substrate and guide and induce new bone migrating and growing in the membranes.

Here, as mentioned above, polymeric membranes are low-cost and easy to fabricate. Therefore, in the next paragraph, we will discuss the principle and possible mechanisms for forming polymeric membranes. The phase inversion methods are most common method to prepare thick membranes. Through the phase inversion methods, the uniformed liquid polymeric solutions can transfer into solid polymeric membranes. In the phase inversion methods, the techniques are divided into precipitation by solvent evaporation, thermal-induced phase separation, precipitation by controlled evaporations, vapor-induced phase separation, and nonsolvent-induced phase separation. First, the precipitation by solvent evaporation method dissolves polymers in a low-boil point solvent. Then, increasing the temperature could evaporate the solvent and form polymeric membranes. The microstructures in membranes could be changed through the different evaporate temperatures. Notably, although the microstructures could be changed, there is no significant macrostructure, so the polymeric membrane made from the precipitation by solvent evaporation can generate skin structures [62]. Similar to the precipitation by the solvent evaporation method, thermal-induced phase separation, precipitation by controlled evaporations is both through the change of temperature to decrease the solubility of the polymer in solutions and promoting the precipitation of polymers, and the formation of solid, thick polymeric membranes. Unlike the former methods, the solvent with a high boil point and affinity with water is usually used in the vapor-induced phase separation method. First, the polymeric solutions cast on substrates are placed in environments with different humidity at the thermo- and humidity-controllable environments. Then, the polymeric solution could absorb the vapor to change the composition of the solution. Subsequently, the solid polymer membranes form in the nonsolvent solution under a phase separation process. An earlier study indicated that this method could be used to generate a thick membrane with high membrane pore connectivity [63].

Next, we will introduce the nonsolvent-induced phase separation method, a major method used to fabricate most asymmetric membranes. In general, the polymeric solutions are first cast or coated on a substrate. Then, the substrate with the polymeric solution is immersed into a nonsolvent bath to progress an exchange between solvents and nonsolvents. Following solvent extraction, the polymer concentration is increased, and a solid thick polymeric membrane forms on the substrates. Therefore, the microstructure of the membrane could be modulated by controlling the types of solvents and nonsolvents and environment temperature. In terms of the nonsolvent-induced phased separation method, the thermodynamic behaviors among polymers, solvents, and nonsolvents, and the mass transfer behaviors between solvents and nonsolvents play an important role in regulating the formation of the membrane. Therefore, we briefly discuss the relationship between microstructure formation and these factors. Generally, the thermodynamic properties of the polymers, solvents, and nonsolvents could be calculated by the Flory-Huggins theory and are shown in a ternary phase diagram (Figure 5-3). In Figure 5-3, it could be observed that the exchange of solvent and nonsolvent enables guiding the compositions of polymeric solutions from the V area into the I, II, III, IV, and VI areas and progressing a phase separation to form the thick polymeric membrane. According to the ternary phase diagram, we can understand that the membrane formation mechanism may go through pathway 1, and the polymeric solutions do not enter any phase separation area and

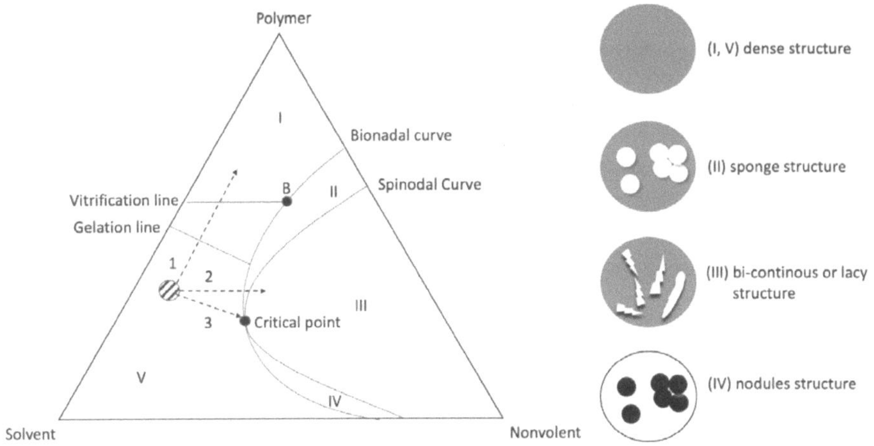

Figure 5-3. Thetenary isothermal phase diagram and the possible microstructure in the different phase areas (B: Berghmans point).

directly enter the vitrification area when the solvents are removed rapidly. Therefore, we can obtain membranes with dense skin-layer microstructures. As compared to pathway 1, if the polymeric membranes are formed via pathway 2, the high polymeric concentration on the surface membrane can easily form an asymmetric structure with a dense skin layer on the surface. Furthermore, if the solvent is slowly removed from the polymeric solution, pathway 3 will be utilized to determine the microstructure of membranes. In this case, the composition of polymer membrane can rapidly pass through the stable area and then enter the spinodal area when the composition is close to the critical point; it may cause high pore connectivity to form membranes with a bi-continuous structure. Therefore, these possible mechanism pathways provide clear information to fabricate the thick membrane with specific microstructures to design the tissue-like cell culture microenvironments.

The above paragraph shows that phase inversion methods have been widely used to fabricate 2D thick membranes for cell culture. Many techniques are developed based on phase inversion methods to produce flat membranes, hollow fiber membranes, and composite membranes. For example, casting or coating processes are the popular methods to fabricate flat membranes [64]. In these processes, polymeric solutions are poured on a substrate, and then a scrapper is used to coat the polymeric solution to form membranes and prevent wrinkle generation on the surface of the membrane. Afterward, if the solvent in the casted membrane is removed by using a solvent evaporation procedure, it can obtain a membrane with a dense skin layer. In contrast, a porous membrane could be obtained when the solvent in the casted membrane is removed via a wet-phase inversion procedure. In addition, compared with flat membranes, hollow fiber membranes are widely used in the industry process because hollow fiber membranes provide a high surface area for use in a limited space. The basic process is that the polymeric solutions are filled into a co-axial nozzle and then extruded from the nozzle. Then, the extruded polymeric hollow fiber forms and is stacked as a membrane structure by using dry- or wet-phase inversion methods to remove the solvents. Therefore, according to the information, it shows that the casting/coating or hollow fiber procedures are two standard techniques for obtaining polymeric membranes to answer the different requirements in stem cell culture.

References

[1] Li, Y., and K.A. Kilian. Bridging the Gap: From 2D cell culture to 3D microengineered extracellular matrices. Adv Healthc Mater, 2015. 4(18): 2780–96.

[2] Altuntas, S. et al. Nanopillared chitosan/gelatin films: a biomimetic approach for improved osteogenesis. ACS Biomater Sci Eng, 2019. 5(9): 4311–4322.

[3] Huang, J. et al. Facile preparation of a strong chitosan-silk biocomposite film. Carbohydr Polym, 2020. 229: 115515.

[4] Ruprai, H. et al. Porous chitosan films support stem cells and facilitate sutureless tissue repair. ACS Appl Mater Interfaces, 2019. 11(36): 32613–32622.

[5] Vedovatto, S. et al. Development of chitosan, gelatin and liposome film and analysis of its biocompatibility *in vitro*. Int J Biol Macromol, 2020. 160: 750–757.

[6] Sechriest, V.F. et al. GAG-augmented polysaccharide hydrogel: a novel biocompatible and biodegradable material to support chondrogenesis. J Biomed Mater Res, 2000. 49(4): 534–41.

[7] Cui, Y.L. et al. Biomimetic surface modification of poly(L-lactic acid) with chitosan and its effects on articular chondrocytes *in vitro*. Biomaterials, 2003. 24(21): 3859–68.

[8] Risbud, M. et al. *In vitro* expression of cartilage-specific markers by chondrocytes on a biocompatible hydrogel: implications for engineering cartilage tissue. Cell Transplant, 2001. 10(8): 755–63.

[9] Lin, Y.H. et al. Keratin/chitosan UV-crosslinked composites promote the osteogenic differentiation of human adipose derived stem cells. J Mater Chem B, 2017. 5(24): 4614–4622.

[10] Kim, S. et al. Chitosan/gelatin-based films crosslinked by proanthocyanidin. J Biomed Mater Res B Appl Biomater, 2005. 75(2): 442–50.

[11] Cheng, N.C. et al. Efficient transfer of human adipose-derived stem cells by chitosan/gelatin blend films. Journal of Biomedical Materials Research Part B: Applied Biomaterials, 2012. 100(5): 1369–1377.

[12] Yuan, B. et al. A strategy for depositing different types of cells in three dimensions to mimic tubular structures in tissues. Adv Mater, 2012. 24(7): 890–6.

[13] Zheng, Y.L. et al. Mesenchymal stem cells obtained from synovial fluid mesenchymal stem cell-derived induced pluripotent stem cells on a matrigel coating exhibited enhanced proliferation and differentiation potential. PLoS One, 2015. 10(12): e0144226.

[14] Kalaskar, D.M. et al. Characterization of the interface between adsorbed fibronectin and human embryonic stem cells. J R Soc Interface, 2013. 10(83): 20130139.

[15] Hung, C.H., and T.H. Young. Differences in the effect on neural stem cells of fetal bovine serum in substrate-coated and soluble form. Biomaterials, 2006. 27(35): 5901–8.

[16] Young, T.H., and C.H. Hung. Behavior of embryonic rat cerebral cortical stem cells on the PVA and EVAL substrates. Biomaterials, 2005. 26(20): 4291–9.

[17] Hung, C.H., Y.L. Lin, and T.H. Young. The effect of chitosan and PVDF substrates on the behavior of embryonic rat cerebral cortical stem cells. Biomaterials, 2006. 27(25): 4461–9.

[18] Zahn, R. et al. Ion-induced cell sheet detachment from standard cell culture surfaces coated with polyelectrolytes. Biomaterials, 2012. 33(12): 3421–7.

[19] Stuart, M.A. et al. Emerging applications of stimuli-responsive polymer materials. Nat Mater, 2010. 9(2): 101–13.

[20] Adusumilli, M., and M.L. Bruening. Variation of ion-exchange capacity, zeta potential, and ion-transport selectivities with the number of layers in a multilayer polyelectrolyte film. Langmuir, 2009. 25(13): 7478–85.

[21] Aytar, B.S., M.R. Prausnitz, and D.M. Lynn. Rapid release of plasmid DNA from surfaces coated with polyelectrolyte multilayers promoted by the application of electrochemical potentials. ACS Appl Mater Interfaces, 2012. 4(5): 2726–34.

[22] Moskowitz, J.S. et al. The effectiveness of the controlled release of gentamicin from polyelectrolyte multilayers in the treatment of *Staphylococcus aureus* infection in a rabbit bone model. Biomaterials, 2010. 31(23): 6019–30.
[23] Sukhishvili, S.A., and S. Granick. Layered, Erasable Polymer Multilayers Formed by Hydrogen-Bonded Sequential Self-Assembly. Macromolecules, 2002. 35(1): 301–310.
[24] Berthelemy, N. et al. Polyelectrolyte films boost progenitor cell differentiation into endothelium-like monolayers. Adv Mater, 2008. 20(14): 2674–8.
[25] Boura, C. et al. Endothelial cells grown on thin polyelectrolyte mutlilayered films: an evaluation of a new versatile surface modification. Biomaterials, 2003. 24(20): 3521–3530.
[26] Rajnicek, A.M., K.R. Robinson, and C.D. McCaig. The direction of neurite growth in a weak DC electric field depends on the substratum: contributions of adhesivity and net surface charge. Dev Biol, 1998. 203(2): 412–23.
[27] Tsai, H.C. et al. A self-assembled layer-by-layer surface modification to fabricate the neuron-rich model from neural stem/precursor cells. J Formos Med Assoc, 2020. 119(1 Pt 3): 430–438.
[28] Cai, S.X. et al. Recent advance in surface modification for regulating cell adhesion and behaviors. Nanotechnology Reviews, 2020. 9(1): 971–989.
[29] Sengupta, P., and B.L.V. Prasad. Surface modification of polymers for tissue engineering applications: Arginine acts as a sticky protein equivalent for viable cell accommodation. ACS Omega, 2018. 3(4): 4242–4251.
[30] Berry, M.F. et al. Mesenchymal stem cell injection after myocardial infarction improves myocardial compliance. Am J Physiol Heart Circ Physiol, 2006. 290(6): H2196–203.
[31] Breitbach, M. et al. Potential risks of bone marrow cell transplantation into infarcted hearts. Blood, 2007. 110(4): 1362–9.
[32] Engler, A.J. et al. Embryonic cardiomyocytes beat best on a matrix with heart-like elasticity: scar-like rigidity inhibits beating. J Cell Sci, 2008. 121(Pt 22): 3794–802.
[33] Xie, J. et al. Substrate stiffness-regulated matrix metalloproteinase output in myocardial cells and cardiac fibroblasts: implications for myocardial fibrosis. Acta Biomater, 2014. 10(6): 2463–72.
[34] Ozturk-Oncel, M.O. et al. Impact of Poly(dimethylsiloxane) surface modification with conventional and amino acid-conjugated self-assembled monolayers on the differentiation of induced pluripotent stem cells into cardiomyocytes. ACS Biomater Sci Eng, 2021. 7(4): 1539–1551.
[35] Ozturk-Oncel, M.O. et al. A facile surface modification of poly(dimethylsiloxane) with amino acid conjugated self-assembled monolayers for enhanced osteoblast cell behavior. Colloids Surf B Biointerfaces, 2020. 196: 111343.
[36] Cobo, I. et al. Smart hybrid materials by conjugation of responsive polymers to biomacromolecules. Nat Mater, 2015. 14(2): 143–59.
[37] Labre, F. et al. DMTMM-mediated amidation of alginate oligosaccharides aimed at modulating their interaction with proteins. Carbohydr Polym, 2018. 184: 427–434.
[38] Golunova, A. et al. Direct and indirect biomimetic peptide modification of alginate: efficiency, side reactions, and cell response. Int J Mol Sci, 2021. 22(11).
[39] Jing, J. et al. Type, density, and presentation of grafted adhesion peptides on polysaccharide-based hydrogels control preosteoblast behavior and differentiation. Biomacromolecules, 2015. 16(3): 715–22.
[40] Preedy, E.C., S. Perni, and P. Prokopovich. Nanomechanical and surface properties of rMSCs post-exposure to CAP treated UHMWPE wear particles. Nanomedicine, 2016. 12(3): 723–734.
[41] Ustyugov, A.A. et al. 3D neuronal cell culture modeling based on highly porous ultra-high molecular weight polyethylene. Molecules, 2022. 27(7).
[42] Zou, J. et al. Evaluation of human mesenchymal stem cell senescence, differentiation and secretion behavior cultured on polycarbonate cell culture inserts. Clin Hemorheol Microcirc, 2018. 70(4): 573–583.
[43] Keirouz, A. et al. Nylon-6/chitosan core/shell antimicrobial nanofibers for the prevention of mesh-associated surgical site infection. J Nanobiotechnology, 2020. 18(1): 51.
[44] Qian, X. et al. Dynamic perfusion bioreactor system for 3D culture of rat bone marrow mesenchymal stem cells on nanohydroxyapatite/polyamide 66 scaffold *in vitro*. J Biomed Mater Res B Appl Biomater, 2013. 101(6): 893–901.

[45] Molyneaux, K. et al. Physically-cross-linked poly(vinyl alcohol) cell culture plate coatings facilitate preservation of cell-cell interactions, spheroid formation, and stemness. J Biomed Mater Res B Appl Biomater, 2021. 109(11): 1744–1753.

[46] Hitscherich, P. et al. The effect of PVDF-TrFE scaffolds on stem cell derived cardiovascular cells. Biotechnol Bioeng, 2016. 113(7): 1577–85.

[47] Carpi, N., and M. Piel. Stretching micropatterned cells on a PDMS membrane. J Vis Exp, 2014(83): e51193.

[48] Etezadi, F. et al. Optimization of a PDMS-based cell culture substrate for high-density human-induced pluripotent stem cell adhesion and long-term differentiation into cardiomyocytes under a Xeno-Free condition. ACS Biomater Sci Eng, 2022. 8(5): 2040–2052.

[49] Mendonca, G. et al. The effects of implant surface nanoscale features on osteoblast-specific gene expression. Biomaterials, 2009. 30(25): 4053–62.

[50] Smieszek, A. et al. Biological effects of sol-gel derived ZrO2 and SiO2/ZrO2 coatings on stainless steel surface—*In vitro* model using mesenchymal stem cells. J Biomater Appl, 2014. 29(5): 699–714.

[51] Premjit Arpornmaeklong, Shelley E. Brown, Zhuo Wang and Paul H. Krebsbach. Phenotypic characterization, osteoblastic differentiation, and bone regeneration capacity of human embryonic stem cell-derived mesenchymal stem cells. Stem Cells Dev 2009. 18(7): 955–68.

[52] Agrawal, A.A. et al. Porous nanocrystalline silicon membranes as highly permeable and molecularly thin substrates for cell culture. Biomaterials, 2010. 31(20): 5408–17.

[53] Carter, R.N. et al. Ultrathin transparent membranes for cellular barrier and co-culture models. Biofabrication, 2017. 9(1): 015019.

[54] Farbod, K. et al. Interactions between inorganic and organic phases in bone tissue as a source of inspiration for design of novel nanocomposites. Tissue Eng Part B Rev, 2014. 20(2): 173–88.

[55] Buonomenna, M.G. Mining critical metals from seawater by subnanostructured membranes: Is it viable? Symmetry, 2022. 14(4): 681.

[56] Bachler, S. et al. Permeation studies across symmetric and asymmetric membranes in microdroplet arrays. Anal Chem, 2021. 93(12): 5137–5144.

[57] Wang, J. et al. Study on the preparation of cellulose acetate separation membrane and new adjusting method of pore size. Membranes (Basel), 2021. 12(1).

[58] Lian, M. et al. A low-temperature-printed hierarchical porous sponge-like scaffold that promotes cell-material interaction and modulates paracrine activity of MSCs for vascularized bone regeneration. Biomaterials, 2021. 274: 120841.

[59] Bottino, M.C. et al. Recent advances in the development of GTR/GBR membranes for periodontal regeneration—a materials perspective. Dent Mater, 2012. 28(7): 703–21.

[60] Rakhmatia, Y.D. et al. Current barrier membranes: titanium mesh and other membranes for guided bone regeneration in dental applications. J Prosthodont Res, 2013. 57(1): 3–14.

[61] Niu, X. et al. Electrospun polyamide-6/chitosan nanofibers reinforced nano-hydroxyapatite/polyamide-6 composite bilayered membranes for guided bone regeneration. Carbohydr Polym, 2021. 260: 117769.

[62] Younga, T.-H. et al. Membranes with a particulate morphology prepared by a dry-wet casting process. Polymer, 1999. 40: 5257–5264.

[63] Tsai, J.T. et al. Retainment of pore connectivity in membranes prepared with vapor-induced phase separation. Journal of Membrane Science, 2010. 362: 360–373.

[64] Lu, Y. et al. A review on 2D porous organic polymers for membrane-based separations: Processing and engineering of transport channels. Advanced Membranes, 2021. 1: 100014.

6

The 3D Culture Technology for Fabrication of Stem Cell-based Biosystem

Yi-Chen Ethan Li,[1,*] *Nai-Chen Cheng*,[2] *I-Chi Lee*,[3,*]
Wen-Yen Huang,[4] *Sung-Jan Lin*,[4] *Chia-Ning Shen*,[5]
and *Min-Huey Chen*[6]

Biomaterials play a groundbreaking role in being artificial substitutes to aid repairing the loss of physiological functions and temporarily or permanently replace the damaged parts of damaged tissues or organs. In the past two decades, stem cells with their potency from pluripotent to unipotent properties have been confirmed to have self-renewal and differentiation abilities that aid the regeneration of injured tissues. Many types of stem cells have been found throughout the embryonic or adult tissues. The niches where stem cells are found in the human body offer a nurturing microenvironment for directly regulating the proliferation and differentiation of stem cells. To create a biomimetic *in vitro* culture environment, using biomaterials as the *in vitro* niches for culturing stem cells has attracted increasing attention, and various new functional biomaterials are springing up with the advancement of material machining techniques in recent years. Furthermore, with the advancement of stem cell biology and decoding genome of cells, the combination of stem cells and biomaterials contribute a new strategy to develop various customized artificial

[1] Department of Chemical Engineering, Feng Chia University, Taiwan.
[2] Department of Surgery, National Taiwan University Hospital, Taiwan.
[3] Department of Biomedical Engineering and Environmental Sciences, National Tsing Hua University, Taiwan.
[4] Department of Biomedical Engineering, College of Medicine and College of Engineering, National Taiwan University, Taiwan.
[5] Genomics Research Center, Academia Sinica, Taiwan.
[6] Graduate Institute of Clinical Dentistry, School of Dentistry, National Taiwan University, Taiwan.
* Corresponding authors: iclee@mx.nthu.edu.tw; yicli@fcu.edu.tw

substitutes with versatile functions for different purposes in the applications of tissue engineering and regenerative medicine.

Stem cells cultured with biomaterials is one of the most important and essential methods in medicine and cell biology. It involves the substrates simulating the microenvironment to enable stem cells to survive, grow, maintain, and reproduce the functions under a sterile *in vitro* system with adequate culture conditions. For translational research and therapies, culturing stem cells on two-dimensional (2D) substrates has been a typical and gold-standard method since the early 1900s [1]. However, most cells are in contact with the extracellular matrix (ECM) and their neighboring cells in the human body to generate a complex crosstalk of mechanical and biological signals for supporting their normal functions. Therefore, the 2D culture systems suffer the drawbacks that they inaccurately represent the real tissue cells and complex cell microenvironment *in vitro* [2]. In addition, the limited space of 2D culture systems may contribute a contact inhibition causing low proliferation rate, changes of cellular morphology and functions, and spontaneous differentiation behaviors. Then, the stemness and specificity of stem cells may be gradually lost in long-term culture [3].

3D *in vitro* cell culture systems

Recently, continuous advancements in technology have improved the limitation of the 2D stem cell culture system. Another strategy known as 3D culture systems has been developed in studies to overcome the limitation of 2D culture systems. Compared with 2D culture systems, 3D culture systems contribute to the designs closer to the structure and functions of the organism in the human body [4]. On one hand, the cells sense the differences of the surrounding ECM and respond by triggering the internal cell signaling pathway, which change the cell behaviors. On the other hand, the changed behaviors of the cells result in different patterns of ECM accumulation. In addition, cells in the 3D culture systems more accurately mimic the cells in the *in vivo* microenvironment vis-à-vis the responses to metabolic waste discharge, nutrients or molecules transition, protein synthesis, and gas exchange through the cell-cell and cell-ECM interactions, which regulate the self-renewal, proliferation, and differentiation of stem cells [5, 6]. All of these make 3D culture methods with various applications model different behaviors of cells *in vivo* when culturing cells *in vitro* [7]. Therefore, 3D culture systems have gradually attracted increasing attention and have become more popular for studies pertaining to different disease models and stem cell research. Based on this, Jensen and co-workers organized the advantages and disadvantages of 2D or 3D culture systems depending on various aspects of experiments being performed (Table 6-1) [8].

Scaffold-free culture systems

Currently, scientists can fabricate 3D cell culture systems by using different 3D culture techniques divided into scaffold-free culture systems, such as liquid overlay culture systems, hanging drop culture systems, rotating bioreactor culture systems, magnetic suspension culture systems, and scaffold-based systems, as shown in

Table 6-1. Comparison of 2D and 3D cell culture. (Reprinted with permission from ref [8]. Copyright (2014) John Wiley & Sons.)

Features	2D Cell Culture System	3D Cell Culture System	References
Cell shape	• Flat and elongation • Monolayer	• Spheroids and natural cell shape • Multiple layers	[2, 7, 9–11]
Cell exposure to medium	• All cells receive the same amount of nutrients and supply components in the medium	• The diffusion of nutrients is unequally divided amongst all cells	
Cell proliferation	• An unnaturally rapid proliferation rate	• Proliferation rates of cell are more realistic	
Cell differentiation	• Poor differentiation	• Cells are well differentiated	
Apoptosis	• Stimulus can easily induce apoptosis in cells	• Higher rates of resistance for drug-induced apoptosis	
Drug sensitivity	• Little resistance	• Cells have more resistance and shown a more accurate representation of the drug's effects	
Response to stimuli	• Inaccurate representation of response to stimulus of cells	• Accurate and realistic representation of response to stimulus of cells	
Usage and analysis	• High replicable and easily interpretable • Better for long-tern cultures	• Difficult to replicate experiments and interpret data	
Cost	• For preliminary testing, it is much cheaper than using the 3D culture system	• More expensive than 2D culture system but can reduce the differences of drug testing results between the *in vitro* and *in vivo* model, and decrease the use of animals	

Figure 6-1 [6]. The scaffold-free systems have no supporting structures for culturing cells, leading to cells forming tissue-like spheroids. In the scaffold-free culture systems, another popular scaffold-free system for 3D cell culture is the hanging drop culture system. The liquid droplets with a small volume (around 10–20 μl) are fixed on the inner lid of culture plates via surface tension, and then the cover is turned upside down. Furthermore, the formation of 3D cellular spheroids relies on gravity, which can condense the cells and induce the aggregation of cells in a liquid drop. The advantage of the hanging drop culture method is that it is easy to obtain the 3D cell spheroids of the same size by adjusting the volume of droplet and the cell density in the liquid drop. However, the limited cell number and size of 3D cell spheroids is the challenge for the hanging drop culture method. Each cell spheroid contains only approximately 50–500 cells in each liquid droplet and the liquid droplet on the lid may fall when the volume of liquid droplet is more than 30 μl. Moreover, vis-a-vis the small liquid droplet, it is difficult to change culture medium and it evaporates easily, so the hanging drop culture system may not be suitable for long-term cell culture.

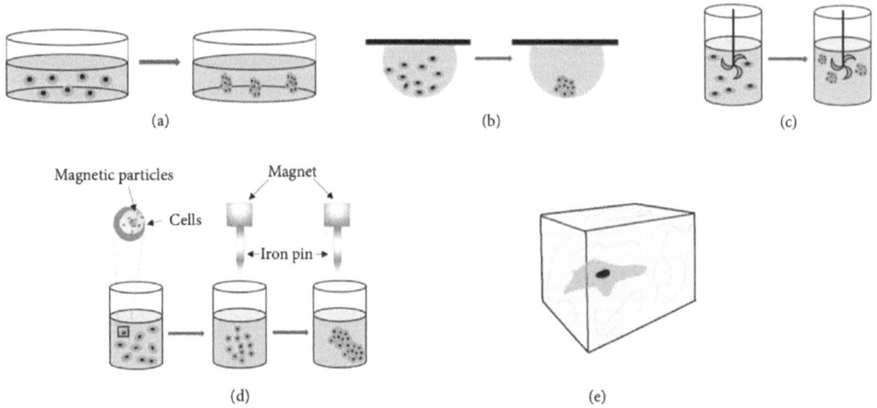

Figure 6-1. The illustration of 3D culture techniques: (a–d) Scaffold-free culture system including liquid overlay culture, hanging drop culture, rotating bioreactor culture, and magnetic suspension culture. (e) Scaffold-based culture system. (Reprinted with permission from [6] Copyright (2021) Hindawi publications.)

Next, the liquid overlay culture method is one of the most economical techniques to fabricate 3D culture systems. The fundamental principle of the liquid overlay culture system is based on inhibiting cell adhesion. Afterward, the cells can spontaneously aggregate to a tissue-like spheroid structure relying on the cell-cell contact or the cell-secreted ECM [12, 13]. To create a non-anchor surface for culturing cells, many hydrophilic materials or commercial culture plates, such as agarose, poly(vinyl alcohol) (PVA), hydroxyethyl-methacrylate (HEMA), and ultralow attachment plates, etc., have been widely used as the low attachment surface for 3D culturing cells. These materials are usually coated or grafted on the surface of culture plates via Van der Waals, ionic, or covalent bonding.

Recently, one of the scaffold-free 3D culture systems, that is, the rotating bioreactor culture method, contributed an efficient strategy to produce a large number of cell spheroids [14]. In general, the rotating bioreactor culture system consists of a continuous stirred-tank reactor (CSTR). The stem cells can be placed in the reactor, and the agitator connected to a motor enables the creation of a turbulent flow-like environment to interfere with the adhesion of cell on the substrates. Afterward, the cells suspend in the medium, increasing the probability of cell contact, and form 3D stem cell-based spheroids. In addition, the rotating bioreactor systems have both continuous input and output flow streams, which can provide the fresh medium to the cells cultured in the rotating bioreactor. Therefore, compared with the hanging drop culture method, the agitator ensures the oxygen and nutrients of the culture medium are uniform distributed in the rotating bioreactor systems where the cells grow in the 3D cell spheroids so that they can obtain enough nutrients to promote the metabolism and enhance the cell viability. Although the rotating bioreactor culture system contributes to a proper nutrient distribution and provides a culture environment similar to the *in vivo* cell system, the non-uniform spheroid size is the major challenge of this method. In addition, to create a well proper distribution environment, the agitator is necessary for the stirring process. However, the stirred fluid easily generates shear stress and foam that may cause unexpected damage to the cells. Therefore, these disadvantages should be overcome and improved by optimizing the speed of the agitator during the stirring process.

Magnetic suspension culture system is a new type of culture method for forming 3D cell spheroids without any scaffold [15]. The magnetic nanoparticles, such as gold or iron oxide nanoparticles, are incubated with cells for several hours to overnight. Then, the cells are magnetized and can easily be gathered into a 3D cell spheroid by regulating the magnetic field. Additionally, some magnetic nanoparticles are modified with the cell-binding domain, which can easily attract the cells to attach to the surface of nanoparticles and aggregate as spheroids [16]. Both methods offer a system to obtain 3D cell spheroids, and the obtained 3D spheroids can be homogenous or heterogeneous spheroids [17]. Furthermore, the purification of 3D spheroids in both methods is easy to execute by using additional magnetic tools or breaking the modified cell-binding domain through enzymes to extract and collect the 3D spheroids. Therefore, the magnetic suspension culture systems contribute to a rapid strategy to form 3D cell spheroids within several minutes and control the size of spheroids. However, a potential concern obviously in the magnetic suspension culture system is that many studies confirm that the cells can uptake the nanoparticles and that the nanoparticles may regulate the cell signaling and affect the original properties of cells [18].

Scaffold-based culture systems

Compared with the scaffold-free culture systems, scaffold-based culture systems provide a novel strategy to develop a 3D biomimetic microenvironment for culturing stem cells. The scaffold-based culture systems are becoming more widely used for culturing stem cells. The possible reason is that the original endogenous ECM from cells in scaffold-free culture systems usually has a slow producing rate and poor mechanical properties [19]. In contrast, scaffolds offer adjustable mechanical properties and plasticity that endow stem cells with a tissue-like microenvironment for growth, proliferation, differentiation, and functionalization. Therefore, various scaffolds are developed for use in scaffold-based culture systems. Basically, the types of scaffolds can be categorized into solid, native tissue-based, and hydrogel scaffolds for 3D stem cell culture, as shown in Figure 6-2 [20].

First, solid scaffolds are an old technique which has been used to fabricate 3D scaffolds in the past several decades. Solid scaffolds provide modellable property and adjustable physical strength that can support stem cells growing, proliferating, and differentiating on the surfaces. In general, stem cells are directly seeded on the solid surface, but a dense surface may cause low cells to migrate into the solid scaffold. In addition, most dense solid scaffolds may have a low water-containing property that results in poor diffusion efficiency of nutrients or oxygen occurring in the centre of dense, solid scaffolds. Therefore, even cells migrate into the dense solid scaffold; the cells easily show low viability when the dense solid scaffold is larger than 500 mm. To overcome the limitation of dense solid scaffolds, scaffolds with porous structures contribute a solution to increase the diffusion efficiency of nutrients or oxygen for thick scaffolds. Porous scaffolds such as sponges or foam have a uniform interconnected structure that endows the scaffold with high porosity. Moreover, the

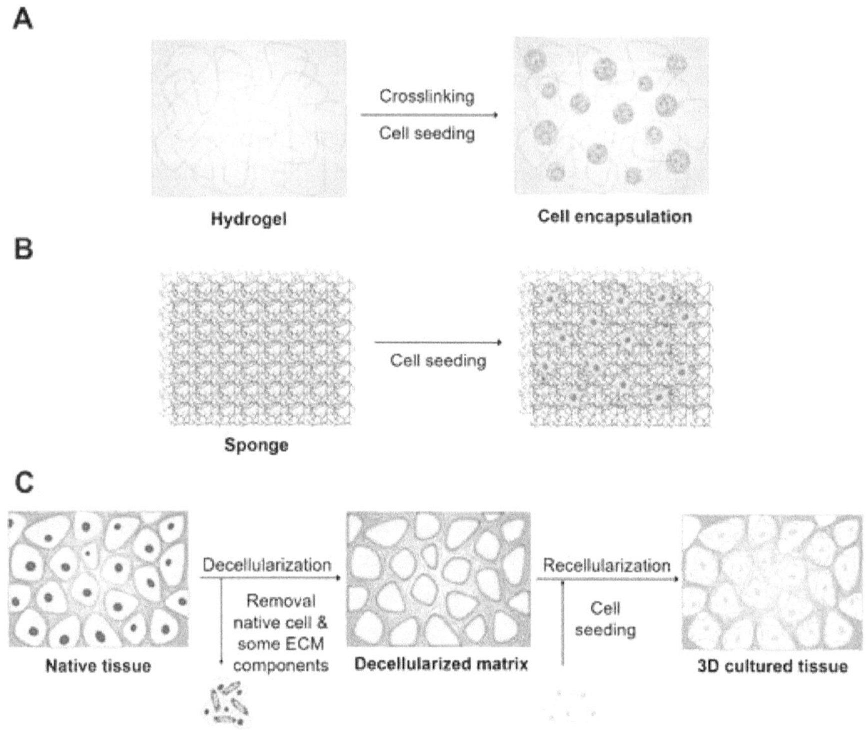

Figure 6-2. The types of scaffolds for 3D scaffold-based culture systems. (Reprinted with permission from [20] Copyright (2021) Hindawi publications.)

interconnected structure enables the diffusion of nutrients or oxygen and allows the cells to migrate into the solid scaffolds easily. Until now, many attempts have been reported to design porous solid scaffolds [21–24]. For example, particulate leaching is a common method in which a material solution contains polymers, and soluble beads form a 3D solid scaffold. Afterward, the soluble beads are leached from the scaffold, and the original site occupied by beads can create a hole [21]. In addition, another similar method, the solvent casting process, is also widely used to combine with the particulate leaching process for creating a porous solid scaffold. Yi and co-workers used a poly(lactic acid) (PLA) solution mixed with sodium chloride particles to fabricate a 3D porous solid scaffold through the solvent casting and particulate leaching (SCPL) process. In their study (Figure 6-3), the results indicated that the porous PLA scaffold provided a 3D cell-anchored area for the attachment of bone marrow stromal cells and cells in the porous scaffold with high cell viability because of the proper nutrient distribution [25]. Besides the SCPL process, other methods, such as gas foaming or emulsion templating, are two common techniques for creating porous solid structures. The gas foaming method is performed by perfusing high-pressure gases (such as carbon dioxide) in the polymer solution and rigidly agitating the solution to obtain the polymeric foam [24]. Through this process, the porosity the polymeric foam can be easily controlled by regulating the flux of gases in the polymer solutions. Noticeably, the scaffolds made from gases foaming may cause unexpected pore interconnectivity. Next, emulsion templating for making the

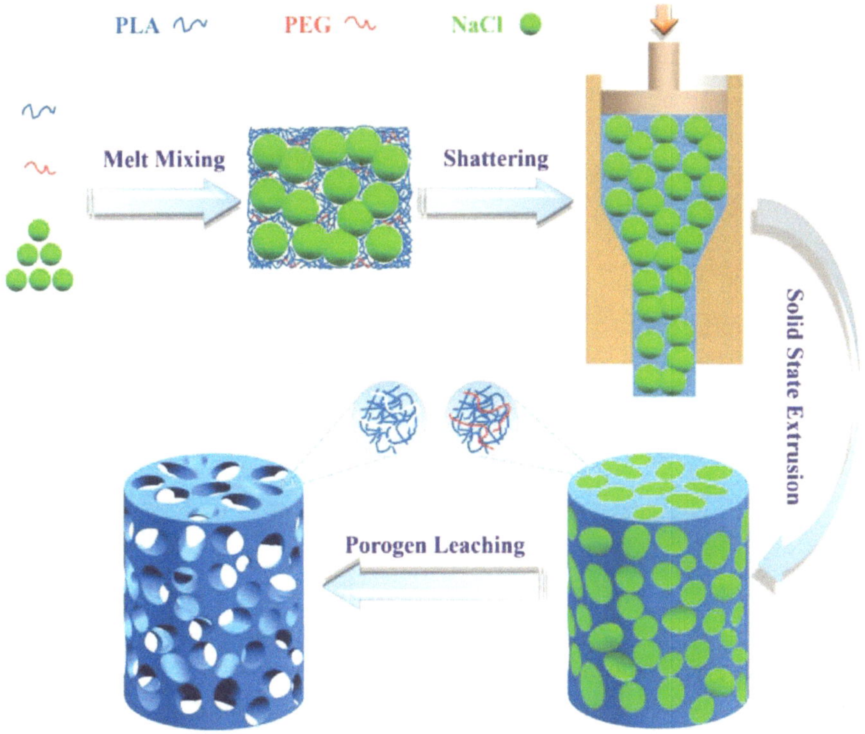

Figure 6-3. The 3D porous solid scaffold made from the PLA-NaCl mixed solution through the solvent casting and particulate leaching process. (Reprinted with permission from [25] Copyright (2016) Hindawi publications.)

porous structure is based on two major steps. The first step is that preparing at least two immiscible components where one component (dispersed phase) is dispersed in the other component (continuous phase). Then the continuous phase of polymeric emulsion is solidified and the droplets in dispersed phase are removed following the solidification of continuous scaffolds. The advantage of the emulsion templating technique is that this technique can generate the scaffolds with high porosity and interconnective pores. In contrast, the disadvantage is that the emulsions are difficult to appropriately monodisperse for crystallization (Figure 6-4) [22].

Although many well-developed methods are used to fabricate 3D porous scaffolds, the problems regarding porosity and the pore size of scaffolds still exist and the abovementioned limitations needed to be overcome for increasing more nutrient transitions and interconnected cellular interactions. To increase the porosity and pore size of scaffolds, fibrous scaffolds provide a new strategy to build 3D porous scaffolds for increasing the surface area and enhancing the gas and nutrient exchange within scaffolds. Many techniques, including fiber bonding, electrospinning, self-assembly, and, etc., have been used to generate 3D fibrous scaffolds [26–28]. The fiber bonding technique is based on binding fibers at the intersections or joints by using crosslinked agents or increasing the temperature over melting points of fibers. After the crosslink, the binded fibers endow the fibrous scaffolds with strong mechanical properties. Moreover, the intricate fiber network also provide a high

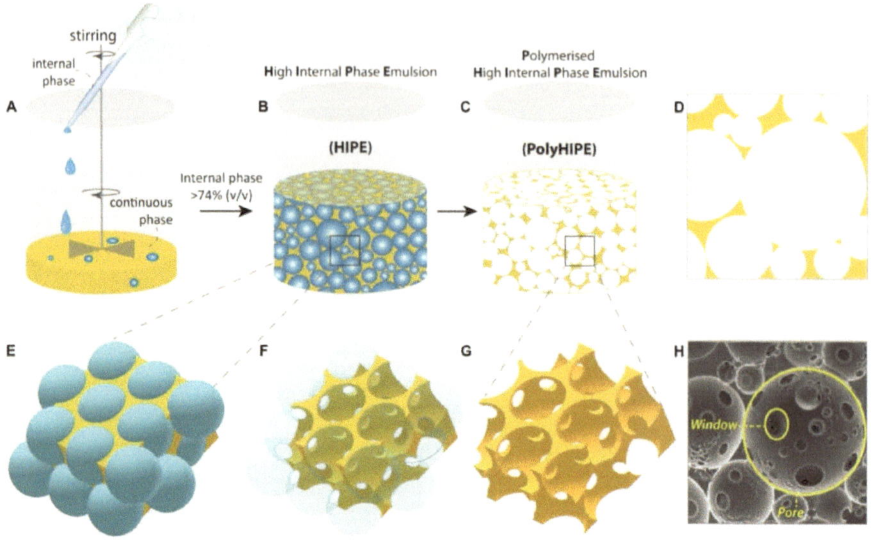

Figure 6-4. Schametic figure in the fabrication steps of the porous scaffold by using the phase emulsion templating technique. (Reprinted with permission from [22] Copyright (2020) Frontiers Media S.A. publications.)

surface to volume ratio and high porosity which allow cells to migrate into the 3D scaffold [26]. But, using chemical agents or thermal method as the manufacturing process for fiber bonding may limit the selection of material sources.

The electrospinning technique uses an electric field as a driving force to generate continuous polymeric fibers, the diameter of which ranges from micrometers to nanometers [27]. The principle of the electrospinning technique is that a polarity polymeric solution is drawn out from a needle placed between two electrodes with opposite electric polarity. The fibers drawn from the needle can be collected via a collector to form different shape of 3D scaffolds such as a mat, a bar, a block, and, etc. [29] for wide applications in stem cell culture. For example, Xue and co-workers utilized the electrospinning technique to make poly(vinyl pyrrolidon) (PVP) and graphene oxide (GO)-PVP nanofibers to fabricate a 3D scaffold for culturing neural stem cells (NSCs) [30]. The results from Figure 6-5 show that the culturing NSCs on the aligned fiber surface of the GO-PVP scaffold can significantly promote NSCs differentiating into neuronal cells. Although the advantage of the electrospinning technique is that the aligned fiber structure in 3D scaffold can induce cells with a directionality, the materials not suitable for culturing cells are used as fibers may require a further modification or functionalization for special purposes, such as proliferation, differentiation, or migration, in stem cell culture [31].

In addition to 3D fibrous scaffolds, some cell cultured systems are now fabricated using a new scaffold fabrication technology, 3D printing, which has been well developed to fabricate engineered cell culture systems with 3D complex structures. To build engineered tissues with 3D configurations more similar to real tissues, people started to employ 3D bioprinting to make biomimetic tissue constructs for stem cell culture systems. In general, inkjet and micro-extrusion printing are two common techniques used in the 3D bioprinting field. Regarding these two printing

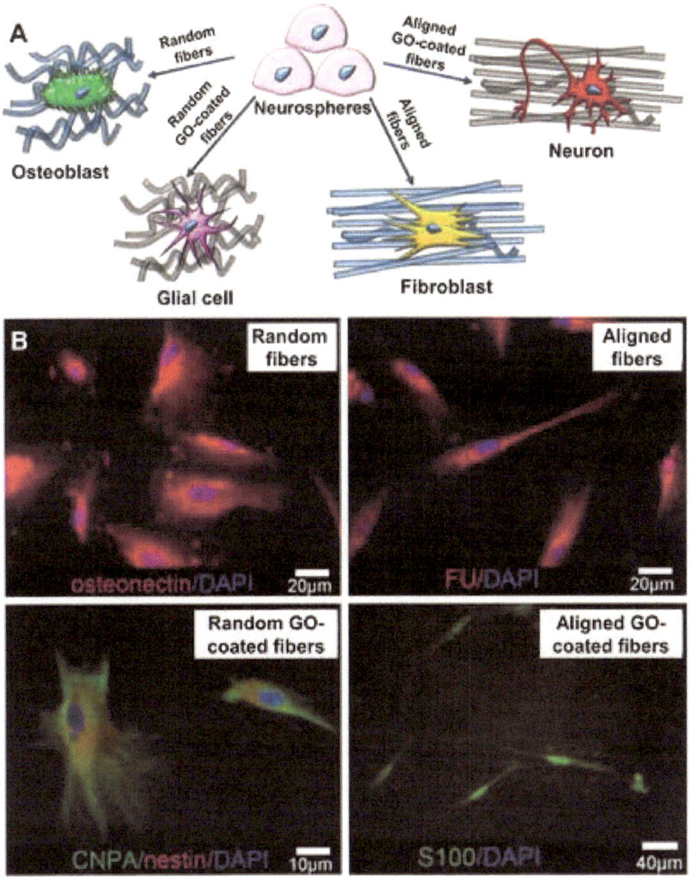

Figure 6-5. Neural stem cell-based neurospheres cultured on an aligned GO-coated fiber. (Reprinted with permission from [30] Copyright (2020) John Wiley & Sons, Inc. publications.)

techniques, injectable biomaterials are crucial in deciding the quality of printed constructs for culture systems. The basic need for biomaterial properties suitable for printing technologies should have injectability, meaning that the rheology properties of materials have to fit the printable requirement of bioinks (i.e., shear thinning behavior) [32]. Additionally, the other physical properties are also essential factors for bioprinting technologies: (i) viscoelasticity, which provides a function to protect the cell from shear stress; (ii) biocompatibility, which decides the cellular viability after the printing process; and (iii) gelation kinetics, which maintains the fidelity of bioprinted structures [33]. This information provides a selection guide for fabricating 3D *in vitro* stem cell culture systems with complex biomimetic configurations. The detail applications will be introduced in the following cases.

Decellularized tissue-based culture systems

To make culture systems close to the components, spatial structures, and chemical/physical properties of native tissues/organs, a new type of scaffold, decellularized tissue, has attracted the scientists' attention since 2010. The basic principle and concept for decellularized tissues is tissues containing rich ECM and specific structure, so it hypothesizes that the *in situ* ECM and tissue structures could be saved after the decellularized process to remove cells. Furthermore, these *in situ* ECM and tissue structures on decellularized tissues have become the new type of scaffold used in culture stem cells to directly induce stem cells to grow *in situ* or differentiate them into the required cells for new tissues. Previously, Ott et al. utilized a perfusion-decellularized heart from a rat to engineer a bioartificial heart; they also proved that, by decellularized process, the architecture and ECM will remain on the decellularized tissue [34]. Following the concept, the decellularized technique applied on different tissues will be a specific tissue-like architecture for reconstructing damaged tissues/organs. Therefore, various strategies have beed developed to produce decellularized tissues/organs according to different purposes. For example, to obtain dense and thick tissues/organs, the perfusion of whole organs is widely used [35]. In contrast, if users want to fabricate thin decellularized tissues/organs, immersing the tissues or organs in the decellularized solution and performing an agilation process is the optimal method recommended for use [36]. In addition to whole or thin tissues, the pressure gradient method is a strategy to obtain tubular or hollow tissues or organs [36]. Moreover, using supercritical fluid is an appropriate method if the decellularized tissues/organs utilized as the decellularized scaffolds can be applied for long-term storage [37]. In general, decellularized scaffolds provide native biomimetic scaffolds with low immunoresponse after transplantation because the cells eliminated from original tissues or organs inhibit the reactions of immunoresponse during the decellularized procedure. Moreover, the decellularized ECM on scaffolds contribute biomedical and anatomical viewpoints to develop an endogenous environment for stem cells. Furthermore, the methods used to recellularize the tissues or organs are divided into two ways. One method is incubating stem cells and decellularized scaffolds together in the culture media. Afterwards, the stem cells can attach on the scaffolds and migrate. However, the efficiency of this method is very low and the cells may not distribute well in the scaffolds. The other method is directly injecting stem cells into the scaffolds for recellularization, and this method is suitable for tubular or hollow tissues or organs. Recently, many reports also show other decellularized scaffolds such as liver [38] and pancreas [39], etc., could be applied in regenerative medicine. For example, Shimoda and co-workers developed a method to evaluate the regeneration of the liver after partial hepatectomy by using a decellularized liver to reconstruct liver tissues. In their study, they used a continuous detergent perfusion procedure for an entire dissected porcine liver to obtain a live-derived ECM scaffold. The perfusion procedure was first used on sodium dodecyl sulfate in deionized water, then the PBS solution containing 0.05% EGTA, 1% Triton X, 4 mM CHAPS, and 0.05% sodium azide was used following the first step. After decellularization, the live-derived ECM scaffold was transplanted and sutured to a left lateral lobe of liver for testing the regenerative effect. Compared with the control lobe without transplantation, the liver sutured with the decellularized liver scaffold shows the clear blood-vessel structures regenerated in the left lateral lobe after 28 days of transplantation (Figure 6-6) [40].

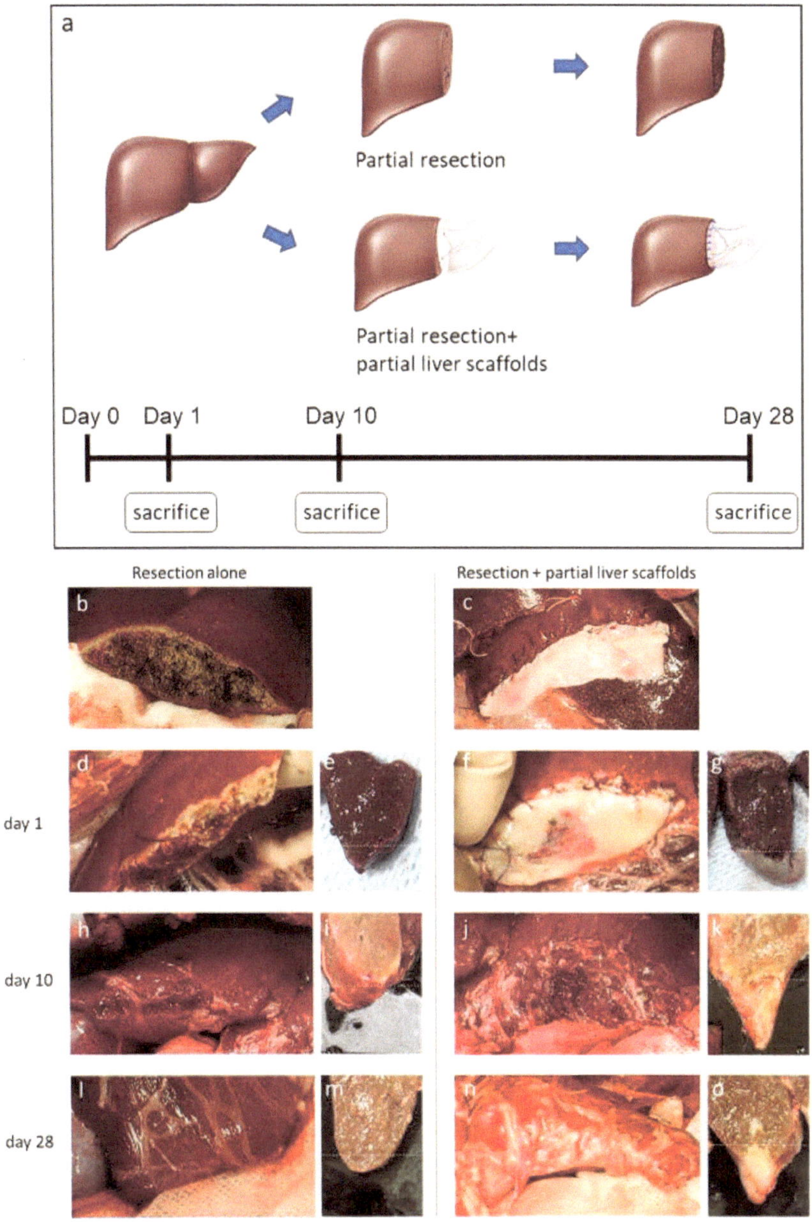

Figure 6-6. The evaluation of a decellularized liver ECM-based scaffold transplanted in the left lateral lobe in a porcine model. (Reprinted with permission from [40] Copyright (2019) Springer Nature Limited.)

Besides the liver-based scaffolds, Guruswamy et al. also created a decellularized pancreata by using perfusion of 0.5% sodium dodecyl sulfate [41]. They found that the decellularized pancreas contributed a suitable scaffold retaining the original ECM and vasculature. After seeding pancreatic cells for 120 days, the decellularized pancreatic scaffold offer the pancreatic cells a biomimetic environment to form pseuoislets.

According to the above information, it can be found that the sodium dodecyl sulfate solution is a common agent used for decellularized tissues or organs. Therefore, Xue and co-workers used a 0.5% sodium dodecyl sulfate solution to decellularize a kidney tissue for culturing adipose tissue-derived stem cells (ADSCs) [42]. After seeding ADSCs in the decellularized kidney organs through an artery or ureter for 10 days, the ADSCs in the decellularized tissue started to differentiate toward tubular or endothelial cells. Based on their results, combining decellularized kidney scaffolds and ADSCs contributes a new alternative strategy regarding a donor kidney for transplantation. Currently, although decellularized tissues have beed used to culture stem cells to form a stem cell-based artificial tissues or organs for regeneration, some challenges still exist in the applications of decellularized tissues/organs [43]. For instance, improving the mechanical properties of a decellularized scaffold is an important topic because the ingredients of tissues such as collagen or proteins loss during the decellularization process. Also, the bioactive site on a decellularized scaffold may deactivate due to the reaction of detergents so that further modification or addition of bioactive factors should be executed for the decellularized scaffolds. Moreover, it is important that some decellularized scaffolds may serve as the host reactions, such as inflammatory and immune response, which negatively impacts the tissue/organ repair process. Therefore, although decellularized scaffolds provide a new strategy for culturing stem cells and creating stem cell-based artificial tissues/ organs, it is necessary to more deeply investigate the relative responses of cells caused via the change of original tissue/organ components and to further design and establish decellularized scaffolds based on the needs of expected cellular responses [43]. In a human body, cell can receive different signals in all three dimensions. Therefore, 3D cell culture systems with better biomimetic environments provide hope for bridging the gap between complex native *in vivo* biological milieu and transitional 2D culture methods. As mentioned above, 3D solid or porous scaffolds and decellularized scaffolds have been offered many benefits for the 3D culture of stem cells. However, lacking the bioactive site in 3D solid/porous scaffolds and unregulable mechanical properties of decellularized scaffolds may limit the development of 3D artificial biomimetic constructs in response to the various requirements of culturing stem cells in tissue engineering and regenerative field.

Gel-based culture systems

In the past decade, scientists were eager to develop a range of biomaterials to overcome the abovementioned limitation. Gel-based biomaterials have been shown to have widely applications and have attracted increasing attention because the properties of gel-based biomaterials enable control of components, structure, and mechanical properties which more accurately replicate features of *in vivo* original tissues. Basically, the gelation of gel-based materials means that polymeric chains in a soluble state change into an insoluble state via the crosslink of polymer chains. The crosslink techniques could be categorized into two mechanisms, physical crosslink and chemical crosslink, and the detail introduction will be mentioned later. Therefore, according to the solvent contained in gel-based biomaterials, the gels can be put into

Table 6-2. The biomedical applications of hydrogels.

Applications	**Examples**
Medical auxiliary appliance	Wound-healing hydrogels, hemostatic hydrogels, heat-insulation hydrogels, and, etc.
Artificial tissues/organs	Contact lens, intraocular lens, wound-dressing hydrogels, and, etc.
Drug carrier	Adhesives for use in oral, intrauterine, or ocular tissues, as well as for pain relief, etc.
Tissue regeneration	Scaffolds, Guided tissue regeneration membrane, and etc.

three categories: lipogel, aerogel, and hydrogel. The lipogels mean polymers dissolve in organic solvents and the hydrogels indicate that polymers dissolve in water.

Water is the most abundant substance in nature; water is the major component in the human body as well. Compared to other components in the human body, the proportion of water in the human body is around 70% to 80%. Therefore, a hydrogel is a hydrophilic and highly water-containing biomaterial with 3D network structure materials [44], and it has been recognized in a variety of biomedical applications (Table 6-2), including drug delivery, regenerative medicine, and other applications [45, 46]. Furthermore, hydrogels are widely used as scaffolds for cell therapy or tissue engineering due to their mechanical properties, multi-tunability, intrinsic biocompatibility, and easy fabrication [47, 48]. Hydrogels used in these applications are typically degradable, can be processed under relatively mild conditions with mechanical and structural properties similar to many tissues and the ECM, and can be delivered in a minimally invasive manner [49].

According to the source of polymers, hydrogels made from polymers can be categorized as natural hydrogels and synthetic hydrogels. In general, natural hydrogels have good biocompatibility because collagen, collagen-derived biopolymers, polysaccharide, etc., are usually used as the source for natural hydrogels. The structures and components of these biopolymers are similar to the ECM so that natural hydrogels can mimic the properties of many native soft tissues. However, although natural hydrogels have been widely used for soft tissues, the mechanical properties of natural hydrogels find it difficult to satisfy the requirement of hard tissues, which limits the applications of natural hydrogels.

Synthetic polymers are the major components for fabricating synthetic hydrogels, as compared to natural hydrogels. Synthetic polymers offer different strategies to produce polymers with versatile functions and regulable mechanical properties according to the design desired from users. Therefore, synthetic polymers can also increase their mechanical properties by crosslinking their own polymer chains and further contribute stronger mechanical properties than natural hydrogels for use in hard tissues. To fabricate synthetic hydrogels, two types of synthetic hydrogel are available based on the bonding of the hydrogel networks; they can be categorized as physical-crosslinked hydrogels and chemical-crosslinked hydrogels [44], as shown in Table 6-3. Physical interactions between polymer chains (e.g., hydrogen bonds, chain entanglements, Van der Waals forces, electrostatic interactions) can result in stable hydrogels [50], such as thermosensitive hydrogels [51, 52]. In contrast, a chemical crosslink means covalent bonding between hydrogel polymer chains to produce

Table 6-3. Synthetic methods and feature of synthetic hydrogels.

Synthetic methods				**Features**
Chemical covalent bonding	Polymerization and crosslinking at the same time		Thermal polymerization	Can combine different monomers
			Photo polymerization	Mild conditions; room temperature
			Radical polymerization	Good polymerization; room temperature
			Inductively coupled as mesorpoperties	Surface polymerization for forming films; high efficiency
	Crosslink after polymerization		Thermal crosslink	Can enable the crosslink of fibers, membrances; or granular powders
			Photo crosslink	
			Radical crosslink	
			Inductively coupled crosslink	Only surface crosslink
Physical force bonding	Crosslink among macromolecules	Hydrogen Bond	Freeze-thraw Freeze-vacuum	Easy to operation; well mechanical properties
		Electrostatic bond	Mix	Easy to operation; mechanical properties change following a change of environment
		Hydrophobic bond	Mix	With ordered structure
	Polyvalent metal bond		Chelation	Easy to operate; a reverable process (unstable)

permanent hydrogel [53]. There are many examples of stem cells cultured within hydrogels for biomedical applications. For example, Kim and co-workers used liquid nitrogen and executed a freeze-throw process to create a cylindrical PVA hydrogel with a stiffness gradient along the longitudinal direction from a soft-top section (~ 1 kPa) to the hard bottom section (~ 24 kPa) [54]. Then, human bone marrow stem cells (hBMSCs) were seeded on the cylindrical PVA hydrogel for inducing the differentiation of hBMSCs. After differentiation, the results confirm that the hBMCs cultured on the PVA hydrogels with a soft-top section, similar to neuronal tissues, towards a neuronal differentiation; in contrast, the bone-like hard bottom section enables the promotion of the osteogenic differentiation of hBMSCs. Next, Chapla and co-workers utilized a photo-crosslink procedure to make poly(ethylene glycol)-dimethacrylate (PEG-dMA) hybrid hydrogels to culture human mesenchymal stem cells (hMSCs). Compared with the single biomaterials, they indicated that mixing PEG-dMA with gelatin-based biomaterials could offer regulable mechanical properties, significantly enhance cell adhesion, and promote the proliferation of hMSCs [55]. Furthermore, a type of non-chemical covalent bonding interaction, chelation, has been widely used as a technique to make hydrogels for culturing stem cells. A classical polymer, alginate, enables crosslinks through the chelation mechanism. Hunt and co-workers first used an RGD-coupled alginate polymer as a scaffold to culture human embryonic stem cells (hESCs) and human-induced pluripotent stem cells (hiPSCs) [56]. In their study, two stem cells were encapsulated into the RGD-alginate hydrogels. Compared with the low-attachment substrate group (i.e., the control group), the RGD-alginate hydrogels provide a tissue-like 3D culture environment to promote the two stem cells with high cellular viability. Moreover, the two stem cells encapsulated in the RGD-alginate hydrogels can form embryoid bodies, and the embryoid bodies enables more differentiation of neural retina tissues than cells cultured on the low-attachment substrate after long-term induction. These results show the potential applications of 3D hydrogel systems for stem cell culture.

Next, we will introduce several common polymers used as hydrogels for culturing stem cells. First, natural polymers are considered as tissue biomimetic polymers suitable for cell culture. Most natural polymers are polysaccharide-based polymers which exist as linear polysaccharide chains; these linear chains will easily extend and move in the solution, contributing entangled networks through the internal molecular interaction and enhancing the gelation ability of natural polymer solutions under modest concentrations.

Cellulose-based culture systems

Cellulose and its derivatives are linear polysaccharide biomass polymers consisting of β-(14)-linked D-glucose units from agricultural raw materials, containing amorphous and crystalline areas [57]. These crystalline areas contribute a high stiffness property considerably similar to cellulose-based biopolymers. Therefore, the crystalline and amorphous areas endow cellulose polymers with tunable mechanical properties for different purposes. Lou and co-workers developed a platform based on cellulose polymers to culture hESCs and hiPSCs and further compared the cellular behaviors in the cellulose hydrogel with that on the 2D Matrigel [58]. The composition of cellulose polymer contains 72.8% glucose, 25.6% xylose, and 1.4% mannose and is prepared as a 1.8% stock solution. Then, a cell density five-times higher than that of the Matrigel group is mixed in the cellulose hydrogel for 3D stem cell culture; final concentration: 0.5% cellulose. One of the major challenges in the 3D hydrogel culture, is how to subculture cells for long-term culture. To overcome this issue, Lou et al. placed the stem cell-laden cellulose hydrogel into the cellulase for 24 hours of incubation. Then the stem cell spheroids could be easily collected, and the collected small stem cell colonies were able to be passaged and subcultured in a new cellulose hydrogel. In their results, it confirms that both hESCs and hiPSCs spheroids cultured from the cellulose hydrogel platform highly express a pluripotency marker, *OCT4*, higher than the marker expressed in the 2D Matrigel group after 24 days of incubation.

Mathieu and co-workers demonstrated a stem cell-laden cellulose hydrogel with a potential application in regenerative medicine [59]. They synthesized a silanized hydroxypropyl methylcellulose by mixing hydroxypropyl methylcellulose with 0.5% silicium. After purification, the 1.5 w/v% of the final concentration was used for the fabrication of a cell-laden hydrogel. Interestingly, the cellulose modified with saline provides sterilizable properties by using steam, which can undergo an autoclaving process. In addition, the silanized hydroxypropyl methylcellulose enables the creation of a water-rich environment that can support the transition of

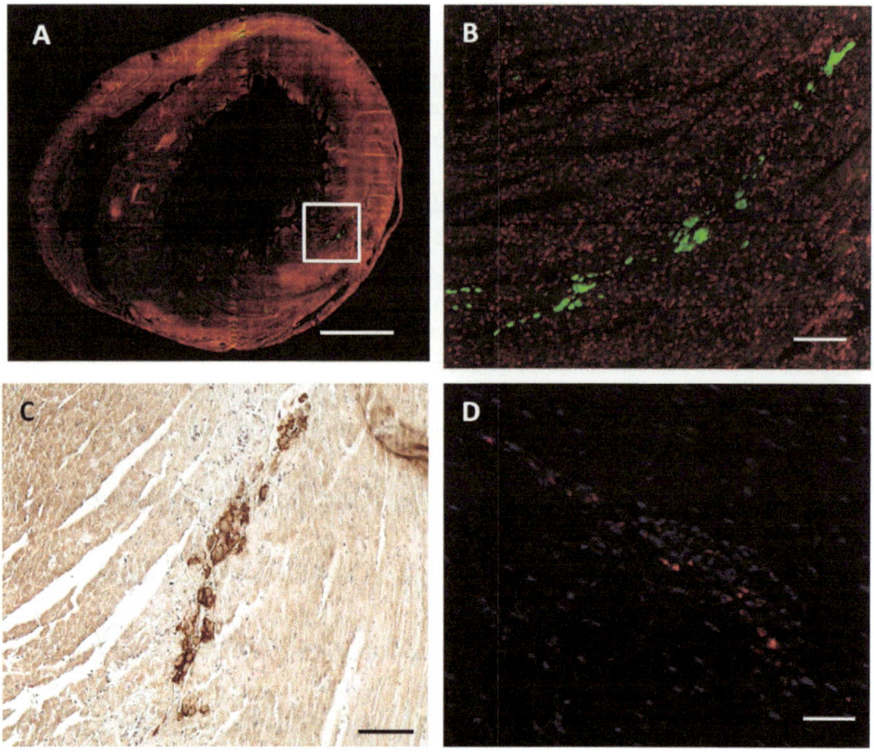

Figure 6-7. Evaluation of mesenchymal stem cell-laden silanized hydroxypropyl cellulose-based hydrogel after *in vivo* injection into the left ventricular cardiac tissue. (A) The nucleus of cells were stained with To-Pro-3 (red color). (B) CFSE-labeled mesenchymal stem cells (green color) can be observed after 14 days of implantation. (C) The engrafted stem cells can be identified with a CD90 marker in the left ventricle after 24 hours of implantation. (D) PKH26 and DAPI staning allowed recognition of implanted stem cells in the heart wall after 14 days in the cardiac tissue. (Reprinted with permission from [59] Copyright (2012) PLOS ONE.)

nutrients and signaling molecules. Moreover, the cellulose-based materials also have an injectable property which can encapsulate mesenchymal stem cells, be injected *in vivo*, and self-crosslink to form a cell-laden scaffold. From the results in this study, as shown in Figure 6-7, the mesenchymal stem cell-laden cellulose ink was injected into the left ventricular wall of cardiac tissue and the cells can still survive at the left ventricular tissue on 14 days after injection. Moreover, the other study also shows that silanized hydroxypropyl methyl cellulose powder was solubilized in NaOH under constant stirring, following which the solution could be mixed with HEPES buffer to form hydrogel [60]. ASCs implanted within Si-HPMC hydrogel were preconditioned in a chondrogenic medium to form a cartilaginous tissue. In addition, the 3D *in vitro* culture of ASC within Si-HPMC hydrogel was found to reinforce the pro-chondrogenic effects of the induction medium [61]. Si-HPCH hydrogel was composed of Si-HMPC and Si-chitosan. It supported the repair of load-bearing osteochondral defects in a canine model in the presence of ASCs [62].

Collagen/Gelatin -based culture systems

Next, we will introduce another popular biopolymer, collagen, for the applications of 3D stem cell culture systems. Collagen is the most abundant protein in the body, which comprises 1/3 of all protein and can be found in skin, cartilage, bone, tendon, connective tissue, etc., in the human body. Collagen can be extracted from a wide range of bioresources, and it is an abundant protein-based biopolymer in all animals. In the human body, collagen is an ECM that contributes various structural support to mechanical properties, such as the elasticity, stiffness, and strength of different tissues/organs. The molecular chain of collagen is built by amino acids, and the major amino acid sequence in collagen consists of glycine-R-hydroxyproline or glycine-proline-R [63]. R can be used as the other seventeen amino acids, and the glycine repeats every three amino acids in the building block of the polymeric chain in collagen. Each collagen polymer contains three chains, and these three chains can further twine together and form an alpha-helix structure. Noticeably, the size of glycine is the smallest among all amino acids, and glycine repeats every three amino acids in the collagen chains so that collagen has a tight structure and a high-stress resistance. Until now, nearly 30 types of collagens have been found according to the types of their structures. Among these collagens, type I collagen through type IV collagen are the common types in the human body. Moreover, compared with these common types of collagens, over 90% is the type I collagen. Therefore, Togo and co-workers cultured ESCs in a three-dimensional type I collagen hydrogel and then induced the ESCs to differentiate into fibroblast-like cells [64]. Fibroblasts are mesenchymal cells with heterogeneity, which play an indispensable role in secreting and maintaining ECM in the human body. However, as the feeder layer to culture ESCs, xenogeneic fibroblasts have several technical and theoretical issues that may lead to the undefined but necessary factors released from xenogeneic fibroblasts affecting the growth or uncontrolled differentiation of ESCs. Although fibroblasts widely exist in our body and many studies use autogenic or syngeneic fibroblasts for co-culture of ESCs [65–67], the mechanism of precisely controlling fibroblast differentiation and the relationship among progenitor and progeny fibroblasts are still unclear. Therefore, Togo and co-workers developed a methodology to address the mechanism for controlling the differentiation of fibroblasts and maturing the functions of a heterogeneous cell population by using type 1 collagen [64]. In their study, they extracted native type I collagen from a rat-tail tendon and used a two-step protocol to induce ESCs differentiating into fibroblast-like cells: (1) Human ESCs were cultured on embryonic fibroblasts with a serum-free maintaining medium containing 80% DMEM/F12 and 20% serum supplements (non-essential amino acid, mercaptoethanol, L-glutamine, and bFGF) for growth. Afterwards, the human ESCs were induced to form embryonic bodies (EBs) by culturing ESC on the plate with a differentiation medium containing 90% DMEM/F12 and 10% serum supplements (non-essential amino acid, L-glutamine, and bFGF) for 4–5 days of incubation to form EBs. Noticeably, compared with maintaining medium, there is no mercaptoethanol in the culture medium; (2) 0.75 mg/l collagen was used to encapsulate EB and cultured in a basal medium (1:1 mixture of DMEM/F12 and differentiation medium) for 21 days of incubation. After 21 days, the spindle fibroblast-like cells could be observed surrounding the EBs. The fibroblast-like cells were collected, re-seeded, and treated in a medium containing 10% fetal calf serum. Following the two-step protocol, the spindle fibroblast-like cells expressing positive staining for fibroblast markers (vimentin and a-SMA) and negative staining for both undifferentiated ESC specific markers (SSEA-41 and SSEA-4) can be obtained from human ESCs.

Type II collagen is a cartilaginous ECM existing in the bone and cartilage tissues, which plays an essential role in developing new bone and regenerating fractured bone tissues. Therefore, type II collagen is a suitable candidate for use in bone tissue engineering. Chiu and co-workers investigated the effect of type II collagen on osteogenic differentiation of bone marrow stem cells [68]. In their study, they coated type I collagen and type II collagen on the culture dish as the substrates for culturing bone marrow stem cells; and a non-coated surface was a control group. After 21 days of incubation, the cells on type II collagen-coated surface showed significant abundant GAG deposition through alcian blue staining. The result indicated that the type II collagen-coated surface enabled bone marrow stem cells with more osteogenic differentiation than bone marrow stem cells with the cells cultured on a type I collagen-coated surface. Furthermore, Chiu and co-workers used integrin-related inhibitors to discuss the possible reasons for the effect of type II collagen on osteogenic differentiation. According to the cell signaling pathway results, the stimulation signals from the activated site on type II collagen-coated surface may pass through the integrin a2b1 on the bone marrow stem cells and then activate the FAK-JNK signaling pathway to induce the osteogenic differentiation [68]. In general, collagen extracted from sources with the retained fibrillar structure is called polymeric collagen, which is widely used for tissue engineering and regenerative medicine applications. Compared with polymeric collagen, the other form of collagen isolated from the original triple helix structure is monomeric collagen. The monomeric collagen extracted from the original structure could be divided into two major groups [69]. One form is the monomeric collagen, whose structure is free of telopeptides, called atelocollagen. The other form of monomeric collagen contains the intact telopeptides, called tropocollagen. The easy and low-cost extraction procedures lead to the current commercial polymeric collagen being widely used in the tissue engineering field. However, polymeric collagen with the naturally crosslinked tropocollagen structure may cause an insoluble property, have variability in each batch, and decrease the reliability of the product. In addition, the sources of collagen are animals that easily evoke immunogenic issues after use in the human body. Therefore, atelocollagen generated from the fibrillar structure by digesting the triple-helix collagen through an enzyme provides a new strategy to obtain a new type of collagen with less native crosslink structures and reduce the issues with immunogenicity. Currently, Tamaddon and co-workers combined monomeric type II collagen and chondroitin sulfate as a strategy to culture human bone marrow mesenchymal stem cells for cartilage repair [70].

The occurrence of osteoarthritis always leads to patients having disabilities and a low quality of life in the adult population. The injury of the cartilage is one of the possible events that leads to osteoarthritis. Therefore, treating the cartilage damage in early stage may prevent osteoarthritis and further prolong the requirement for join replacement. In fact, the cartilage suffers from a limit self-healing ability so that the patients with osteoarthritis require mosaicplasty or autologous implantation [71, 72]. Therefore, Tamaddon extracted type II collagen from fetal bovine cartilage. Then the collagen was dissolved in diluted acetic acid with a pH of 3.2, and the final concentration of collagen was 0.5% (w/v). Afterwards, 0.5 g chondroitin sulfate (CS) was dissolved in the above collagen solution. Then the type II collagen-CS solution was placed in the 48 culture as well as the substrate, and the primary human mesenchymal stem cells were seeded with 1.35×10^6 cell/substrate cell density on the top of the collagen-based substrate. In their results, the cells cultured on type II collagen substrate in the presence of CS had a high chondrogenesis behavior compared to the pure type II collagen substrate in the absence of CS. This is because that the bioactive CS provides a chondro-favorable environment and contributes to a high porosity for condensation and differentiation of human mesenchymal stem cells. In addition, the results also show that the human mesenchymal stem cells on the type I collagen substrate expressed low proteoglycan deposition compared to cells on the type II collagen substrate, indicating that the composition of type II collagen could affect the ECM secretion of cells and stimulate mesenchymal stem cells toward chondrogenesis.

Type III collagen is a major ECM component in skin and organs, which endows many tissues or organs with tensile strength. Type III collagen is one of the fibrillar collagen family that consists of 5–20% of the whole collagen in the human body, and the structure of type III collagen comprises three alpha 1 chains. Compared with other types of collagens, type III collagen is typically produced by cells as pre-procollagen, which can then form procollagen and combine with other collagen chains to create a right-handed triple helix structure. So, previous studies indicated that type III collagen colocalizes with type I collagen and plays a role in regulating the size of type I collagen fibrils [73, 74]. In addition, type III collagen is also a modifier in cartilage to modulate the fibril network through the crosslink of type II collagen fibril. Therefore, type III collagen has been confirmed as having a role in the structural integrity of hollow organs such as the bowel, uterus, and cardiovascular system [75]. Moreover, it is also considered type III collagen has a function involving platelet aggregation in a hemostatic cascade [76] and a wound healing signaling pathway [75, 77]. Based on this information, type III collagen shows potential applications in skin and cardiovascular tissue engineering.

Type IV collagen is a basement membrane protein involving self-renewal and differentiation of stem or progenitor cells in the stem cell biology field. For example, Sillat and co-workers reported that human mesenchymal stem cells generate type IV collagen during the differentiation in the adipogenic process [78]. In general, type IV collagen is a network structure composed of six different α chains with heterotrimers such as α1α1α2, α3α4α5, and α5α5α6 [79]. These α chains could link together, form a sheet-like networks, and further constitute a 2D basement membrane. Notably, the α chain-formed network provides a necessary tensile strength to the basement membrane, which not only contributes regulatory, structural support, and flexibility to the development and maturation of the adipogenic process, but also influences the maturation of adipocytes. The possible mechanism of type IV collagen inducing adipogenesis is attributed to the α chains integrating the integrin and non-integrin matrix receptors on stem cells [80]. In addition to adipocyte tissues, type IV collagen also exists in different tissues and is required to regulate cellular behaviors [81]. For example, the basement membranes of the human liver and skin type IV collagen are in the presence of type IV collagen [82, 83]. Therefore, several reports have shown that decellularized liver or derma tissues were used as biological scaffolds to culture stem cells for liver and cardiac repairing and regeneration [83–85]. In these studies, the results indicated that the cells cultured in type IV collagen-contained decellularized scaffolds allowed the stem cells with the homing and targeting effects to reach the correct location and then expressed many of the *in situ* native features. Besides the decellularized type IV collagen-based scaffolds, one of the famous commercial products, Matrigel, also enriched the basement membrane components such as laminin and type IV collagen, which is widely used for 3D culturing stem cells, including embryonic stem cells and iPSCs. For example, commercial Matrigel contains 8% entactin, 31% type IV collagen, and 56% laminin, which is similar to the basement membrane components [86]. Therefore, Lam and co-workers used Matrigel to culture three human stem cells (i.e., hESCs, hiPSCs, and hASCs) and found that Matrigel enabled hESCs to proliferate compared to the other two stem cells [87]. In addition, Kim and co-workers also reported that Matrigel could be a bridge for the migration of ASCs, enhance the internal cellular communication, and induce the angiogenesis of ASCs to form a homogenous endothelial cellular network when they co-cultured HUVECs and ASCs in the Matrigel [88]. Therefore, these results showed type IV collagen with the potential application as a basement membrane-like scaffold for 3D culturing stem cells.

Next, we will introduce a collagen-derived biomaterial, gelatin, which is popular and widely used in the applications of tissue engineering. Gelatin is a polypeptide structure from the parent protein-based collagen, which contains a heterogeneous mixture of peptides [89]. In general, enzymatic hydrolysis is a common strategy for modifying the functional properties of proteins [90]. Therefore, the production of gelatin from collagen is a procedure using enzymatic hydrolysis involving the breakage and destruction of the bonding among the polypeptide chains. Furthermore, regulating the degree of hydrolysis could control the number of small peptides in gelatin chains during the hydrolysis process of collagen; and the molecular weight distribution of gelatin also could be determined according to the enzyme types, gelatin source, and hydrolysis conditions. For example, the molecular weight distribution of squid gelatin is below 1.4 kDa while producing squid gelatin by using neutrase or alcalase [91]. In contrast, preparing squid gelatin through the hydrolysis of trypsin can obtain the gelatin chain with high proportions of peptides of more than 1.4 kDa [91]. Generally, repeating glycine-based amino sequences (called, Gly-X-Y repeating triplets) are the main structures in the gelatin structures, and X and Y are usually proline and hydroxyproline. For example, the peptide sequences in the Alaska pollock skin gelatin present the typical sequence with Gly-Glu-Hyp-Gly-Pro-Hyp-Gly-Pro-Hyp [92], and the repeating sequence of peptide in giant squid gelatin is identified as Gly-Pro-X-Gly-X-X-Gly-PheX-Gly-Pro-X-Gly-X-Ser [91]. Notably, the peptide sequences in the gelatin chain have been reported for their benefits to human health or biological functions [93, 94]. These peptide sequences may provide cells or tissues with several specific functional properties such as anticancer, antihypertensive, antioxidant, cholesterol-lowering, and antiphotoaging. For example, Mendis and co-workers reported that both Gly and Pro and His (N-terminus site) and Leu (C-terminus site) in the sequence, His-Gly-Pro-Leu-Gly-Pro-Leu prepared from Hoki skin, play a role in radical scavenging activity [95]. In addition, Giri and co-workers stated that the N-terminal site in the presence of hydrophobic amino acid (Ala, Leu, Pro, and Val) and the C-terminal site with Glu, Ile, Met, Leu, Trp, Tyr, Val could provide a strong radical scavenging activity in the related peptides [96]. Therefore, many studies also indicated that the associated sequences in gelatin from different sources enable removing free radicals [97, 98]. In addition to antioxidative ability, gelatin is also abundant in an Arg-Gly-Asp sequence (i.e., RGD sequence), a well-known peptide that is considered a focal adhesion sequence for promoting the adhesion of cells on substrates. An earlier study indicated the mechanism associated with the integrin on the cell surface, which can activate the cell adhesion pathway when the RGD sequence binds with the integrin receptor [99]. Besides, because the RGD sequence binds with the integrin receptor, the RGD sequence on gelatin also involves other integrin-mediated cell behaviors such as proliferation, differentiation, and migration [100]. Therefore, gelatin has attracted increasing attention for its application in cell culture to regulate cellular behaviors.

To culture stem cells in a 3D gelatin-based culture system, Sulaiman and co-workers developed a gelatin microsphere-based method to culture bone marrow-derived MSCs [101]. Microsphere-based scaffolds are 3D culture environments that offer a high surface-to-volume ratio for shortening the cell expansion process and growing [102]. Previous studies have confirmed that the surface of the gelatin microsphere enables the promotion of the cell adhesion and growth [103, 104]. So, Sulaiman et al. combined gelatin microspheres and a dynamic system to culture MSC to investigate the potential therapeutic methods for cartilage damage. In their study, gelatin (molecular weight = 100 kDa) was dissolved in water to prepare a 10% gelatin solution. Then, through a water-in-oil emulsion method, a 20 ml 10% gelatin solution was drop-wise slowly added to 600 ml olive oil at 40 °C. Then, the mixture solution was stirred at 400 rpm for 10 min. After the water-in-oil emulsion, the residual oil was removed by washing with cold acetone in a centrifugation process (4 °C and 5000 rpm for 5 min). Furthermore, the microspheres were fractionated via the sieves with different apertures and dried by air at 4 °C. Afterward, the dried gelatin microspheres were crosslinked by using the dehydrothermal crosslinking method in the vacuum oven at an environment of 0.1 torrs and 140°C. Subsequently, MSCs were seeded on the sterilized gelatin microsphere (cell density: 5×10^4 cells/mg microsphere) and the cell-laden microspheres were further cultured in a static and a dynamic culture environment for chondrogenic differentiation experiments. Compared with the static culture environments, they found the proliferation and chondrogenic differentiation of MSCs in the dynamic system with the growth medium and chondrogenic induction medium are both significantly higher than with the cells culture in the static system. Especially the chondrogenic differentiation, combination of 3D gelatin microspheres and the dynamic system significantly enhanced the cell-matrix and cell-cell interaction, leading to MSC differentiation and a higher production of type II collagen. The results indicated that the 3D gelatin microsphere system is a candidate for the cartilage repair in tissue engineering.

Besides the microsphere carriers, pure gelatin has been used as the other types of 3D porous scaffold for culturing stem cells. For example, electrospinning technology provides a strategy to easily fabricate scaffolds with high porosity. Therefore, Vardiani and co-workers develop a gelatin-based 3D scaffold to make a co-culture system for the induction of embryonic stem cell differentiation [105]. First, the pure gelatin was dissolved in acetic acid to prepare a 15% gelatin solution. Then, the 15% gelatin solution was electrospun under 0.5 ml/hour injection rate and 12 kV voltage. The distance between the collector and needle was 16 cm, and the rotation speed of the collector was 28-RCF. After 5 hours of the electrospinning process, the electrospun gelatin scaffold was treated with glutaraldehyde vapor for two days to endow the scaffold with a water-stable property. Afterward, embryonic stem cell-formed embryoid bodies (EB) and Sertoli cells were seeded on the crosslink gelatin scaffold to induce EB differentiation. Compared with the cell culture on the gelatin-coated culture plate, Vardiani and co-workers found that only the EB-laden gelatin scaffold could increase the gene expressions of the ESC and germline markers (*c-Kit*) and reduce the gene of expression of Nanog. Furthermore, compared with the co-culture of EB and Sertoli cells on a gelatin-coated culture plate, co-culturing cells on the electrospun gelatin scaffold significantly up-regulated the gene expression level of germline markers. They had no significant difference in the gene expression of pluripotency. The results revealed the 3D electrospun gelatin scaffolds with potential applications in the Sertoli-EB co-cultured system for creating a biomimetic seminiferous tubule structure and promoting the ESCs to differentiate into germ-like cells.

Although gelatin-based scaffolds enable the culture of stem cells and offer a cell-friendly environment for the proliferation and differentiation of stem cells, the fast enzymatic degradation and poor mechanical properties of gelatin-based scaffolds are the major challenge and limit the applications of gelatin. Therefore, many strategies, such as blend or covalent bonding, are developed as the standard method to reinforce the gelatin properties for a broad spectrum of applications [106, 107]. First, polysaccharides are natural polymers widely produced as hybrid biomaterials with gelatin to overcome the above shortcoming via the blend strategy. For example, cellulose is a typical polysaccharide that is the most natural polymer from abundant sources. Various sources in nature, such as tunicates, algae, bacteria, and others, endow cellulose with versatile derivatives, including macro-, micro-, and nano-fibrillated cellulose [108]. Xing and co-workers developed high porous hybrid gelatin-cellulose scaffolds to culture human mesenchymal stem cells [109]. Based on the advantages of cellulose, the glucan functional group contributes cellulose fibers with a high packed chain structure, which can provide a desired mechanical property for supporting the cell adhesive structure. In addition, cellulose and its derivates have poor degradability, which offers a stable environment for long-term culture applications under dynamic stress conditions. To fabricate a gelatin-cellulose hybrid scaffold, Xing et al. prepared a 1% gelatin solution using gelatin type B. Then, 1-ethyl-3-(3-dimethyl aminopropyl) carbodiimide (EDC) and N-hydroxysuccinimide (NHS) (molar ration 1:1, final concentration 5mM) were added with the 1% gelatin solution [109]. Afterward, the cellulose polymer was added to about 1% gelatin solution as a mixture, and the mixture was put through a reaction process in a refrigerator at 4°C overnight. Furthermore, the mixture was kept at –20°C to freeze. The freeze gelatin-cellulose mixture was placed in a lyophilized freeze-dryer for more than 24 h. Compared with the pure gelatin scaffold, with the addition of cellulose, a high cellulose concentration contributed more fiber content to significantly reinforce the strength of gelatin-cellulose hydrogels whose Young's modulus can achieve 3 MPa. The result was attributed to good fiber-matrix adhesion. Following the results, the human mesenchymal stem cells were seeded with 2.5×10^6 cells/ml on the gelatin-cellulose 3D scaffold for 28 days. After 28 days of incubation, the fibers in the gelatin-cellulose scaffolds provided cells with both a cell-laden area and an aligned cell morphology, and the high strength of hybrid scaffolds also shows that the stem cells cultured on the scaffold still possessed a multilineage of differentiation ability to execute the osteogenesis and adipogenesis processes.

In the polymer blending systems, electrostatic interaction is one plan used to develop hybrid biomaterials with desired mechanical properties. In general, gelatin polymers contains functional groups with negative and positive charges [110]. Therefore, blending the other charged polymers may provide an electrostatic effect to gelatin that allows for the fabrication of hybrid gels via ionic crosslink or the formation of a polyelectrolyte complex [111]. Recently, one of the polysaccharides, chitosan, is usually used as a polycation polymer for developing various polyelectrolyte complexes. Chitosan is a natural polymer from marine resources. The chitosan polymer contains an *N*-deacetylation functional group which contributes a positive charge to the chitosan backbone. Therefore, the positive charge of chitosan can provide an electrostatic effect to gelatin and further form the gelatin-chitosan polyelectrolyte complex via the polycation and polyanion interaction. The complex type may be a new strategy to reinforce the mechanical properties of the gelatin-based hydrogel. An earlier study has developed a gelatin-chitosan hybrid scaffold for 3D culture mesenchymal stem cells [112]. To fabricate a 3D hybrid scaffold, Miranda and co-workers dissolved gelatin and chitosan in 0.1 M acetic acid, respectively, to prepare the stock solutions, and the concentration of stock solutions was 0.7%. Then, two stock solutions were mixed to prepare the gelatin-chitosan solution at a 1:3 (gelatin/chitosan) ratio. Afterwards, the gelatin-chitosan solution was crosslinked using a glutaraldehyde solution (final concentration = 0.1%). After 1 hour of agitation, the hybrid polymer solution was added to a 24-well cell culture plate at 1 ml/well to make a cylindric gelatin-chitosan sponge. Subsequently, the sponge was dried at 37°C for 24 hours, and the dried sponge was frozen and performed with a lyophilization process. Furthermore, the 25% glutaraldehyde solution at 0.1% concentration was used to re-crosslink the dried sponge for another 30 mins. Furthermore, the gelatin-chitosan sponge was re-treated with a procedure containing heat, freezing, and lyophilization processes. Afterward, the sponge was placed in 100% ethanol for sterilization and dried completely for further cell culture. To investigate the stem cell behaviors in the 3D gelatin-chitosan sponge, primary rat bone marrow mesenchymal stem cells were seeded on the sponge for 14 days of incubation under treatment with an osteogenesis inductive medium, including 50 mg/mL ascorbic acid, 10 mM b-glycerophosphate, and 0.1 mM dexamethasone. After 14 days of incubation, the stem cells could bridge together within the gelatin-chitosan layers in the sponge and significantly express the calcium deposition in the

3D sponge. The possible inductive mechanism is attributed to the similar structural similarities of gelatin and chitosan, offering a favorable biological and physiochemical environment with both collagen- and GAGs-like regions for regulating stem cells. Furthermore, through the evaluation of the *in vivo* experiments, the sponge placed in the tooth sockets of rats for 35 days shows that the gelatin mixed with chitosan could increase the strength and endowed the sponge with a slow degradation behavior. In addition, hybridizing gelatin and chitosan also forms an interconnected network for promoting cell adhesion. Moreover, the degradation of chitosan through lysozyme-catalyzed hydrolysis within 35 days enabled the generation of oligosaccharides from chitosan to induce the migration, growth, and new bone formation of bone cells in the tooth sockets. The results indicated the bone healing function of hybrid gelatin-chitosan biomaterials in bone tissue engineering.

Next, we will introduce another method, crosslink, widely used for reinforcing the strength of gelatin-based hydrogel. In general, the major crosslink mechanisms, including physical and chemical crosslinks, are usually used to strengthen hydrogels. Ionic crosslink, enzymatic crosslink, temperature-stimuli crosslink, and photo-crosslink methods are common strategies for designing various crosslinkable polymers [113]. The UV light-based photo-crosslinked method is a popular method that enables the modification of natural polymers and endows natural polymers with regulable mechanical properties. Since a decade ago, UV-triggered photo-crosslinkable gelatin has been developed and has been widely applied in different biomedical applications. In general, a popular synthetic method for photo-crosslinkable gelatin is modifying gelatin by using methacrylic anhydride. In the synthetic procedure [114], methacrylic anhydride was dissolved in the 10% w/v type A gelatin solution and reacted under stirred conditions for one hour at 50°C. After the reaction, the solution was diluted five times in DPBS, filled in a dialysis membrane (12–14 kDa), and then put through a purified dialysis process in distilled water for 1 week at 40°C. During the dialysis process, the unreacted residual methacrylic acid and salts in the mixture were removed to obtain the pure methacrlated gelatin (i.e., GelMA). Furthermore, the purified GelMA solution was lyophilized for 1 week to remove the water and generate the GelMA polymer. Basically, the reaction mechanism between gelatin and methacrylic anhydride is that the gelatin provides the lysine functional group to react with methacrylic anhydride to form an amide group. Then, the carbon-carbon double bond on the GelMA backbone can create a new covalent bond and crosslink with the other carbon-carbon double bond on the neighbor GelMA chains through the treatment of UV light when the GelMA solution is in the presence of a photoinitiator. Importantly, the degree of amide group formation from the lysine fraction can be controlled by the initial amount of methacrylic anhydride added, and the mechanical properties of GelMA polymer could be regulated via the change of lysine substitution degree and UV-triggered time. Subsequently, the synthetic GelMA polymer has been confirmed to have regulable mechanical properties and biocompatibility for the culture of 3T3 fibroblasts, indicating that the GelMA polymer is a candidate as a cell-laden scaffold [114]. With the advancement of new technology, the combination of GelMA polymer and different engineering means providing a variety of applications for culturing stem cells. For example, embryonic stem cells can form a 3D embryoid bodies (EBs) during the early stage of the differentiation process. In the 3D aggregates, embryonic stem cells pack together and form a 3D structure. Many common methods for creation of EBs aggregates are widely used, such as hanging drop and liquid suspension culture in a dish [115, 116]. Currently, microscale technologies are a useful tool for mimicking the native biological structure. Samad and co-workers used dielectrophoresis to prepare 3D EB aggregates through GelMA particles. The dielectrophoresis provides a rapid method to manipulate the aggregation of particles through an AC electric field. They encapsulate embryonic stem cells in the GelMA hydrogel, and the hydrogel was sandwiched into two ITO electrodes. When the cell-laden GelMA is more polarizable than the surrounding culture medium, the dielectrophoresis force enables the stem cell in the gel to head towards the high electric field region. Furthermore, under applying an electric field at 12 V and 1 MHz, the embryonic stem cells can rapidly move and accumulate within 15 s. Compared with the standard hanging drop culture system, the dielectrophoresis allows the formation of EB aggregates in less then 6 mins. Notably, the EB size has been confirmed as a parameter to regulate the fate and differentiation of stem cells. So, the size of 3D EB aggregates can be controlled by regulating the distance between two ITO distance and the distribution of EB aggregate diameter were arranged from 50 μm to 300 μm. Moreover, Samad et al. also demonstrated that a differentiation-related transcription factor in stem cells, Nagnog, has a significant down-regulation expression in the patterned EB aggregates generated from the dielectrophoresis system than from unpatterned system, indicating that the GelMA-based culture system combine with other engineering technology enables maintain the stemness of embryonic stem cells and provides a high chance of compatibility for use in other technology.

Beside embryonic stem cells or EB spheroids, GelMA-based 3D scaffolds are also applied for culturing mesenchymal stem cells from different sources. For example, endothelial cells have increasingly attracted attention for their applications in basic science or clinical research. However, the applications of endothelial cells suffered from the limited proliferation rates, leading to operate endothelial cells for autologous vascular therapies or re-constructed injured vascular networks. Currently, mesenchymal stem cells show their potential in generation of endothelial cells [117]. Therefore, Lin and co-workers used a GelMA hydrogel to culture human mesenchymal stem cells and create an inductive environment for endothelial differentiation [118]. They prepared 7.5%, 10%, and 15% GelMA solutions and then added the solutions in a mold with 6.2 mm diameter and 8 mm height to form a cylindrical shape under blue light at intensity 10 mW/cm^2. To fabricate a mesenchymal stem cell-laden hydrogel, Lin and co-workers mixed mesenchymal stem cells in the abovementioned three GelMA solutions. After one min of blue light exposure, the mesenchymal stem cells were encapsulated in the hydrogel and further cultured the hydrogel with EGM-2 medium for 10 days of incubation for the induction endothelial and osteogenic differentiation. In their study, the results showed they could provide a range of stiffness of GelMA hydrogel (25 kPa – 41 kPa) for use in the mesenchymal stem cell differentiation. In the 7.5% GelMA hydrogel, the hydrogel with lowest strength provides a suitable environment for the cell proliferation compared with the other groups. In contrast, 10% GelMA hydrogel offers an endothelial- and osteocyte-favoring stiffness so that endothelial marker

(VEGFR-2) and osteogenic marker (OPN) expressions were significantly increased. The reason for this is that a low concentration of GelMA hydrogel enhances the cell proliferation because the larger pore sizes in 7.5% GelMA contribute more nutrient transportation. In the high concentration groups (10% and 15%), the 10% GelMA hydrogel can achieve a stiffness around 33 kPa which is close to the native bone tissue (25 to 40 kPa), however, the stiffness of 15% GelMA hydrogel is over that of 40 kPA. Therefore, this is the possible reason that the stem cells can show higher endothelial and osteogenic differentiation, spindle shape, and capillary-like structure in the 10% GelMA hydrogel. Based on this information, Lin and co-workers demonstrated that GelMA polymers can serve as a culture system with adjustable mechanical properties, which can regulate the proliferation and differentiation of mesenchymal stem cells and affect their functionalities.

There are many sources where mesenchymal stem cells could be isolated. Adipose tissues have been considered as a source for isolating those cells with the characteristics of mesenchymal stem cells. Donnell and co-workers used GelMA hydrogel as a scaffold to culture adipose tissue-derived mesenchymal stem cells (ASCs) in a 3D bioreactor [119]. Similarly, they synthesized GelMA polymer by using methacrylic anhydride to modify gelatin polymers, prepared 15% (w/v) GelMA in Hank's balanced salt solution, and added 0.15% (w/v) lithium phenyl-2,4,6-trimethylbenzoylphosphinate in the GelMA solution. Afterward, ASCs at p3–p4 stages were mixed with the GelMa solution at 2×10^7 cells/ml, and a dental curing light (395 nm wavelength) was used to activate the photo-initiator for 2 minutes and trigger the formation of ASCs-laden GelMA hydrogels. Furthermore, the stem cell-laden hydrogels were cultured in a custom-designed 3D printing bioreactor with dual streams. The bioreactor was designed such that the ASCs-laden GelMA hydrogel was placed in a culture well sandwiched between the two perfusion channels and the growth or the differentiation-inductive medium were introduced into the channels both above and below the stem cell-laden hydrogel. Compared with the monolayer culture, GelMA-based hydrogel provide an adipose tissue -like spatial structure and mechanical properties and the dynamic system can enhance the expression of the adipocyte gene, promote the secretion of adipocyte-required cytokines/proteins, and mature the adipogenic differentiation. These results indicated that the 3D GelMA-based engineered adipose model combining a dynamic perfusion system allowed for culturing stem cells derived from adipose-relative tissues.

Currently, GelMA-based scaffolds are also used for culturing iPSCs. For example, the diseased or traumatic issues usually cause damage to the spinal cord.

Furthermore, the devastating situation may result in the dysfunction of motor and sensory neurons at the injury site. Unfortunately, no effective therapeutic methods have yet been developed so that the harsh microenvironment in the injury site can vitiate the regeneration of neurons and the recovery of body function after spinal cord injury. With the advancement of stem cell therapy, iPSCs for cell therapy have achieved promising results and offer a promising approach to repairing the injured spinal cord [120, 121]. Therefore, Fan and co-workers use GelMA as a scaffold to photo-encapsulate the neural stem cells derived from iPSCs (i.e., iNSCs) [122]. To create a neural stem cell-like microenvironment, the stiffness of the scaffold can modulate the fate of neural stem cells [123]. In general, scaffolds with very low pr high stiffness (< 0.1 kPa or > 100 kPa) are unsuitable for the survival of iNSCs. In addition, if scaffolds with a stiff modulus (7–10 kPa), the high stiffness can induce the astrocyte differentiation of neural stem cells. In contrast, neural stem cells are prone to neuronal differentiation in a scaffold with a soft modulus (0.1–1 kPa). In the above cases, we know GelMA polymer with tunable stiffness can be suitable for the needs of different cultural environments. Therefore, Fan et al. prepared a 3% GelMA solution in the presence of 0.5% Irgacure 2959 and harvested the iPSC-derived iNSCs by treating iPSCs with 1 μM retinoic acid. Then, the iNSCs-laden GelMA hydrogel with cell density at 107 cells/ml is formed by using UV light (6.9 mW/cm2) at 15, 25, and 40 s. Under the different UV exposure times, three different GelMA hydrogels (soft, medium, and stiff, respectively) displayed an increasing Young's modulus (0.68, 1.23, and 2.03 kPa, respectively) following the increase of UV exposure times. Furthermore, they also found that iNSCs in the soft GelMA hydrogel observed higher neuronal marker expression after seven days of incubation compared to cells in the medium and stiff hydrogels. Furthermore, the 2 mm spinal cord segment of mice was transected at the T9–T10 level, and the iNSCs-laden GelMA hydrogel was implanted into the transected site and connected the rostral and caudal sites of the spinal cord. After 6 weeks, the iNSCs-laden GelMA hydrogel group significantly reduced the inflammation, offered high mature neuronal expression, and promoted locomotor recovery, which indicates the potent applications of GelMA hydrogel with tunable modulus for culturing neural stem cells.

The above case shows the application of 3D GelMA-based neural stem cell culture system in spinal cord regeneration. Here, we will introduce the GelMA-based culture system, which is applied for the applications of fabricating *in vitro* neural tissue- and brain-like models. *In vitro* neural tissue or brain models are very useful for investigating therapeutical methods of neuronal diseases, mechanisms of neuronal development, and effects of pharmaceutical medicine on neural tissues or brain. Over the past five years, over 1 million patients worldwide have suffered from brain diseases such as Alzheimer's, ischemic stroke, brain tumor diseases, and others [124–126]. Brain disorders cost almost $1.5 trillion per year worldwide and usually cause a high mortality rate for patients [12–131]. This is largely attributed to the inability of current *in vitro* cell cultures and pre-clinical animal models to predict human responses toward treatment. Despite some advances in 2D neuronal cell cultures and animal models, the existing 2D cultures find it difficult to necessarily recapitulate the important complex features of the 3D brain environments. It leads to the current 2D neural models as brain disease models lacking accurate, specific 3D configurations, cell-cell interactions, and functions such as the directionality of the neurites [132, 133] and the 3D glutamate transition between astrocyte and neuron cells [134, 135]. Therefore, creating a 3D biomimetic *in vitro* brain model with the features of the brain is an important requirement for testing the effects of drugs in the pre-clinical process. To fabricate a 3D brain model, many technologies can offer effective methods for creating 3D complex structures; bioprinting has emerged as a versatile technology that allows the fabrication of 3D tissue constructs of high complexity at extremely high spatial precision and reproducibility. To date, a variety of different techniques for bioprinting have been proposed, such as extrusion [136–139], injection [140–143], magnetic [144], and laser-based approaches [145–147].

However, current strategies are limited in their abilities to bioprint soft and self-supporting materials, such as the anisotropic directional nerve fiber tracts in the brain that extend across a large scale in the 3D space. Recently, a novel 3D printing strategy (i.e., embedded 3D printing) based on direct extrusion of a shear-thinning ink into a self-healing supporting hydrogel matrix has allowed for the embedded printing of freeform structures [148–150]. By taking advantage of the embedded 3D printing technology, it can overcome gravitational force to control anisotropy and microstructures in the 3D space and print very soft materials and soft gel structures *in situ* during the fabrication process to prevent the structures collapsing. The current embedded printing technologies are made with polydimethylsiloxane (PDMS)-, elastomer-, carbopol polymer-, pluronic-based materials as the supervisors for fabricating sensors and vascularized constructs after removing the supervisors [149–152]. These reports provided the new concepts to apply embedded bioprinting technology on the construction of biological tissues. Based on the above information, Li and co-workers developed a 3D biomimetic brain construct by executing a 3D embedded bioprinting technique, as shown in Figure 6-8 [153].

Through the embedded printing technique, they hypothesized that a neuron-laden bioink can be used to fabricate the free-standing neuronal fibers in an astrocyte-laden supporting bath for aligning the growth of neurite and mimicking the nerve fiber tracts in the brain. The astrocyte-laden supporting bath was designed with brain-like stiffness to support astrocyte cell growth and extension. In addition, the co-culture of astrocyte and neuron cells was designed to recapture the astrocyte-neuron interaction in the brain. In the results, the brain-like co-culture construct was fabricated by the neural stem cell-laden bioink embedded printing in the astrocyte-laden bath. After seven days of incubation, neural cells can spread well in the printed nerve fibers. Then, via incubating the neural stem cells for 14 days, the cells with elongated morphologies could be observed in the confluent fibers, which could be attributed to the ECM properties of the bioink suitable for the growth and spread of neural stem cells. Additionally, the encapsulated GFAP-expressed astrocyte in the bath revealed a large network of astrocytes after 14 days of culture due to the gradual degradation of the bath, providing the astrocytes with spatial vacancies to grow and spread throughout the bath. Furthermore, through inducing differentiation in the co-culture constructs, interconnected networks could be formed within the printed fibers, leading to the formation of a patterned self-organizing brain-like structure with astrocytes surrounding the printed neuronal fibers. Moreover, high TUJ1 expression of cells within the printed niche in the co-culture system indicates that astrocytes may promote the neuronal differentiation of neural stem cells, similar to the case observed here [154]. In the brain, it is well known that neurons in the central nervous system can release and transport neurotransmitters such as glutamate to regulate physiological functions; however, excess glutamate release from the neurons results in the overstimulation and destruction of neural cells, a phenomenon called excitotoxicity. Averting this condition, astrocytes uptake and recycle the excess levels of glutamate into glutamine. The translated glutamine can be released from the astrocytes and reabsorbed by the neurons. This process is known as the glutamate-glutamine cycle. Using this mechanism, the glutamate concentration can remain in equilibrium in the brain. In the study, the printed 3D culture system with

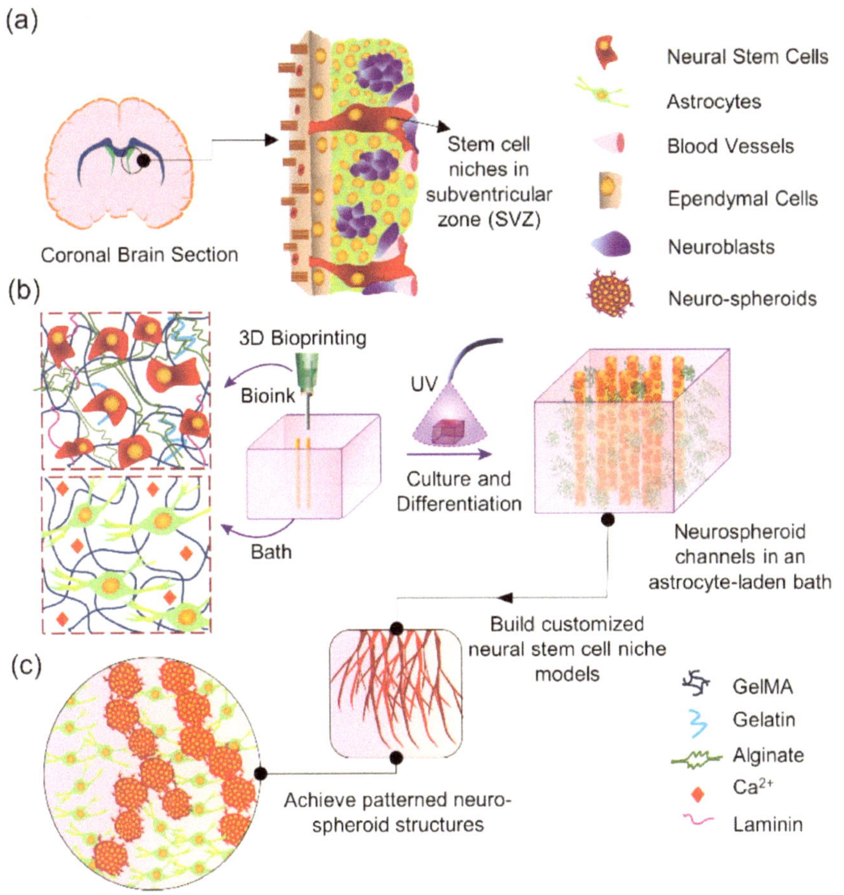

Figure 6-8. (a–c) Schematic diagram of the stem cell niches in the subventricular zone and the fabrication process of embedded printing a 3D tissue-like microenvironment, which consists of the neural stem cell-laden bioink inside an astrocyte-laden support bath. (Reprinted with permission from [153] Copyright (2021) IOP Publishing.)

neuron mono-culture showed higher amounts of released glutamate compared to the printed systems with the co-culture and astrocyte mono-culture. Notably, compared with the neuron mono-culture group, the concentration of glutamate in the co-culture group experienced a reduction in the presence of astrocytes in the brain-like construct. This difference may be attributed to the role of adjacent astrocytes in the uptake of the glutamate produced by the neurons in the printed fibers and the bath. Therefore, combining the GelMA polymers and 3D bioprinting technology can contribute to variations in the mechanical properties by optimizing various compound concentrations, UV exposure time, or chemical crosslinking ratios to achieve an *in vitro* neural stem cell co-culture with the desired biomimetic brain geometry. By adjusting the stiffness of the sink and bath, the developed brain-like co-culture system could effectively interface neurons with glial populations and thus provide a pathway toward modeling the 3D neural stem cell niche. Moreover, the stem cell-based co-culture system demonstrated that the optimized system could support the growth and differentiation of neural stem cells and astrocytes. Also, the results showed the feasibility of fabricating complex free-standing structures by patterning self-organizing neural stem cells and the flexibility of a bioprinted stem cell culture system in creating diverse neuronal populations within heterogeneous cell cultures. As such, spatial gradients of neural stem cells could be generated by adjusting the printing parameters. Therefore, the stem cell-based co-culture brain-like system can provide a means to study and engineer a heterogenous neural stem cell niche.

Alginate-based culture systems

Next, a natural biomass polymer, alginate, derived from abundant resources, will be introduced to the potential applications in stem cell culture. In general, alginate is typically a naturally occurring anionic polymer, which is also a polysaccharide polymer mainly obtained from brown seaweed. Alginate polymer is a popular candidate for cell culture that has been extensively investigated and used for many biomedical applications due to its biocompatibility, low toxicity, and relatively low cost [155, 156]. Additionally, the alginate polymer chain contains unbranched biomass polysaccharides, which consist of an α-(14)-linked L-guluronic acid (G) or 1,4 linked residues of β-d-mannuronic acid (M) structure such as -GGG-, -MMM-, or alternating -MGMG- repeat units. Notably, this α-(14)-linked chain configuration is similar to the structure of an egg-box. Therefore, the configuration provides an "egg-box" effect on alginate chains conjugating with metal ions. For example, calcium chloride is usually added to the alginate solution to create an alginate-based hydrogel because the calcium ions with divalent charge enable the formation of a physical crosslinked ionic bond with the alginate polymeric chain when calcium chloride is dissolved in the solution. Furthermore, via the addition of different divalent cation ions such as Sr^{2+} or Ba^{2+}, the different affinity between alginate polymers and divalent cation ions contributes a selectivity to the amount and type of ions for hydrogel formation, which can tune the properties of hydrogels such as swelling, strength, elasticity, and stability [157]. Therefore, alginate-based biomaterials have versatile applications in the biomedical field. For example, Gao and co-workers recently developed a novel hybrid cell-laden alginate bioink via a mixture of alginate, vascular-tissue derived extracellular matrix (ECM), and endothelial progenitor cells [158]. Then, this hybrid bioink enables a direct fabrication of hollow tubular vessels by using a 3D coaxial bioprinting technique, indicating that the alginate scaffold is suitable for cell culture.

A previous earlier study reported that the molecular weight of alginate might affect the formation of hydrogel and its applications in cell culture [159]. To investigate the formation and differentiation of EBs, Fannon and co-workers used alginate polymers with two different molecular weights (i.e., low molecular weight (high G/M ratio) and high molecular weight (low G/M ratio)) to culture embryonic stem cells or EBs [160]. In the study, the low molecular weight alginate polymer was defined as 100,000–200,000 g/mole, and the G/M ratio was around 3; and the high molecular weight alginate polymer was defined as 450,000–550,000 g/mol, and the G/M ratio was around 0.67. Subsequently, they prepared 1% alginate solutions, and the alginate solutions were mixed with mouse embryonic stem cells (cell concentration at 3 × 10^6 cells/ml). Furthermore, the cell-mixed alginate solutions were extruded into a crosslinking bath consisting of 10 mM 4-(2-hydroxyethyl)-1-piperazineethanesulfonic acid (HEPES) and 100 mM calcium chloride to generate ESC-laden alginate beads by using a syringe connected with 21 g needle. Then, the cell-laden beads were immersed in the crosslinking bath for 5 mins to ensure the beads with fully crosslinked. After crosslink, the cell density per each bead were evaluated: the low molecular weight alginate-based beads with 6 × 10^4 cells/bead and the high molecular weight alginate-based beads with 5 × 10^4 cells/bead. After six days of incubation, the results showed no significance of the EBs formation between low molecular weight and high molecular weight alginate groups, indicating that both types of alginate polymers were suitable for culturing ESCs and forming EBs. Furthermore, the EBs were confirmed that the cells in the EB could maintain their stemness and enable differentiation and express the makers from three germ layers, such as Tuj1, Nestin, SMA, and AFP, after treatment of differentiation condition.

Vis-à-vis retinal diseases, producing retinal-like tissue through tissue engineering technology benefits both the *in vivo* tissue transplantation and the *in vitro* fundamental study of diseases. According to the anatomy of retinal tissue, the normal retinal tissue is a multiple-layer structure, which includes supportive retinal pigmented epithelium (RPE) and neural retina. However, retinal diseases, such as retinitis pigmentosa, age-related macular degeneration, glaucoma, and vascular retinopathies may cause permanent blindness. The current gold standard for 70% of patients is by transplanting foetal neural retina and retinal pigmented epithelium, but the limited foetal tissue source and ethical issues can reduce the options of therapeutic methods. Therefore, developing suitable protocols through a tissue engineering strategy may be useful for studying the retinal disease. Human ESCs or iPSCs have been shown their ability to generate the RPE for transplantation in clinical [161]. However, despite previous reports confirming the successful use of neural retinal sheets derived from hiPSCs or hESCs in tissue transplantation for clinical cases, the major challenge remains that the protocol requires a over 200 days lengthy process for cell differentiation [162]. To overcome the challenge, the combination of ECM-like scaffolds and stem cells is recognized as a novel strategy to provide the various microenvironments for the efficient differentiation of the neural retina with RPE. In the previous study, Hunt and co-workers developed an ECM-like alginate-based hydrogel to culture hiPSCs for improving retinal tissue development [56]. To synthesize the ECM-like alginate polymer, Hunt et al. coupled a peptide sequence, GRGDSP, from ECM with high molecular weight alginate. The RGD-based sequence was added to the alginate solution by using the concentration at 0.5% or 1% (called 0.5% RGD-alginate and 1% RGD-alginate). Then, hiPSC-derived EBs were mixed with the RGD-alginate solutions, which were further immersed into a 0.1 mM calcium chloride for crosslinking and encapsulating cells to form hiPSC-laden alginate hydrogels. The results show that the hiPSC cells in both types of RGD-alginate hydrogel express a low level of cleaved-caspase-3 (CASP-3), an apoptosis marker, indicating that there is no adverse effect on the cells encapsulated in the 0.5% or 1% RGD-alginate hydrogels. Additionally, the hiPSCs were found to form embryoid body (EB)-like structures within the hydrogel, which exhibited an increase in size over time. The expressed Ki-67-positive cells also frequently represented in the hydrogel confirmed that the stem cells encapsulated in the hydrogel enabled them to maintain their proliferation ability. Afterward, the cells maintained in the RGD-alginate hydrogels allowed the cells to differentiate into pigmented RPE and optic vesicles (OVs) after 45 days of incubation. Interestingly, 0.5% RGD-alginate hydrogel increased the expressions of both pigmented RPE, OVs, and neural retinal markers, and 1% RGD-alginate hydrogel only enhanced the cells with OVs or RPE expressions. Therefore, the results indicated that the 0.5% RGD-alginate hydrogel as a scaffold may be useful to culture hiPSCs and could significantly decrease the culture period to 45 days for generating the RPE and neural retinal tissues.

As mentioned above, it is well known that ECM plays an indispensable role in the microenvironment for cells, which can contribute to the biomimetic spatial structure and mechanical properties and activate the cellular function or regulate cell behaviors. The content of the alginate polymer chain consists of natural polysaccharides, which can provide both physical- and chemical-crosslinked mechanisms to form hydrogels. The hydrogels have internal 3D networks and are oxygen/nutrient-permeable scaffolds that enable the survival of cells. Moreover, alginate hydrogels are versatile and their biophysical and biochemical properties can be modified. Therefore, by using alginate polymers, Bidarra and a co-worker established some key steps for creating a 3D culture microenvironment for mesenchymal stem cells [163]. Here, they describe an ionic crosslink method to fabricate an MSC-laden alginate hydrogel. First, they prepared a RGD-modified alginate polymer to endow the alginate with more bioactive sites for cells. The 1% alginate solution was mixed with N-hydroxy-sulfosuccinimide (Sulfo-NHS), and then 1-ethyl-(dimethylaminopropyl carbodiimide) (EDC) was added to the solution at a ratio of 1:2 (Sulfo-NHS/EDC) so that the EDC could react with the carboxylic group on the alginate polymer chains to form an O-acylisourea intermediate. After adding EDC to the alginate solution for 5 mins, the GRGDY-peptide was added, and the N-terminal of GRGDY could react with the intermediate to form an RGD-modified alginate. Afterwards, the synthesized RGD-modified alginate polymer was further purified through a dialysis process (MWCP = 3500). Subsequently, the mesenchymal stem cells were added to the RGD-modified alginate solution, and the calcium carbonate was added to an adequate amount in the cell-mixed alginate solution following a 0.288 Ca_2/COO- ratio to perform a crosslinking process to fabricate an MSC-laden alginate hydrogel. Furthermore, the MSC-laden alginate hydrogel was treated with a basal medium (DMEM + 10% FBS) for proliferation or an osteogenic inductive media (DMEM + 10% FBS + 100 nM dexamethasone + 10 mM ß-glycerophosphate + 0.05 mM 2-phosphpo-L-ascorbic acid) for osteogenesis. Moreover, Bidarra et al. provided a strategy to release the MSC from the cell-laden alginate. The calcium-crosslinked MSC-laden alginate hydrogel was immersed in a solution including 50 mM EDTA for 5 mins. Afterward, the MSC could be released from the hydrogel because the EDTA could be used a chelator to catch the calcium ions from the hydrogel and induce the occurrence of disintegration. This phenomenon implied that the alginate polymer as a stem cell culture system might offer a variety of applications for encapsulating or releasing cells.

Adipose-derived mesenchymal stem cells are a type of mesenchymal stem cell candidate with immunomodulatory and regeneration potentials for stem cell therapy and tissue engineering. Especially, the self-assemble morphology of MSC-based spheroid has been confirmed to enhance the stemness, expansion, differentiation, and other properties to contribute a supplemental role to making up the cell-cell interactions in the *in vitro* culture systems [164]. In addition, stem cell therapy typically requires a large number of stem cells, at least more than 1,000,000 cells per kilogram of body weight, in order to achieve clinically relevant outcomes. However, the scaling -up of stem cell numbers in the *in vitro* culture systems is majorly obstructed by the requirement of space, sophisticated control systems, and complex gas/nutrient pipelines to set up bioreactor systems. Therefore, developing a simple culture system for harvesting stem cell-based spheroid benefits extending stem cell applications in the cell therapy field. Thus, Nebel and co-workers used the alginate polymer to create core-shell capsules for 3D culture adipose-derived mesenchymal stem cells [165]. To create the core-shell capsules, two different materials, xanthan gum, and carboxymethylcellulose, were used as thickening agents by increasing the viscosity of suspensions. Furthermore, the thickening agents would be mixed with calcium chloride and cells; and the final concentrations of calcium chloride in the agents were 1.3%. In Figure 6-9, the schematic figure detail shows the procedure for fabricating the stem cell core-alginate shell capsules.

Firstly, xanthan gum and carboxymethyl cellulose were UV-sterilized and mixed with the culture medium to prepare a 0.3% w/v xanthan gum solution and a 1.1% w/v carboxymethyl cellulose solution containing 1.3% calcium chloride and 1×10^6 cells/mL human adipose-derived mesenchymal stem cells. Subsequently, the cell-laden solutions were gently extruded through a 30 G needle and added to the 0.5% alginate bath drop by drop at 400 RPM stirring speed for well dispersion. Once the cell-laden droplets were submersed in the alginate solution, the calcium ions provide an ionic crosslinking interaction to link alginate polymers and form a shell on the cell-laden droplet surfaces instantly. After 5 mins of crosslink and stirring,

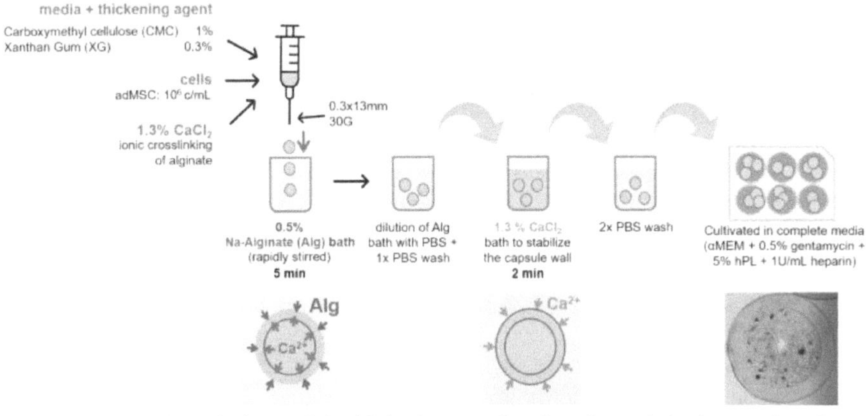

Figure 6-9. The schematic figure of the fabrication procedure for adipose-derived mesenchymal stem cells-laden core-shell alginate capsules for 3D culture strategy (Reprinted with permission from [165]. Copyright (2022) MDPI).

the PBS solution was used to dilute the capsule-containing alginate solution to further disperse the capsule to prevent them from sticking. Afterward, the cell-laden capsules were gently collected and incubated in the calcium bath for another 2 mins to stabilize the outer alginate shell. Then, the capsules were washed with PBS and cultured for further applications. Through the mechanical properties, the results show that two different materials could offer different mechanical properties even through the same procedure to form the core-shell capsules. The xanthan gum-based capsules can achieve 5 kPa and the carboxymethyl cellulose-based capsules can reach higher than 10 kPa, indicating that the mechanical properties of core-shell capsules can be regulated by the properties of core-materials. Notably, the stem cells in the core-shell capsules could be released quickly for other characterizations of cells when the medium in the presence of 0.1 M Na citrate due to the enzymatic digestion effect on the alginate shell. Furthermore, during the three days of incubation, the cell numbers of stem cells in the spheroids were increased more than 2.5-fold, and no significant difference in both core structures by evaluating the cell viability, implying that the stem cell appeared to have a high recovery rate and could maintain the proliferation ability in the core-shell capsules. With these results, Nebel and co-workers offer a simple and low-cost-effective procedure to fabricate reproducible 3D culture systems for adipose-derived mesenchymal stem cells. Based on this system, the established system paves the way for producing large numbers of MSCs and scaling up the next generation of dynamic 3D culture systems.

Human neural tissues such as the central nervous system contain a complex spatial tissue network to provide cell-cell and cell-matrix interactions, playing a crucial role in the maturation of neural stem cells. Hydrogels have attracted increasing attention for their role in creating 3D culture systems for neural stem cells from the beginning of this century. Hydrogels possess a versatile property that enables them to easily make various sophisticated structures through different manufacturing technologies. Moreover, modulating the self-organization of neural tissues could be achieved by harnessing the 3D hydrogel properties. For example, hydrogels contribute to 3D spatial structures to encapsulate neural stem cells and drive the formation of neurospheres to mimic the neural organoid structure and enhance the function of neural stem cells [166]. Kapr and co-workers developed an alginate-based hybrid hydrogel for the culture of iNSCs to produce 3D *in vitro* neural models. They prepared a bioink by mixing an alginate polymer with another natural polysaccharide, gellan gum, which can also crosslink with divalent cations. In their study, they prepared two polymer solutions based on 1.5% and 0.3% alginates. One of the solutions is 1.5% alginate/0.5% gellan gum/0,01% laminin, and the other solution is 0.3% alginate/0.8% gellan gum/0,01% laminin. Notably, the presence of laminin in the alginate/gellan gum blends contributes more biofunctions to the blends despite alginate and gellan gum polymers being biologically inert [166]. In the mechanical properties testing, 1.5% alginate/0.5% gellan gum/0,01% laminin contribute to a high modulus (around 35 kPa) and 0.3% alginate/0.8% gellan gum/0,01% laminin provided a lower stiffness (around 20 kPa). Furthermore, the hiNSCs were used to prepare the two types of cell-laden alginate-based hydrogels (cell density ~ 4.9×10^3 neurospheres/ml gel) for evaluating neural differentiation. The hiNSCs show no significant difference in neuronal differentiation between both two groups. But, interestingly, when a Cockayne Syndrome B disease-based cell line was encapsulated in the alginate hydrogels, the cell appeared to have a quicker neurite outgrowth within the 0.3% alginate/0.8% gellan gum/0,01% laminin in comparison to 1.5% alginate/0.5% gellan gum/0,01% laminin composition. It suggests that the 0.3% alginate/0.8% gellan gum/0,01% laminin with softer mechanical properties exhibited a fast stress relaxation effect on the slowly reducing alginate content benefiting spontaneous neural differentiation, migration, and neurite outgrowth. Furthermore, Kapr et al. also confirmed that the alginate solution mixed with gellan gum provided printability, enabling to fabrication of a hiNSC-laden construct for 3D culture neural stem cells in a specific spatial structure. This is because gellan gum can increase the viscosity of alginate solution to support the printed structure and prevent the occurrence of collapse, showing that alginate and gellan gum hybrid hydrogels have potential as a bioink. With these results, the alginate-based hybrid hydrogels provide an opportunity to develop 3D neural stem cell culture system for long-term neural differentiation and *in vitro* biomimetic neural models in the future. Overall, the alginate polymer is a natural polysaccharide with ionic crosslink ability via conjugating with calcium ions. In addition to calcium ion,s other divalent cations such as manganese ion, barium ion, strontium ion, also can trigger the ionic crosslink with alginate polymer chains [167], and strontium and barium ions have stronger affinity to alginate polymer chains in comparison to calcium ions. Moreover, some trivalent cations such as iron ion can offer more affinity than divalent ions to provide an ionic interaction with alginate polymer chains [168]. Beside ionic interactions, alginate polymers could also modify their residue functional groups to create other crosslink strategies. For example, Hou and co-workers conjugated dopamine with the carboxylic groups through EDC/NHS reaction [169]. The modified alginate-dopamine polymers provided the catechol group which allowed for performing an enzymatic crosslink to form hydrogels by reacting with horseradish peroxidase and hydrogen peroxide. These applications endow alginates with various feasible mechanisms to fabricate cell-based 3D culture systems.

Chitin-Chitosan-based culture systems

Next, we will introduce a natural polysaccharide, chitin polymer, and its derivates, which are majorly obtained from marine bioresources. Chitin polymer is considered as the second most abundant biopolymer in the world. In general, chitin polymer can be harvested for over 1000 tons yearly. The block-based structure of the chitin polymer chain consists of β-(1-4) linked *N*-acetyl-d-glucosamine and *N*-glucosamine, which is similar to the structure of cellulose polymer. In addition, chitin polymers possess two polymorphs, α and β types, which lead to the chitin polymers containing α and β chains in the polymeric structures. The α polymeric chain provides a high degree of internal hydrogen bonding, stabilizing the polymeric structures, and the β polymeric chain offers a high chemical reactivity due to the weaker internal interactions among the chains. Following the internal interactions through the α and β chains in the chitin polymeric structure, chitin polymers have high insolubility and poor biodegradable properties for long-term use in tissue engineering. In addition, the two types of the chain further endow chitin polymers with different functions and use for cell culture.

For example, Shou and co-workers used α-type chitin polymer to prepare an MSC-laden hydrogel [170]. First, they minorly modified the α-type chitin polymer by dissolving chitin in a 33% sodium hydroxide solution containing 0.03 g $NaBH_4$ as a 4% chitin solution, and then the solution was heated at 90°C for 3 hours to generate a partially deacetylated chitin. Afterward, the modified chitin was washed by using deionized water and dissolved in 0.02 M acetic acid to prepare a chitin nanofiber solution with a 4 mg/ml concentration. Subsequently, the chitin nanofiber solution with 4×10^6 MSCs was further mixed with a serum-free medium by using the 1:5 volume ratio of medium : cell-laden chitin solution. Their results demonstrated that the chitin nanofiber-based hydrogel could provide dermis and granulation tissue-like elastic properties, which can induce the fibroblast differentiation of MSCs in comparison to 2D monolayer culture, as well as the wound-healing effect of stem cell-encapsulated chitin hydrogel on the rats. Besides α-type chitin polymer, Liu and co-workers investigated the effect of β-type chitin polymer on the culture of stem cells. To prepare a β-type chitin polymer, they prepare the β-type chitin polymers from a squid pen. First, the squid pen was pre-treated through a procedure of washing and drying, and then a demineralization process was operated by using a 50 g squid pen dissolving in 750 ml of 100 mM hydrochloric acid for 20 hours. Afterward, the sample was washed and deproteinized by using 750 ml of 4% sodium hydroxide solution at 80°C for 10 hours. Subsequently, the β-type chitin polymer sample was washed, purified, and ready for use. Furthermore, the purified β-type chitin polymer was dissolved in a pH 3 acetic acid to obtain a 0.3% β-type chitin polymer solution. The solution was further ultrasonicated for 6 min at the power of 300 W and with a pulse of 19.5 Hz for homogenization. Then, the homogenized chitin solution was mixed with the culture medium at a 1:3 ratio of medium :chitin solution to drive the gelation of the chitin solution. Next, the adipose-derived mesenchymal stem cells were obtained from a mouse, and the medium with 1×10^6 ADSCs was mixed with the β-type chitin/medium solution by using a 1:2 volume ratio of ADSC: β-type chitin/medium to prepare the stem cell-laden chitin hydrogel. Similar to the function of an α-type chitin polymer, the ADSC in the β-type chitin hydrogel enables the secretion of more exosomes and accelerates the wound-healing rate. As opposed to the stem cells-laden α-type chitin hydrogel, ADSC encapsulated in the β-type chitin hydrogel can further activate the TGFb/smad3 pathway signaling pathway to significantly promote the deposition of collagen I and III and expression of VEGF, which can improve the epithelialization behavior of skin and enhance angiogenesis. These results implied that the stem cell-laden β-type chitin hydrogel may have a potential use in clinical applications for the treatment of chronic and acute wounds.

Although the chitin polymer as a stem cell scaffold has been demonstrated to have potential applications in tissue engineering, one of the features of chitin polymer, an antiparallel arrangement structure, endows it with high insolubility in water and organic solvents. This issue majorly limits the use of chitin polymer in cell culture applications. Currently, to overcome the issue of chitin polymer, we want to introduce one of the chitin derivates, chitosan polymer, for the applications of stem cell culture. Chitosan is a linear polysaccharide consisting of (1–4) linked *N*-acetyl-d-glucosamine and (1–4) linked *N*-glucosamine, which is close to chitin polymer. The synthesis of chitosan polymer is generated from deacetylating the amide functional group of chitin polymer via an alkaline treatment process. In general, the degree of deacetylation of chitin, polymer means the ratio of D-glucosamine residue on the chitin backbone to the sum of D-glucosamine and *N*-acetyl D-glucosamine residues on the chitin backbone. The definition of the chitosan polymer corresponds to the degree of deacetylation of the chitin polymer, which should be over at least 60% after the alkaline treatment process. In comparison to the chitin polymer, the deacetylation in the chitosan backbone enables raising the solubility of chitosan, especially when the degree of deacetylation is achieved over 85%. To the best of our knowledge, chitosan is a popular biomass polymer used in biomedical applications. The structure of chitosan has one of the ECM components—glycosaminoglycan—considering that chitosan can interact with other ECM to regulate the cell-cell and cell-substrate adhesions [171]. ECM regulates the cell-cell and cell-substrate adhesions [166]. Additionally, the positive charge of chitosan-based polymers also provides an antimicrobial property by disrupting the integrity of the bacterial cellular membrane. This antimicrobial feature has increasingly attracted people to use chitosan as a candidate for implants in tissue engineering [172]. As mentioned above, the degree of deacetylation of the chitosan polymer could provide the different contents of the D-glucosamine group, which could modulate the proliferation, adhesion, or differentiation of cells and regulate the functional, mechanical, or other physiochemical properties of scaffolds. Therefore, Thein-Han and co-workers investigated the effect of the degree of deacetylation of chitosan on the regulation of embryonic stem cell behaviors [173]. First, they purchased three different chitosan polymers with 70%, 88%, and 95% degrees of deacetylation. Notably, to precisely control the effect of the degree of deacetylation, the three chitosan polymers have the same molecule weight (around 810 kDa). Then, three chitosan polymers were dissolved in 1% acetic acid to prepare 2% chitosan solutions. The 2% chitosan polymer solutions were added to the culture dish and formed a 2 cm^2 area and 3–3.5 mm height 3D chitosan scaffolds through a freezing and lyophilization process. Afterwards, the chitosan scaffolds were further dehydrated and sterilized in an ethanol series immersed in absolute ethanol for 1 hour, and then 70% and 50% ethanol for 30 mins, respectively. Subsequently, the scaffolds were stored in PBS solution at 4°C. Next, they isolated the buffalo embryonic stem cells from the inner cell mass and were seeded on the top of three different chitosan scaffolds at the cell concentration 1×10^5 embryonic stem cells/100 mL for 4 hours. Notably, the four-hour pre-culturing cells allowed the cell to attach to the chitosan scaffold, and the medium would be refilled to cover the cell-laden chitosan scaffolds for incubation after the pre-seeding process. Thein-Han and co-workers found that around 85–90% of cells could adhere to the three types of chitosan scaffolds the day after the pre-seeding process. Interestingly, the embryonic stem cells cultured on 88% and 95% degree-deacetylation chitosan scaffolds were significantly higher in proliferation behavior than those on the scaffold with a 70% degree of deacetylation. In comparison to the 70% degree of deacetylation, the degree of deacetylation of chitosan scaffolds higher than 88% contributes to appropriate mechanical properties (~ 6 kPa) for cell proliferation. In this study, the scaffold with a 70% degree of deacetylation had a rapid degradation rate, leading to the scaffold easily transferring as a gel-like structure, which may not be useful for embryonic stem cells. Importantly, the stem cells cultured on the chitosan scaffolds with over

88% degree of deacetylation possess a high proliferation ability enabling a 10-fold increase in cell numbers after 14 days of incubation. Furthermore, these cells on the chitosan scaffolds with 88% and 95% degrees of deacetylation still maintained their stemness, which was observed in the high pluripotency expression of SSEA-1, Oct-4, TRA-1, and SSEA-4. According to these results, the report found that the chitosan with a higher degree of deacetylation offers a stronger internal molecular hydrogen bonding interaction, providing scaffolds with high mechanical properties for cell growth. Additionally, the pore size in chitosan scaffolds was large enough, providing the appropriate space for cell expansion. Therefore, the chitosan scaffolds are a promising candidate as a 3D culture system for embryonic stem cells.

In the past decade, stem cells have shown their potential applications in cell therapy. Therefore, iPSCs with pluripotency are promising transplantation tools to regenerate damaged tissues. For example, the damage of limbal stem cells usually causes chronic inflammation and vascular-related lesion of the cornea, cornea opacification, and the unnormal growth of fibroblast [174]. A traditional gold standard therapy is using an amniotic membrane to transplant and co-culture with limbal stem cells to induce the self-renewal and differentiation of limbal stem cells. This method provides the epithelial cells on the corneal surface with an effective regeneration. Although the gold standard amniotic membrane transplantation is commonly used in the therapy of clinical damage to limbal stem cells, but the long-term preparation and cultivation may be the major issues, and the amniotic membrane from the donor may cause inflammation and amniotic membrane-related infection. Therefore, Chien and co-workers developed a possible option for corneal repair by combining chitosan hydrogel and iPSCs to overcome the above issues [175]. In their study, carboxymethyl-hexanoyl chitosan polymer was dissolved in distilled water and formed a sponge-like powder through a freeze-dried procedure. Afterward, 0.6 g freeze-dried chitosan powder was dissolved in 20 ml medium in an ice bath to prepare a viscosity solution, and 50 mg β-glycerol phosphate disodium salt hydrate (β-GP) was added to 1ml chitosan solution. Furthermore, the chitosan/β-GP solution was induced to form a solid gel through a thermally sol-gel transition process, which increases the temperature from 4°C to over 37°C. The possible mechanism is that the β-GP could neutralize the positive charge of chitosan nanocapsules in the sol-gel solution and decrease the repulsive force to induce the aggregation of chitosan nanocapsules to form a chitosan hydrogel. Then, different from other research projects, Chien and co-workers used the corneal keratocytes derived from a human patient and reprogrammed them to corneal-derived iPSCs by transfecting the factors, Klf4, c-Myc, Sox2, and Oct-4, via pMxs vectors. Subsequently, the corneal-derived iPSCs were seeded in the chitosan hydrogel at a cell density 2×10^4 cells/ml. The chitosan hydrogel formed via the aggregation of 50–100 nm nanocapsules enables to provide a continuous network as a 3D scaffold for the adhesion and growth of seeded iPSCs and maintained iPSC stemness. Additionally, due to the chitosan hydrogel with a thermal-responsive property, the iPSC-laden chitosan hydrogel was injected and *ex vivo* transplanted in an alkaline-induced rat model to investigate the effect of cell-laden hydrogel on the corneal wound healing. After injecting the cell-laden hydrogel into the corneal wound, the gel was formed at a warm ambient temperature in the rat. Compared with the control (without treatment) and hydrogel alone groups, the iPSC-laden chitosan 3D hydrogel offered a stem cell niche for the proliferation of limbal stem cells, significantly promoting the growth of epithelial in the damaged wound and increasing the thickness of the corneal epithelium. The results appeared that the chitosan hydrogel with potential as a 3D scaffold to culture iPSCs and the cell-laden chitosan hydrogel possess the feasibility to reconstruct the damaged corneal tissue.

From the above case, we know that chitosan polymer as a 3D culture system is suitable to culture iPSCs and a combination of chitosan and β-GP enables obtaining a hydrogel with the sol-gel transition behavior. Here, we will introduce the applications of chitosan hydrogel for culturing mesenchymal stem cells. As we know, stem cells play an important role in regenerative medicine because the stem cells not only provide an ability to directly reconstruct the damaged tissues but also contribute to protein or cytokine secretion to modulate the behaviors of themselves or the neighbor cells, called autocrine and paracrine [176]. The secretion factors/ molecules from stem cells, including cytokines, growth factors, and ECM, were considered useful to mediate the proliferation, angiogenesis, immunosuppressive, anti-apoptosis, and other behavior of neighbor cells to regulate the tissue repair or regeneration, indicating that the paracrine behavior plays a crucial role in the regulation of cell-cell interactions. Based on this concept, Boido and co-workers developed a chitosan-based hydrogel to culture mesenchymal stem cells and further used the chitosan hydrogel to activate the paracrine behavior from mesenchymal stem cells in the regeneration of injured neural tissues [177]. First, they dissolved a chitosan polymer with a higher than 92% deacetylation degree in 0.2M HCl for 24 hours to obtain a 3.6 (w/v) % chitosan solution. Afterwards, similar to another study [175], they prepared the gelling agent, β-GP solution, and generated a chitosan/ β-GP hydrogel by adding the gelling agent drop-by-drop to the chitosan solution in a low-temperature environment for forming a chitosan-based hydrogel via the thermal-responsive properties of chitosan/β-GP hybrid hydrogels. The chitosan/β-GP hybrid hydrogels have confirmed that the gel contains high porosity, which provides permeability and stability in long-term use. Furthermore, 2×10^4 MSCs MSCs isolated from C57BL/6J mice were directly seeded on the chitosan/β-GP hydrogel. After 10 mins of incubation, the MSCs would permeate the hydrogel because the chitosan hydrogel contains a high porosity structure, and then the culture medium would be added to cover the hydrogel. After encapsulating and culturing MSCs in the hydrogel for five days, the *in vitro* experiment provided evidence to show that MSCs cultured in the hydrogel enabled releasing more extracellular vesicles, which can eliminate the ROS level expression in comparison to both MSC alone and chitosan hydrogel alone groups. Furthermore, the MSC-laden chitosan hydrogel was injected into a mice model with SCI injury. After one week of transplantation, the MSC could be observed around and inside the injury sites, indicating that the MSC-laden chitosan hydrogel as a 3D culture system not only avoided MSCs with an uncontrol migration but also offered cells with high cell viability after transplantation. Moreover, the *in vivo* results also appeared that the transplantation of MSC-laden chitosan hydrogel did not affect the other normal cellular behaviors and activated the astrogliosis, can integrate into the host tissue, and scavenge the inflammation process and immune response around the injury site. The results benefit the development of a 3D stem cell culture system with a better biointegration for applications in regeneration medicine.

Many previous studies have demonstrated that the chitosan polymer is promising for culturing MSCs and regulating their cellular behaviors. Since its first identification in 2001 [178], adipose-derived stem cells are a popular type of mesenchymal stem cells that have been extensively applied for inducing adipogenesis, cardiogenesis, osteogenesis, and chondrogenesis because ADSC could be easily isolated from adult human adipose tissue. To endow ADSCs with more applications in the tissue engineering field, combining different culture materials and methods as culture systems have been well developed. Culturing systems are well known to play an indispensable role in regulating stem cells. In general, 2D culture systems provide an easy operation and low-cost environment for observing the response of stem cells while the system added a new stimulation factor, and 3D culture systems offer a strategy to culture stem cells in a 3D-designed spatial structure that could mimic the cells in the human body. Therefore, to understand the effects between 2D and 3D culture systems on the response of ADSCs after exogenous stimulation, Yang and co-workers cultured ADSCs through a 2D and 3D culture system and combined a direct-current electrical field to investigate the orientation and motility of ADSCs [179]. In their study, a 2D culture system was fabricated by using 0.1% gelatin to coat the six-well culture tissue plate, and ADSCs were cultured on the gelatin-coated culture dish at the cell density 1.0 × 104 cells/cm^2. Compared to the 2D culture system, a 3D culture system used the chitosan/β-GP hydrogel system. The detailed protocol was that a chitosan polymer with 93% deacetylation was dissolved in 0.1 M acetic acid solution to prepare a 2% chitosan polymeric solution. The β-GP solution was added to the 2% chitosan solution, and the final concentration of β-GP was 5%. Then, the mixture was stirred to obtain a homogeneous solution, and 1.0 × 10^4 cells/cm^2. Compared to the 2D culture system, a 3D culture system used the chitosan/β-GP hydrogel system. The detailed protocol was that a chitosan polymer with 93% deacetylation was dissolved in 0.1 M acetic acid solution to prepare a 2% chitosan polymeric solution. The β-GP solution was added to the 2% chitosan solution, and the final concentration of β-GP was 5%. Then, the mixture was stirred to obtain a homogeneous solution, and 1 × 10^6 cells/mL ADSCs were mixed with the above solution. Subsequently, an ADSC-laden hydrogel could gradually form while the ADSC-chitosan solution was incubated at 37°C for 30 mins. Then, the 2D and 3D culture systems were treated with continuous electrical stimulation (2, 4, 6, and 8 V/cm) and lasted for 6 hours. Afterwards, they found that the various electrical field strengths significantly can change the morphology and behaviors of ADSCs. For example, under the 6 V/cm strength of electrical stimulation, ADSCs cultured on the 2D system showed a regular morphology that aligned vertically to the orientation of the electrical field. Additionally, the ADSC on the 2D culture system appeared to have a high survival rate when the cells were treated with the 6 V/cm strength of electrical stimulation. Compared to the 2D culture system, the chitosan hydrogel contributed a 3D spatial structure that not only provides a structure effect on ADSCs but also leads to the galvanotaxis of ADSCs in the hydrogel response the stimulation from 3D ways. Therefore, the results indicated that the synergistic effect of electrical stimulation and chitosan-based hydrogels provided high cellular mobility and intercellular calcium activity. This study created a 3D culture system combining the electrical field stimulation to be an *in vitro* model for investigating the research related to the cell galvanotaxis response of stem cells after electrical stimulation.

Until now, chitosan polymers have been shown to have many applications in stem cell culture. Next, we will introduce the chitosan polymers as a 3D culture system for neural stem cells. As we know, tissue engineering provides versatile strategies to supply cells for injury tissue or induce endogenous cell regeneration. Recently, neural stem cells have attracted increasing attention as a source of stem cells for replacing the injured cells in the damaged tissues because neural stem cells enable self-renewal and differentiation into all types of cells, such as oligodendrocyte, astrocyte, and neurons, in the principal central nervous system. However, the downstream cell from spontaneous differentiation of neural stem cells is major astrocyte that diminishes the effect of neural stem cell transplantation [180]. Therefore, it is important to find an effective method to enhance the proliferation of neural stem cells and prevent differentiation into astrocytes. An earlier study also indicated that a multi-cell environment enhances the cell-cell interactions for neural regeneration [181]. For example, co-transplanting olfactory ensheathing cells and neural stem cells could provide a synergistic effect on promoting the recovery of spinal cord injury [181]. Therefore, Wand and co-workers want to investigate the chitosan-based hydrogel, whether it contributes to a neural cell favorite in a *in vitro* culture system with a high proliferation effect and transplant the neural stem cells in the body for differentiation by the interactions from the other cells in the body [182]. Here, as opposed to the chitosan/β-GP hybrid hydrogel system, they designed a chitosan-based hydrogel system by mixing chitosan and alginate polymers. Due to the chitosan polymer with the positive charge amino functional group and the alginate polymer containing the negative charge carboxyl functional group, the opposite charge functional group offers an ionic interaction between the chitosan and alginate backbones, which enables the formation of a hydrogel. To prepare the chitosan-alginate hydrogel, the chitosan polymeric powder with 93% degree of deacetylation was dissolved in 1.2% acetic acid, and the polymeric alginate powder was dissolved as well, but the alginate solution contained 0.03 M sodium chloride. Subsequently, the chitosan solution was added to the alginate/sodium solution drop-by-drop and well stirred to form a homogeneous hybrid polymeric solution consisting of 1% chitosan and 1% alginate. Afterward, the pH value of the chitosan/alginate solution was further adjusted to 7.4, incubated at 60°C for 1 hour, and placed in the 4°C refrigerators for 5 hours to obtain a chitosan/alginate hydrogel. Furthermore, the primary neural stem cells isolated from the cortex of E12–E14 fetal SD rat were cultured with the chitosan/alginate hydrogel. The remaining positive charge of the chitosan backbone could attract the neural cells attached on the scaffold. Compared to the Matrigel control group, the expression of cell proliferation is not significantly different between Matrigel and the chitosan/alginate hydrogel, indicating that the chitosan/alginate hybrid hydrogel enabled achieving the same effect as commercial Matrigel. This might be because the chitosan/alginate hydrogel contributes to a high porosity structure with highly hydrophilic properties, allowing oxygen and nutrients to transition well. The properties endow the seeded neural stem cells with a culture environment with enough space and nutrients to enhance the proliferation of neural stem cells. Therefore, the encouraging results implied that the chitosan/alginate hybrid system possesses a promising potential as a low-cost alternative for providing a 3D *in vitro* culture system for use in neural stem cell research.

In the case of chitosan hydrogel for neural stem cells, the positive charge of chitosan backbone may provide a cell-matrix interaction to regulate the behaviors of neural stem cells. This is because the amine groups on the chitosan polymer contribute a pH-sensitive property at the natural pH conditions which can partially transfer from a–NH_2 group to a –NH_3^+ group. Nonetheless, several challenge existed in using chitosan polymers and the user should develop suitable strategies to overcome them. For example, the pKa of chitosan (around 6.5) limits its solubility at natural pH conditions. Additionally, the abundance of lysozyme in the body cleaving the β-(1,4) glycosidic linkages of chitosan chains leads to occurrence of chitosan degradation *in vivo* quickly and easily [183]. These issues inevitably constrict the applications of chitosan in tissue engineering. Fortunately, chitosan contains amine- and carboxymethyl-side groups which provide the active sites for modification with other materials to overcome the abovementioned limitations, even endowing chitosan with new functions. For instances, Hu et al. developed a photo-crosslinkable chitosan by modifying glycol chitosan with a methacrylate functional group. Through the cured process, the chitosan hydrogel decreases the degradation rate, supporting the proliferation of encapsulated chondrocytes and deposition of ECM, enhancing the repair of the damaged cartilage [184]. Besides small molecules, chitosan was further modified by other polymers such as poly(ethylene glycol) to obtain an injectable hydrogel for repairing the deficits of the central nervous system [123], or by collagen as an biomimetic skin for healing of the defects in skin [185]. These strategies provide chitosan polymers with various modifications, thereby expanding the potential applications of chitosan polymers for fabricating 3D *in vitro* culture systems for stem cells.

Hyaluronic acid-based culture systems

In this paragraph, we will discuss the applications of hyaluronic acid polymers, also called hyaluronate or hyaluronan, which is a biomass polymer and widely exists in many parts of the extracellular matrix in the body. The structure of the hyaluronic polymer is a linear, non-sulfated, anionic glycosaminoglycan. This type of glycosaminoglycan is a polysaccharide that consists of the β-1,4- D-glucuronic acid and β-1,3-N-acetyl-D-glucosamide repeating disaccharide units and hyaluronic acid contained in around 5k~30k sugar-based repeating units in the backbone of hyaluronic acid so that the molecular weight of hyaluronic acid could achieve 100 kDa [186]. Hyaluronic acid possesses many hydroxyl and carboxyl groups, which endows hyaluronic acid with a highly hydrophilic property and viscoelasticity that enables it to generate a hydrogel structure via the interactions with the intermolecular macromolecules. Therefore, based on the concentration and molecular weight of hyaluronic acid, hyaluronic acid hydrogels could produce versatile properties, such as ionic exchange, permeability, and the water-binding ability for different uses in tissue engineering applications. For example, the hyaluronic acid in the different tissues appears with various degrees of viscosity according to the mutual crowding degree of hyaluronic acid and other ECMs in the human body [187]. Therefore, through the regulation of hyaluronic acid concentration such as 2% to 20% [186], the degradation behavior, compressive modulus, and volumetric swelling ratio of hyaluronic acid hydrogels could range from 1 to 38 days in the presence of hyaluronidase, 2 to over 100 kPa, and 8 to 42, respectively. As mentioned above, hyaluronic acid is a glycosaminoglycan, which can be synthesized via hyaluronic acid synthesis on the cell membrane. Therefore, hyaluronic acid provides well biocompatibility, which may generate interactions with internal molecules, ECM, or cells. For example, hyaluronic acid weakly binds to interleukin-8 (IL-8) secreted from the cells when the body is inflamed [188]. This interaction contributes to a strategy to protect the exogenous implant and reduce the risk of an undesired immune response [189]. Additionally, hyaluronic acid as a scaffold could also offer a cell-matrix interaction or culturing cell because hyaluronic acid with pharmaceutical and biological effects enables it to bind and modulate the molecules of stem cells, such as the intercellular adhesion molecule 1 (ICAM-1), receptor for HA-mediated motility (RHAMM), and cluster determinant 44 (CD44) [190]. Therefore, hyaluronic acid polys play an essential role in regulating and activating the survival, migration, differentiation, and proliferation pathways of cells. For example, the CD 44 receptor offers an internalization function for HA and its derivates during the inflammation response, recruits fibroblasts aggregating around the wound, and triggers the growth and migration of cells in the wound healing process.

In terms of embryonic stem cells, the cells could be derived from the inner cells mass of a developing blastocyte encapsulated in a 3D environment, which could modulate the cell proliferation, self-renewal, and differentiation. The current standard protocol is culturing embryonic stem cells on an embryonic fibroblast feeder layer or the Matrigel containing various growth factors, laminin, and fibronectin [191]. However, these 2D culture systems may limit the need for cell transfer and differentiation because the human embryonic stem cells inhabit a 3D culture environment during early development. Additionally, previous studies have shown that inner cell mass cultured in a 3D matrix could modulate the stemness such as self-renewal or differentiation during the period of embryogenesis [192, 193]. Therefore, creating a controllable and unique 3D culture system could provide a biomimetic environment allowing embryonic stem cells to maintain an undifferentiation stage and induce differentiation in response to the stimulation. Hydrogel possesses a high-water content ability and native tissue ECM-like mechanical and structural properties for enhancing the proliferation of cells and regulating cellular behaviors. Therefore, Gerecht and co-workers developed an HA-based scaffold to culture embryonic stem cells and further investigate the effect of the HA scaffold on embryonic stem cells [194]. The reasons for HA being selected as the scaffold are that (1) HA plays a role as the feeder layer for the culture of ESC, and (2) HA possesses controllable degradation, mechanics, and regulable hydrogel architecture through the different machining technologies. In their study, they prepared 2% HA hydrogel dissolved in PBS solution, and the HA hydrogel solution contained 10^7 cell/ml ESC. During the culture period, the MEF medium was used to maintain the growth of the ESC, and the endothelial growth medium in the presence of 100 ng/ml VEGF was used for the vascular differentiation of ESCs. Compared with a dextran hydrogel as a control group, the low concentration (10 μl/ml) of HA hydrogel contributed to a cell-favoring environment for the growth of ESCs, and the high concentration of HA hydrogel (50 μl/ml) slightly induced p53 protein expression, meaning the high concentration of HA hydrogel had a toxic effect on ESCs after four days of incubation. Furthermore, the ESCs could form EB-like tissue when releasing the ESCs from the HA hydrogel after treating hyaluronidase, and the EB-like tissue could spontaneously undergo the differentiation of all three germ layers. Additionally, if specifically inducing the ESCs in the HA hydrogel by treating the endothelial culture medium, the vasculogenic sprouting of ESCs could be observed in the HA, and the differentiated endothelial cells from ESCs could further modulate their sprouting, proliferation, and migration behaviors, implying that the HA also offers a tissue-like environment to modulate the angiogenesis and vascular cellular functions of cells from ESCs.

Next, the effect of HA as scaffolds for culturing BMSCs and its application in tissue regeneration will be introduced in this paragraph. For example, the articular cartilage tissue with mechanical injuries or degenerative diseases can cause a large population of people, regardless of age, to suffer from the disability affecting their motor skills. To deal with articular cartilage injuries, the currently major gold standard clinical approaches for repairing the injured tissues are abrasion and microfracture, which are intended to induce the growth of cartilage and then stimulate the articular cartilage regeneration [195]. Previous studies have confirmed that the implantation of autologous chondrocyte contributed to a breakthrough method for full-thickness treatment, which allowed for restoring hyaline-like cartilage [196, 197]. However, the autologous chondrocyte in bodies can rapidly differentiate in the expansion stage leading to reducing the deposition of *in situ* type II collagen [198]. In the past decade, stem cell therapy has brought a promising strategy to stimulate tissue regeneration. Compared with the injection of cell suspension, stem cell-laden scaffold-based systems have more practical value for implantation because the scaffold-based systems could prevent the cells from moving out of the target site. In general, cartilage tissues majorly consist of proteoglycans and type II collagen; hyaluronic acid is the second component in articular cartilage. Therefore, HA is a candidate for the regeneration of articular cartilage. To understand the effect of HA on stem cell therapy, Deszcz and co-workers investigated the effect of HA on the chondrogenic differentiation of BMSC by culturing BMSCs and chondrocytes on a 2D and a 3D culture model [199]. In their study, the culture flask was used as a 2D culture system, and a HA fiber-based thin membrane was used as a 3D culture system. First, they cultured only chondrocytes on the 2D and 3D culture systems. The chondrocytes were passaged three times, and the cells on the 3D HA fibrous scaffold were removed from the scaffold after 14 days of incubation for the observation of chondrogenesis. Compared with the chondrocyte on the 2D culture flask, the type II collagen expression of the cell cultured in the 3D HA fibrous membrane could be increased nine-fold as compared to that on the 2D culture flask. Afterward, the chondrocyte-induced MSCs by using a chondrogenic induction medium (called 3D standard culture group) and the direct co-cultured MSCs/chondrocytes (1:1 cell ratio) were cultured on the 2D and 3D culture systems, respectively. According to the results from 17 days of incubation, similar to the results from chondrocytes, the type II collagen expressions of both 3D culture systems were significantly higher than that of the 2D culture system. Moreover, in comparison to the 3D standard culture group, the direct co-culture of MSC and chondrocyte in the 3D fibrous scaffold possessed a better ability to secrete type II collagen. The possible reason is that the direct co-culture of MSC and chondrocyte in the HA fibrous scaffold is similar to the cell distribution and spatial structure of articular cartilage tissue, leading to stimulating the cells in the co-culture group to express the components of cartilage. These results indicated that the 3D HA fibrous scaffold combining the co-culture MSCs and cells provided a new type of 3D stem cell culture system for the regeneration of articular cartilage tissues with defects.

From the above cases, we can understand that HA is a natural ECM from tissues and is widely used for tissue regeneration. However, its major disadvantages, such as rapid degradation and poor mechanical properties, limit HA's application *in vivo* in tissue engineering. Fortunately, HA contains large number of functional groups, which permit HA to be used as scaffold and which could overcome the its disadvantages through chemical modification or crosslinking methods. Therefore, the degradation, solubility, mechanical and biological properties, and viscosity of HA could be improved after modification and then the modified HA can offer more versatile applications.

In general, to modify HA, the functional groups on HA backbone, including carboxyl groups, hydroxyl groups, and acetanilide groups, offer the targeted sites for modification. For example, carbonyldiimidazole or carbodiimides could be used as a mediate agent to covalently modify the carboxyl group to generate amide bones. Moreover, the hydroxyl group in HA enable hemiacetal formation, ester formation, oxidation, and ether formation; the acetanilide group provides the reaction site to perform the reactions for deacetylation, hemiacetylation, amidation. For example, HA-modified thermoresponsive poly(N-isopropylacrylamide) (HA-PNIPAAm-CL) hydrogels also enhanced rabbit ASC chondrogenesis during articular cartilage tissue engineering applications [200]. In a study, ASCs were encapsulated with PEG-HA-RGD hydrogel. Combining with RGD peptide, as a cell adhesion motif, significantly altered the cellular morphology, enhanced cell proliferation, and increased the paracrine activity of angiogenesis and tissue remodeling growth factors. The hydrogel protected the implanted cells from the harmful wound environment in a burn model. Hydrogel-ASC treatment significantly enhanced neovascularization, accelerated wound closure and reduced the scar formation [201]. In addition, as we know, iPSCs possess the ability to self-renew and differentiate into any cells in the three germ layers, which offers a new strategy to decrease the need for organ donation and reduce immune rejection after transplantation. Furthermore, iPSCs also provide a benefit for a complement to the study of animal models in the drug screening of toxicology and pharmacology. Although these benefits from iPSCs are used in biomedical applications, the applications of iPSCs are still constrained by the gap between the cell number at the laboratory level and clinical studies [202]. Therefore, developing a culture strategy to expand the iPSC number by achieving more than 1×10^9 cells is required for the application for further clinical uses. 3D culture system for cell culture is highly promising in providing a high surface area for cell adhesion and cell expansion. Therefore, Ekerdt and co-workers hypothesized that establishing a 3D cell culture system to encapsulate iPSCs in 3D scaffolds could efficiently increase the yield of cell numbers related to 2D culture systems [203]. In the study, they prepared a thermoreversible HA-PNIPAAM hydrogel to 3D culture hiPSCs. First, 0.1 M NaOH solution was used to dissolve HA, and then the hydroxyl groups on HA backbone deprotonated a proton. Then, the deprotonated HA solution was mixed with divinyl sulfone and allowed to react for 20 min before adding 1 M HCl to stop the reaction. Notably, the molar concentration of divinyl sulfone was used in excess in the reaction, which prevented crosslinking. Afterwards, the functionalized HA was purified via a dialysis process in DI water to remove unreacted divinyl sulfone. Besides HA, PNIPAAM with terminal thiol functional groups was synthesized by using 2-(Dodecylthiocarbonothioylthio)-2-methyl propionic acid (DMP), a cleavable chain transfer agent (CTA), and an initiator (azobisisobutyronitrile (AIBN)) through reversible addition-fragmentation chain transfer (RAFT) polymerization. The NIPAAM monomer was mixed with DMP and AIBN in dioxane at 60°C for 24 h of reactions. Then, the thiol group at the terminal end of the PNIPAAM chain was revealed by the cleaving effect of CTA following the reduction of NaBH4 when PNIPAAM was treated with CTA in the presence of NaBH4. After the reaction, the PNIPPAM-SH solution was further treated with 10 mM EDTA and 20 mM TCEP to reduce disulfide bonds. Subsequently, the solution was performed in another dialyzed process and lyophilized to obtain the PNIPAAM-SH polymers. Furthermore, to synthesize the HA-PNIPAAM polymer, the functionalized HA and PNIPAAM were dissolved in 0.3 M triethanolamine (TEOA) buffer at 4°C and pH = 8 environment. Notably, the weight ratio of PNIPAAM : functionalized HA is 5:1. Afterward, a temperature-sensitive HA-PNIPAAM polymer could be obtained after the dialysis procedure. Due to the HA-PNIPAAM polymer with a liquid-gel transition property, iPSCs could be mixed with the polymer solution at 4°C and form an iPSC-encapsulated HA hydrogel at 37°C. Therefore, the thermo-reversible HA systems provide a gentle method to execute the passage of iPSCs without any chemical agent such as trypsin. In this study, the results confirmed that the iPSCs culture in the HA-PNIPAAM hydrogel could increase the cell number 10-fold compared with a control 2D culture method and maintain the high expression of iPSCs specific markers Nanog and Oct4 after five continuous passages. Significantly, the iPSCs harvested from the hydrogel also still have the ability to differentiate into cells in all three germ layers after injecting the iPSCs into a NOD/SCID mice. These results indicated that the functionalized thermo-reversible HA-PNIPAAM polymers with the potential to be a 3D culture system to scale up the cell number and maintain the pluripotency of iPSCs.

In addition to the synthetic HA-based copolymers, Jiang and co-workers also developed another functionalized HA polymer, methacrylated HA (called HA-MA), to overcome the poor mechanical properties and rapid degradation issues of native HA [204]. In their study, 2% HA solution was prepared, and a 20-fold excess methacrylic anhydride was added to the 2% HA solution at pH value = 8. The mixture was reacted at 4°C for 24 hours under continuous stirring. Then, the modified HA-MA polymers were precipitated and washed by using ethanol to remove unreacted methacrylic anhydride and methacrylic acid. To create the HA-MA hydrogels, 0.75 %w/v modified HA-MA polymers were dissolved in the cell medium containing 0.05 %w/v photoinitiator (2-hydroxy-1(4-(hydroxyethox)phenyl)-2-methyl-1-propanone, Irgacure 2959). Then the HA-MA solution was treated with

UV light to form a photo-crosslinked HA-MA hydrogel. Compared with native HA and other functionalized HA-based copolymers, the HA-MA polymer could form a hydrogel triggered by UV light, which provides a simple and easy operation process to fabricate a hydrogel-based culture system. In addition, the other advantage of HA-MA polymer is that the stiffness of HA-MA hydrogel could be regulated according to the UV exposure time, indicating that HA-MA-based culture systems with a tailorable mechanical property for use in different applications. Furthermore, Jiang and co-workers validated the effect of HA-MA polymers on the culture of hiPSC-NSCs. Stem cell replacement therapy offers a great attraction for the regeneration and function recovery of neural tissues when there is occurrence of disease or injury. Despite the fact that cell replacement therapy is a promising method, it is necessary to understand and predict the possible risk and behaviors of stem cells after transplantation. To reduce the risk and raise the probability of the success of stem cell transplantation, developing 3D *in vitro* models for culturing stem cells plays an indispensable role in investigating stem cells. In the development of neuronal tissues, previous studies have shown that neuronal growth, migration, or neurite extension may be in response to the rigidity of culture environments [205, 206]. For example, a stiff culture environment could enhance the adhesion of neuronal cells, and a soft culture environment is friendly for neuronal maturation, differentiation, and neurite outgrowth [207, 208]. Based on this information, Jiang and co-workers encapsulated hiPSCs in the HA-MA hydrogels with different stiffness [204]. They mixed hiPSCs in HA-MA solution and generated the hiPSC-laden scaffold by treating UV light for the 30s and 60s. Via the different UV exposure times, the hiPSC could be encapsulated in the soft HA-MA hydrogel (from the 30s) and the stiff HA-MA hydrogel (from the 60s). Compared with the 2D culture groups, the hiPSCs could form an organoid morphology which contributes better differentiation and cell growth because the cellular morphology could mimic the 3D CNS structure in a physiological environment. Additionally, after 28 days of incubation in 3D culture systems, the hiPSCs at the hydrogel-free group and the stiff HA-MA hydrogel group showed a spheroid shape, but fewer cells sprouted out from the spheroid and had low expression of neuronal specific markers (TUBB3 and MAP2). Compared with the hydrogel-free and stiffness hydrogel groups, hiPSC-based spheroid cultured in the soft HA-MA hydrogel presented the highest TUBB3 and MAP2 expression, indicating that the soft HA-MA hydrogel (0.5 kPa) contribute to a CNS-like culture environment for culturing stem cells and inducing the neuronal differentiation. Furthermore, to validate the results, Jiang and co-workers further encapsulated Down Syndrome (DS) patient-specific hiPSCs-derived NPCs (DS-NPCs) in the soft and stiff HA-MA hydrogels. The results were similar to the groups cultured with normal hiPSC that presented the highest neuronal differentiation behaviors in the soft HA-MA hydrogels. The exciting results show that the HA-MA hydrogel as a 3D culture system is suitably used for culturing stem cells and regulating the behaviors of stem cells by the superior tailorable mechanical properties of HA-MA hydrogels.

The above example has demonstrated the potential applications of HA-MA hydrogels used in the 3D *in vitro* stem cell culture systems. Additionally, HA-based scaffolds were widely fabricated as circular disks or cylinder shapes. However, it has been reported that fabricating constructs with biomimetic spatial structures could provide cells with a tissue-like structure for promoting cell-matrix interactions [209]. Therefore, Highley and co-workers developed a 3D bioprinting technology to build complex and multiscale structures by using HA-MA as ink [148]. Currently, hydrogel-based inks are valuable and have attracted increasing attention vis-à-vis fabricating scaffolds to mimic or recapture features of tissue environment for generating *in vitro* biomimetic tissues or models in biomedical applications. Therefore, the design from Highley and co-workers was that of an HA-based shear-thinning hydrogel as an ink directly printed into another HA-based self-healing supporting hydrogels. Notably, both HA polymers used for inks or supporting hydrogels were specifically designed to endow the two HA-based polymers with functionalized residuals for supramolecular assembly via the interaction of guest-host complexes. To prepare the HA-based guest-host complexes, the HA polymers were functionalized by using adamantane (Ad) or β-cyclodextrin (β-CD). To prepare adamantane modified HA (Ad-HA), the Ad-HA polymer was reacted by 1-adamantane acetic acid through bicarbonate (BOC2O)/4-dimethylaminopyridine (DMAP) esterification. Similarly, to functionalized HA polymers, β-CD was grafted on the HA backbone via (benzotriazol-1-yloxy)tris(dimethylamino) phosphoniumhexafl uorophosphate (BOP) amidation to obtain CD-HA. Moreover, the synthesized Ad-HA and CD-HA executed a methacrylate process to obtain Ad-HAMA and CD-HAMA polymers. The advantages of Ad-HAMA and CD-HAMA polymers contribute to the grafted Ad and CD moieties on the HA backbone and provide a supramolecular interaction to rapidly generate assemblies via guest-host bonds. Therefore, during the printing process, the printed structure could be rapidly fixed in the Ad-HAMA/CD-HAMA supporting bath when modified HA ink was directly writing structures in the supporting bath. Afterward, the methacrylate functional groups on the Ad-HAMA and CD-HAMA polymers could offer a permanent crosslink to the printed structure, which could possess tailored mechanical properties according to the UV exposure time. Moreover, this technology indicated that the free-standing scaffolds with specific patterns or configurations could be printed in any direction of 3D space through the guest-host interaction of Ad-HAMA and CD-HAMA polymers. In addition to the fabrication of scaffolds, Ad-HAMA and CD-HAMA polymers were further validated for their potential applications in stem cell culture. Therefore, MSCs were added to the ink, and 3T3 fibroblasts were added to the supporting bath. Subsequently, the MSC-laden bioink was patterned in a specific structure in the fibroblast-laden supporting bath to fabricate a multicellular structure. Interestingly, both MSC and 3T3 fibroblasts in the multicellular structure were observed to have nontoxic behavior and high cell viability. These results revealed that the modified HA polymers opened a new strategy to fabricate cell-laden constructs with more tissue-like configurations and have the feasibility of creating 3D *in vitro* stem cell culture systems with multicellular co-culture interactions.

Until now, we have introduced some of popular synthetic and natural polymers which were commonly used as 3D *in vitro* stem cell systems by various fabrication processes such as chemical crosslink, physical crosslink, 3D bioprinting, and others. However, it is difficult to introduce all of biomaterials used as 3D *in vitro* culture systems for stem cell applications. Therefore, the other synthetic or natural polymers as hydrogel culture systems will be briefly mentioned in the following paragraphs.

Other synthetic and polysaccharide-based culture systems

In this section, we will briefly discuss other synthetic and natural polymers, such as polyethylene glycol (PEG), Pluronic F-127 (PF-127), gellan gum, aloe vera, glycosyl-nucleoside-amphiphiles (GNA), and pullulan for the fabrication of 3D stem cell culture systems and the regulation of stem cell behaviors.

Polyethylene glycol monomethyl ether (mPEG), was mixed with N,N-carbonyl diimidazole (CDI)-activated Poly(L-glutamic acid) (PLGA), followed by adding crosslinker Cys, which contains disulfide bonds. After dialysis and lyophilization, the PLGA-Cys-mPEG porous hydrogel was achieved. MSCs were seeded into the PLGA-Cys-mPEG porous hydrogel to generate spheroids. In a hindlimb ischemia model of nude mice, the enhanced paracrine secretion of stem cell spheroids resulted in promoted angiogenesis and muscle regeneration, exhibiting obvious therapeutic effects after 21 days [210]. In addition, another photo-crosslinkable PEG-based hyperbranched multifunctional homopolymers were developed via RAFT homopolymerization of the divinyl monomer of poly(ethylene glycol) diacrylate (PEGDA). Due to its high degree of multi-acrylate functionality, the hyperbranched polyPEGDA (HP-PEG) can rapidly crosslink with a thiolated hyaluronic acid under physiological conditions and form an injectable hydrogel for cell delivery [211]. HP-PEG hydrogels encapsulating stem cells have demonstrated promising regenerative capabilities, such as the maintenance of stemness and secretion abilities of stem cells. For example, ASCs embedded in such hydrogels were tested in a diabetic murine animal model and showed enhanced wound healing [212].

Pluronic F-127 (PF-127) is an amphiphilic copolymer consisting of units of polypropylene oxide and ethylene oxide, which provides injectability and a thermo-reversible gelation process [213]. PF-127 not only has biodegradability, biocompatibility, and non-toxicity but also possesses a thermosensitive characteristic that enables a rapid liquid-gel phase transition from the liquid state at 4°C to the gel state at 37°C. The properties of PF-127 are similar to PNIPPAM, so the PF-127 polymer is also widely used in the controlled release and drug delivery field [214]. In addition, an earlier study demonstrated that PF-127 is a promising polymer for encapsulating MSC, yielding high proliferation ability, and maintaining the pluripotency of MSC, which can proceed with osteogenesis and adipogenesis [213]. Moreover, another study also showed that allogenic extracellular vesicles (EVs) from ASCs were administered in a porcine fistula model through a thermoresponsive PF-127 gel, injected locally at 4°C and gelling at body temperature to retain EVs in the entire fistula tract. Complete fistula healing was reported to be 100% for the gel plus EVs group, 67% for the gel group, and 0% for the control, supporting the therapeutic use of Pluronic F-127 gel combined with ASC-EVs [215].

A popular natural polymer, agarose, is extracted from ocean plants such as seaweed and red algae. The structure of agarose polymers includes 1,4-linked 3,6-anhydro-α-l-galactopyranose and 1,3-linked-d-galactopyranose. The agarose polymer possesses a thermo-sensitivity that exists a gel state at room temperature and can transit a liquid state while the temperature is increased. This property endows agarose polymers with a favorable application for mixing cells in the polymer solution at 37°C and then forming cell-laden hydrogel at room temperature. In addition, agarose polymers also have tunable mechanical properties according to controlling the parameters in the synthesis process. Previous studies showed that if immobilizing vascular VEGF on agarose as a scaffold to encapsulate ESCs, the ESC during the early stages of development can differentiate toward blood progenitor cells [216]. Furthermore, agarose polymers mixed with human stem cells enable a mechanically tunable bioink to fabricate stem cell-laden constructs with various specific tissue-like architectures, indicating that agarose-based stem cell culture systems offer a versatile tissue-like platform for investigating stem cell therapy [217].

Gellan gum is a polysaccharide manufactured by microbial fermentation of the *Sphingomonas paucimobilis* microorganism, being commonly used in the food and pharmaceutical industry [218]. Gellan gum can form gels in the following way: at high temperatures, Gellan gum is in the coil form; upon temperature decrease, a thermally reversible coil to double-helix transition occurs, which is a prerequisite for gel formation [219]. The gellan gum encapsulated chondrogenic pre-differentiated rabbit MSCs have shown significantly higher type II collagen and aggrecan gene expression after eight weeks, and histological analyses further suggested better overall cartilage tissue structure [220].

Aloe vera was noted to have antioxidant [221], antimicrobial [222], and immunomodulatory activities [223, 224]. Aloe vera contains a polysaccharide component that can serve as a trigger factor for cell proliferation, differentiation, mineralization, and dentin formation [225]. In a rat burn wound model, applying injectable aloe vera hydrogel encapsulated with stem cells resulted in more cornification of the stratum corneum, more complete mature epidermis, higher fibroblasts/fibrocytes counts, lower inflammatory cell counts, and a more organized arrangement of collagen fibers in the dermis, suggesting enhanced cutaneous regeneration [226].

Glycosyl-nucleoside-amphiphiles are a new class of low molecular weight gelators (LMWG) possessing a sugar, nucleoside, and lipid covalently linked by triazole bridges. The Glycosyl-Nucleoside Fluorinated amphiphiles (GNFs) used in a study featured three building blocks: a hydrophobic fluorinated carbon backbone, a central thymidine group and a carbohydrate moiety, the fluorinated aliphatic chain and the carbohydrate being linked to the central thymidine. ASCs mixed with GNF hydrogel displayed a strong sustained signal at the site of implantation after 30 days [227].

Next, a fungal exopolysaccharide, pullulan polymer, from *Aureobasidium pullulans* is a neutral linear polysaccharide with α-1,6-linked maltotriose residues [228]. In addition, pullulan polymers have a wide range of molecular weights from several thousand to 2 million Da [229]. Furthermore, the linear polysaccharide structure contributes to a unique linkage pattern for pullulan, which provides well water-solubility, degradable, and tunable mechanical properties. Moreover, because pullulan is generated from fungi, pullulan polymers have confirmed their hemocompatibility and non-immunogenic response [230]. Especially pullulan polymers are FDA-approved biomaterials that have been used in drug and gene delivery, wound healing, and tissue engineering [231]. For example, a salt-induced phase inversion technique was employed to ensure that the pullulan-based hydrogel recapitulated the structure of the skin while remaining soft for skin repair [232, 233].

Seeding MSCs within a soft pullulan–collagen hydrogel enhanced ASC survival and improved wound healing [234]. Kosaraju et al. demonstrate that the application of ASC-seeded hydrogels to wounds, when compared with injected ASCs or a noncell control, increased the recruitment of provascular circulating bone marrow-derived mesenchymal progenitor cells (BM-MPCs). When exposed to a hydrogel-seeded ASC-conditioned medium, a statistically significant increase in BM-MPC migration, proliferation, and tubulization was noted relative to the control ASC-conditioned medium [235].

In this chapter, we introduced various commonly synthetic and natural polymers and their unique properties. Furthermore, the combination of multiple types of scaffolds, microfabrication technologies, and types of stem cells have shown their potential applications in developing a 3D *in vitro* culture system for regulating the survival, proliferation, migration, and differentiation of stem cells. Despite the fact that not all biomaterials are being introduced in this chapter, this chapter provides a brief concept for the selection of materials and design of versatile stem cell-based microsystems with variable cell activities to promote the applicability and therapeutic activity of stem cells in regenerative medicine.

References

[1] Ferreira, L.P., V.M. Gaspar, and J.F. Mano. Design of spherically structured 3D *in vitro* tumor models -Advances and prospects. Acta Biomater, 2018. 75: 11–34.

[2] Costa, E.C., A.F. Moreira, D. de Melo-Diogo, V.M. Gaspar, M.P. Carvalho, and I.J. Correia. 3D tumor spheroids: an overview on the tools and techniques used for their analysis. Biotechnol Adv, 2016. 34(8): 1427–1441.

[3] Kim, J., J.W. Kang, J.H. Park, Y. Choi, K.S. Choi, K.D. Park, D.H. Baek, S.K. Seong, H.K. Min, and H.S. Kim. Biological characterization of long-term cultured human mesenchymal stem cells. Arch Pharm Res, 2009. 32(1): 117–26.

[4] Su, G., Y. Zhao, J. Wei, J. Han, L. Chen, Z. Xiao, B. Chen, and J. Dai. The effect of forced growth of cells into 3D spheres using low attachment surfaces on the acquisition of stemness properties, 2013. Biomaterials 34(13): 3215–22.

[5] Ylostalo, J.H. 3D Stem Cell Culture, Cells, 2020. 9(10).

[6] Wu, X., J. Su, J. Wei, N. Jiang, and X. Ge. Recent advances in three-dimensional stem cell culture systems and applications. Stem Cells Int, 2021. 9477332.

[7] Ravi, M., V. Paramesh, S.R. Kaviya, E. Anuradha, and F.D. Solomon. 3D cell culture systems: advantages and applications. J Cell Physiol, 2015. 230(1): 16–26.

[8] Jensen, C., and Y. Teng. Is it time to start transitioning from 2D to 3D Cell Culture? Front Mol Biosci, 2020. 7: 33.

[9] Langhans, S.A. Three-dimensional *in vitro* cell culture models in drug discovery and drug repositioning. Front Pharmacol, 2018. 9: 6.

[10] Imamura, Y., T. Mukohara, Y. Shimono, Y. Funakoshi, N. Chayahara, M. Toyoda, N. Kiyota, S. Takao, S. Kono, T. Nakatsura, and H. Minami. Comparison of 2D- and 3D-culture models as drug-testing platforms in breast cancer. Oncol Rep, 2015. 33(4): 1837–43.

[11] Kapalczynska, M., T. Kolenda, W. Przybyla, M. Zajaczkowska, A. Teresiak, V. Filas, M. Ibbs, R. Blizniak, L. Luczewski, and K. Lamperska. 2D and 3D cell cultures—a comparison of different types of cancer cell cultures. Arch Med Sci, 2018. 14(4): 910–919.

[12] Costa, E.C., V.M. Gaspar, P. Coutinho, and I.J. Correia. Optimization of liquid overlay technique to formulate heterogenic 3D co-cultures models. Biotechnol Bioeng, 2014. 111(8): 1672–85.

[13] Chen, H., H. Fu, X. Wu, Y. Duan, S. Zhang, H. Hu, Y. Liao, T. Wang, Y. Yang, G. Chen, Z. Li, and W. Tian. Regeneration of pulpo-dentinal-like complex by a group of unique multipotent CD24a(+) stem cells. Sci Adv, 2020. 6(15): eaay1514.

[14] Achilli, T.M., and J. Meyer, J.R. Morgan, Advances in the formation, use and understanding of multi-cellular spheroids, Expert Opin Biol Ther, 2012. 12(10): 1347–60.
[15] Desai, P.K., H. Tseng, and G.R. Souza. Assembly of hepatocyte spheroids using magnetic 3D cell culture for CYP450 inhibition/induction. Int J Mol Sci, 2017. 18(5).
[16] Kim, J.A., J.H. Choi, M. Kim, W.J. Rhee, B. Son, H.K. Jung, and T.H. Park. High-throughput generation of spheroids using magnetic nanoparticles for three-dimensional cell culture. Biomaterials, 2013. 34(34): 8555–63.
[17] Adine, C., K.K. Ng, S. Rungarunlert, G.R. and Souza, J.N. Ferreira, engineering innervated secretory epithelial organoids by magnetic three-dimensional bioprinting for stimulating epithelial growth in salivary glands. Biomaterials, 2018. 180: 52–66.
[18] Yaman, S., M. Anil-Inevi, E. Ozcivici, and H.C. Tekin. Magnetic force-based microfluidic techniques for cellular and tissue bioengineering. Front Bioeng Biotechnol, 2018. 6: 192.
[19] Evans, N.D., J. EileenGentleman and M. Polak. Scaffolds for stem cells. Materialsoday, 2006. 9(12): 26–33.
[20] Park, Y., K.M. Huh, and S.W. Kang. Applications of biomaterials in 3D cell culture and contributions of 3D cell culture to drug development and basic biomedical research. Int J Mol Sci, 2021. 22(5).
[21] Wu, L., D. Jing, and J. Ding. A “room-temperature” injection molding/particulate leaching approach for fabrication of biodegradable three-dimensional porous scaffolds. Biomaterials, 2006. 27(2): 185–91.
[22] Aldemir Dikici, B., and F. Claeyssens. Basic principles of emulsion templating and its use as an emerging manufacturing method of tissue engineering scaffolds. Front Bioeng Biotechnol, 2020. 8: 875.
[23] Thadavirul, N., P. Pavasant, and P. Supaphol. Development of polycaprolactone porous scaffolds by combining solvent casting, particulate leaching, and polymer leaching techniques for bone tissue engineering. J Biomed Mater Res A, 2014. 102(10): 3379–92.
[24] Salerno, A., M. Oliviero, E. Di Maio, S. Iannace, and P.A. Netti. Design of porous polymeric scaffolds by gas foaming of heterogeneous blends. J Mater Sci Mater Med, 2009. 20(10): 2043–51.
[25] Yin, H.M., J. Qian, J. Zhang, Z.F. Lin, J.S. Li, J.Z. Xu, and Z.M. Li. Engineering porous poly(lactic acid) scaffolds with high mechanical performance via a solid state extrusion/porogen leaching approach. Polymers (Basel), 2016. 8(6).
[26] Loh, Q.L., and C. Choong. Three-dimensional scaffolds for tissue engineering applications: role of porosity and pore size. Tissue Eng Part B Rev, 2013. 19(6): 485–502.
[27] Gizaw, M., A. Faglie, M. Pieper, S. Poudel, and S.F. Chou. The role of electrospun fiber scaffolds in stem cell therapy for skin tissue regeneration. Med One, 2019. 4: e190002.
[28] Blum, A.S., C.M. Soto, C.D. Wilson, T.L. Brower, S.K. Pollack, T.L. Schull, A. Chatterji, T. Lin, J.E. Johnson, C. Amsinck, P. Franzon, R. Shashidhar, and B.R. Ratna. An engineered virus as a scaffold for three-dimensional self-assembly on the nanoscale. Small, 2005. 1(7): 702–6.
[29] Chen, Y., M. Shafiq, M. Liu, Y. Morsi, and X. Mo. Advanced fabrication for electrospun three-dimensional nanofiber aerogels and scaffolds. Bioact Mater, 2020. 5(4): 963–979.
[30] Xue, J., D. Pisignano, and Y. Xia. Maneuvering the migration and differentiation of stem cells with electrospun nanofibers. Adv Sci (Weinh), 2020. 7(15): 2000735.
[31] Amores de Sousa, M.C., C.A.V. Rodrigues, I.A.F. Ferreira, M.M. Diogo, R.J. Linhardt, J.M.S., and Cabral, F.C. Ferreira. Functionalization of electrospun nanofibers and fiber alignment enhance neural stem cell proliferation and neuronal differentiation. Front Bioeng Biotechnol, 2020. 8: 580135.
[32] Chimene, D., K.K. Lennox, R.R. Kaunas, and A.K. Gaharwar. Advanced bioinks for 3D printing: A materials science perspective. Ann Biomed Eng, 2016. 44(6): 2090–102.
[33] Bertassoni, L.E., J.C. Cardoso, V. Manoharan, A.L. Cristino, N.S. Bhise, W.A. Araujo, P. Zorlutuna, N.E. Vrana, A.M. Ghaemmaghami, M.R. Dokmeci, and A. Khademhosseini. Direct-write bioprinting of cell-laden methacrylated gelatin hydrogels. Biofabrication, 2014. 6(2): 024105.
[34] Ott, H.C., T.S. Matthiesen, S.K. Goh, L.D. Black, S.M. Kren, T.I. Netoff, and D.A. Taylor. Perfusion-decellularized matrix: using nature’s platform to engineer a bioartificial heart. Nat Med, 2008. 14(2): 213–221.
[35] Guyette, J.P., S.E. Gilpin, J.M. Charest, L.F. Tapias, X. Ren, and H.C. Ott. Perfusion decellularization of whole organs. Nat Protoc, 2014. 9(6): 1451–68.

[36] Syed, O., N.J. Walters, R.M. Day, H.W., and Kim, J.C. Knowles. Evaluation of decellularization protocols for production of tubular small intestine submucosa scaffolds for use in oesophageal tissue engineering. Acta Biomater, 2014. 10(12): 5043–5054.

[37] Casali, D.M., R.M. Handleton, T. Shazly, and M.A. Matthews. A novel supercritical CO2-based decellularization method for maintaining scaffold hydration and mechanical properties. J. Supercrit. Fluids, 2018. 131: 72–81.

[38] Kim, S.J., M.H. Park, H.J. Moon, J.H. Park, D.Y. Ko, and B. Jeong. Polypeptide thermogels as a 3D culture scaffold for hepatogenic differentiation of human tonsil-derived mesenchymal stem cells. ACS Appl Mater Interfaces, 2014.

[39] Mirmalek-Sani, S.H., G. Orlando, J.P. McQuilling, R. Pareta, D.L. Mack, M. Salvatori, A.C. Farney, R.J. Stratta, A. Atala, E.C. Opara, and S. Soker. Porcine pancreas extracellular matrix as a platform for endocrine pancreas bioengineering. Biomaterials, 2013. 34(22): 5488–95.

[40] Shimoda, H., H. Yagi, H. Higashi, K. Tajima, K. Kuroda, Y. Abe, M. Kitago, M. Shinoda, Y. Kitagawa. Decellularized liver scaffolds promote liver regeneration after partial hepatectomy. Sci Rep, 2019. 9(1): 12543.

[41] Guruswamy Damodaran, R., and P. Vermette. Decellularized pancreas as a native extracellular matrix scaffold for pancreatic islet seeding and culture. J Tissue Eng Regen Med, 2018. 12(5): 1230–1237.

[42] Xue, A., G. Niu, Y. Chen, K. Li, Z. Xiao, Y. Luan, C. Sun, X. Xie, D. Zhang, X. Du, F. Kong, Y. Guo, H. Zhang, G. Cheng, Q. Xin, Y. Guan, and S. Zhao. Recellularization of well-preserved decellularized kidney scaffold using adipose tissue-derived stem cells. J Biomed Mater Res A, 2018. 106(3): 805–814.

[43] Young, B.M., K. Shankar, C.K. Tho, A.R. Pellegrino, and R.L. Heise. Laminin-driven Epac/Rap1 regulation of epithelial barriers on decellularized matrix. Acta Biomater, 2019. 100: 223– 34.

[44] Liu, C., F. Lei, P. Li, K. Wang, and J. Jiang. A review on preparations, properties, and applications of cis-ortho-hydroxyl polysaccharides hydrogels crosslinked with borax. International Journal of Biological Macromolecules, 2021. 182: 1179–1191.

[45] Thanos, C.G., and D.F. Emerich. On the use of hydrogels in cell encapsulation and tissue engineering system. Recent Pat Drug Deliv Formul, 2008. 2(1): 19–24.

[46] Drury, J.L., and D.J. Mooney. Hydrogels for tissue engineering: scaffold design variables and applications, Biomaterials, 2003. 24(24): 4337–51.

[47] Slaughter, B.V., S.S. Khurshid, O.Z. Fisher, A. Khademhosseini, and N.A. Peppas. Hydrogels in regenerative medicine. Adv Mater, 2009. 21(32-33): 3307–29.

[48] Gwon, K., E. Kim, and G. Tae. Heparin-hyaluronic acid hydrogel in support of cellular activities of 3D encapsulated adipose derived stem cells. Acta Biomaterialia, 2017. 49: 284–295.

[49] Lee, K.Y., and D.J. Mooney. Hydrogels for tissue engineering. Chemical Reviews, 2001. 101(7): 1869–79.

[50] Raemdonck, K., J. Demeester, and S. De Smedt. Advanced nanogel engineering for drug delivery. Soft Matter, 2009. 5(4): 707–715.

[51] Cui, Z., B.H. Lee, C. Pauken, and B.L. Vernon. Degradation, cytotoxicity, and biocompatibility of NIPAAm-based thermosensitive, injectable, and bioresorbable polymer hydrogels. Journal of Biomedical Materials Research. Part A, 2011. 98(2): 159–66.

[52] Cui, Z., B.H., and Lee, B.L. Vernon. New hydrolysis-dependent thermosensitive polymer for an injectable degradable system. Biomacromolecules, 2007. 8(4): 1280–6.

[53] Parhi, R. Cross-linked hydrogel for pharmaceutical applications: A review. Adv Pharm Bull, 2017. 7(4): 515–530.

[54] Kim, T.H., D.B. An, S.H. Oh, M.K. Kang, H.H. Song, and J.H. Lee. Creating stiffness gradient polyvinyl alcohol hydrogel using a simple gradual freezing-thawing method to investigate stem cell differentiation behaviors. Biomaterials, 2015. 40: 51–60.

[55] Liang, J., Z. Guo, A. Timmerman, D. Grijpma, and A. Poot. Enhanced mechanical and cell adhesive properties of photo-crosslinked PEG hydrogels by incorporation of gelatin in the networks, Biomed Mater, 2019. 14(2): 024102.

[56] Hunt, N.C., D. Hallam, A. Karimi, C.B. Mellough, J. Chen, D.H.W. Steel, and M. Lako. 3D culture of human pluripotent stem cells in RGD-alginate hydrogel improves retinal tissue development. Acta Biomater, 2017. 49: 329–343.

[57] Klemm, D., F. Kramer, S. Moritz, T. Lindstrom, M. Ankerfors, D. Gray, and A. Dorris. Nanocelluloses: a new family of nature-based materials. Angew Chem Int Ed Engl, 2011. 50(24): 5438–66.

[58] Lou, Y.R., L. Kanninen, T. Kuisma, J. Niklander, L.A. Noon, D. Burks, A. Urtti, and M. Yliperttula. The use of nanofibrillar cellulose hydrogel as a flexible three-dimensional model to culture human pluripotent stem cells. Stem Cells Dev, 2014. 23(4): 380–92.

[59] Mathieu, E., G. Lamirault, C. Toquet, P. Lhommet, E. Rederstorff, S. Sourice, K. Biteau, P. Hulin, V. Forest, P. Weiss, J. Guicheux, and P. Lemarchand. Intramyocardial delivery of mesenchymal stem cell-seeded hydrogel preserves cardiac function and attenuates ventricular remodeling after myocardial infarction. PLoS One, 2012. 7(12): e51991.

[60] Merceron, C., S. Portron, M. Masson, J. Lesoeur, B.H. Fellah, O. Gauthier, O. Geffroy, P. Weiss, J. Guicheux, and C. Vinatier. The effect of two- and three-dimensional cell culture on the chondrogenic potential of human adipose-derived mesenchymal stem cells after subcutaneous transplantation with an injectable hydrogel. Cell Transplantation, 2011. 20(10): 1575–88.

[61] Portron, S., C. Merceron, O. Gauthier, J. Lesoeur, S. Sourice, M. Masson, B.H. Fellah, O. Geffroy, E. Lallemand, P. Weiss, J. Guicheux, and C. Vinatier. Effects of *in vitro* low oxygen tension preconditioning of adipose stromal cells on their *in vivo* chondrogenic potential: application in cartilage tissue repair. PloS one, 2013. 8(4): e62368.

[62] Boyer, C., G. Réthoré, P. Weiss, C. d'Arros, J. Lesoeur, C. Vinatier, B. Halgand, O. Geffroy, M. Fusellier, G. Vaillant, P. Roy, O. Gauthier, and J. Guicheux. A self-setting hydrogel of silylated chitosan and cellulose for the repair of osteochondral defects: from *in vitro* characterization to preclinical evaluation in dogs. Frontiers in Bioengineering and Biotechnology, 2020. 8: 23.

[63] Szulc, P. Bone turnover: Biology and assessment tools. Best Pract Res Clin Endocrinol Metab, 2018. 32(5): 725–738.

[64] Togo, S., T. Sato, H. Sugiura, X. Wang, H. Basma, A. Nelson, X. Liu, T.W. Bargar, J.G. Sharp, and S.I. Rennard. Differentiation of embryonic stem cells into fibroblast-like cells in three-dimensional type I collagen gel cultures. *In Vitro* Cell Dev Biol Anim, 2011. 47(2): 114–24.

[65] Chen, H.F., C.Y. Chuang, Y.K. Shieh, H.W. Chang, H.N. Ho, and H.C. Kuo. Novel autogenic feeders derived from human embryonic stem cells (hESCs) support an undifferentiated status of hESCs in xeno-free culture conditions. Hum Reprod, 2009. 24(5): 1114–25.

[66] Yoo, S.J., B.S. Yoon, J.M. Kim, J.M. Song, S. Roh, S. You, and H.S. Yoon. Efficient culture system for human embryonic stem cells using autologous human embryonic stem cell-derived feeder cells. Exp Mol Med, 2005. 37(5): 399–407.

[67] Stojkovic, P., M. Lako, R. Stewart, S. Przyborski, L. Armstrong, J. Evans, A. Murdoch, T. Strachan, and M. Stojkovic. An autogeneic feeder cell system that efficiently supports growth of undifferentiated human embryonic stem cells. Stem Cells, 2005. 23(3): 306–14.

[68] Chiu, L.H., W.F. Lai, S.F. Chang, C.C. Wong, C.Y. Fan, C.L. Fang, and Y.H. Tsai. The effect of type II collagen on MSC osteogenic differentiation and bone defect repair. Biomaterials, 2014. 35(9): 2680–91.

[69] Tamaddon, M., R.S. Walton, D.D. Brand, and J.T. Czernuszka. Characterisation of freeze-dried type II collagen and chondroitin sulfate scaffolds.J Mater Sci Mater Med, 2013. 24(5): 1153–65.

[70] Tamaddon, M., M. Burrows, S.A. Ferreira, F. Dazzi, J.F. Apperley, A. Bradshaw, D.D. Brand, J. Czernuszka, and E. Gentleman. Monomeric, porous type II collagen scaffolds promote chondrogenic differentiation of human bone marrow mesenchymal stem cells *in vitro*. Sci Rep, 2017. 7: 43519.

[71] Newman, A.P. Articular cartilage repair. Am J Sports Med, 1998. 26(2): 309–24.

[72] Lynn, A.K., R.A. Brooks, W. Bonfield, and N. Rushton. Repair of defects in articular joints. Prospects for material-based solutions in tissue engineering. J Bone Joint Surg Br, 2004. 86(8): 1093–9.

[73] Vuorio, E., and B. de Crombrugghe. The family of collagen genes. Annu Rev Biochem, 1990. 59: 837–72.

[74] Fleischmajer, R., B.R. Olsen, R. Timpl, J.S. Perlish, and O. Lovelace. Collagen fibril formation during embryogenesis. Proc Natl Acad Sci U S A, 1983. 80(11): 3354–8.

[75] Kuivaniemi, H., and G. Tromp. Type III collagen (COL3A1): Gene and protein structure, tissue distribution, and associated diseases. Gene, 2019. 707: 151–171.

[76] Balleisen, L., S. Gay, R. Marx, and K. Kuhn. Comparative investigation on the influence of human and bovine collagen types I, II and III on the aggregation of human platelets. Klin Wochenschr, 1975. 53(19): 903–5.
[77] Wu, J.J., M.A. Weis, L.S. Kim, and D.R. Eyre. Type III collagen, a fibril network modifier in articular cartilage. J Biol Chem, 2010. 285(24): 18537-44.
[78] Sillat, T., R. Saat, R. Pollanen, M. Hukkanen, M. Takagi, and Y.T. Konttinen. Basement membrane collagen type IV expression by human mesenchymal stem cells during adipogenic differentiation. J Cell Mol Med, 2012. 16(7): 1485–95.
[79] Khoshnoodi, J., V. Pedchenko, and B.G. Hudson. Mammalian collagen IV. Microsc Res Tech, 2008. 71(5): 357–70.
[80] Aratani, Y., and Y. Kitagawa. Enhanced synthesis and secretion of type IV collagen and entactin during adipose conversion of 3T3-L1 cells and production of unorthodox laminin complex. J Biol Chem, 1988. 263(31): 16163–9.
[81] Chun, T.H., K.B. Hotary, F. Sabeh, A.R. Saltiel, E.D. Allen, and S.J. Weiss. A pericellular collagenase directs the 3-dimensional development of white adipose tissue. Cell, 2006. 125(3): 577–91.
[82] Sierra-Sanchez, A., K.H. Kim, G. Blasco-Morente, and S. Arias-Santiago. Cellular human tissue-engineered skin substitutes investigated for deep and difficult to heal injuries. NPJ Regen Med, 2021. 6(1): 35.
[83] Mazza, G., K. Rombouts, A. Rennie Hall, L. Urbani, T. Vinh Luong, W. Al-Akkad, L. Longato, D. Brown, P. Maghsoudlou, A.P. Dhillon, B. Fuller, B. Davidson, K. Moore, D. Dhar, P. De Coppi, M. Malago, and M. Pinzani. Decellularized human liver as a natural 3D-scaffold for liver bioengineering and transplantation. Sci Rep, 2015. 5: 13079.
[84] Barakat, O., S. Abbasi, G. Rodriguez, J. Rios, R.P. Wood, C. Ozaki, L.S. Holley, and P.K. Gauthier. Use of decellularized porcine liver for engineering humanized liver organ. J Surg Res, 2012. 173(1): e11–25.
[85] Belviso, I., V. Romano, A.M. Sacco, G. Ricci, D. Massai, M. Cammarota, A. Catizone, C. Schiraldi, D. Nurzynska, M. Terzini, A. Aldieri, G. Serino, F. Schonauer, F. Sirico, F. D'Andrea, S. Montagnani, F. Di Meglio, and C. Castaldo. Decellularized human dermal matrix as a biological scaffold for cardiac repair and regeneration. Front Bioeng Biotechnol, 2020. 8: 229.
[86] Kleinman, H.K., M.L. McGarvey, L.A. Liotta, P.G. Robey, K. Tryggvason, and G.R. Martin. Isolation and characterization of type IV procollagen, laminin, and heparan sulfate proteoglycan from the EHS sarcoma. Biochemistry, 1982. 21(24): 6188–93.
[87] Lam, M.T., and M.T. Longaker. Comparison of several attachment methods for human iPS, embryonic and adipose-derived stem cells for tissue engineering. J Tissue Eng Regen Med, 2012. 6 Suppl 3: s80–6.
[88] Kim, S.J., H. Byun, S. Lee, E. Kim, G.M. Lee, S.J. Huh, J. Joo, and H. Shin. Spatially arranged encapsulation of stem cell spheroids within hydrogels for the regulation of spheroid fusion and cell migration. Acta Biomater, 2022. 142: 60–72.
[89] Liu, D., M. Nikoo, G. Boran, P. Zhou, and J.M. Regenstein. Collagen and gelatin. Annu Rev Food Sci Technol, 2015. 6: 527–57.
[90] Klompong, V., S. Benjakul, D. Kantachote, K.D. Hayes, and F. Shahidi. Comparative study on antioxidative activity of yellow stripe trevally protein hydrolysate produced from Alcalase and Flavourzyme. Int J Food Sci Tech, 2008. 43(6): 1019–1026.
[91] Aleman, A., B. Gimenez, E. Perez-Santin, M.C. Gomez-Guillen, and P. Montero. Contribution of Leu and Hyp residues to antioxidant and ACE-inhibitory activities of peptide sequences isolated from squid gelatin hydrolysate. Food Chem, 2011. 125(2): 334–341.
[92] Kim, S.K., Y.T. Kim, H.G. Byun, K.S. Nam, D.S. Joo, and F. Shahidi. Isolation and characterization of antioxidative peptides from gelatin hydrolysate of Alaska pollack skin. J Agric Food Chem, 2001. 49(4): 1984–9.
[93] Mills, S., C. Stanton, C. Hill, and R.P. Ross. New developments and applications of bacteriocins and peptides in foods, Annu Rev Food Sci Technol, 2011. 2: 299–329.
[94] Hamzeh, A., S. Benjakul, and T. Senphan. Comparative study on antioxidant activity of hydrolysates from splendid squid (*Loligo formosana*) gelatin and protein isolate prepared using protease from hepatopancreas of Pacific white shrimp (*Litopenaeus vannamei*). J Food Sci Technol, 2016. 53(9): 3615–3623.

[95] Mendis, E., N. Rajapakse, and S.K. Kim. Antioxidant properties of a radical-scavenging peptide purified from enzymatically prepared fish skin gelatin hydrolysate. J Agric Food Chem, 2005. 53(3): 581–7.
[96] Giri, A., and T. Ohshima. Bioactive marine peptides: nutraceutical value and novel approaches. Adv Food Nutr Res, 2012. 65: 73–105.
[97] Nikoo, M., S. Benjakul, A. Ehsani, J. Li, F. Wu, N. Yang, B. Xue, Z. Jin, and X. Xu. Antioxidant and cryoprotective effects of a tetrapeptide isolated from Amur sturgeon skin gelatin. Journal of Functional Foods, 2014. 7: 609–620.
[98] Zhang, Y., X. Duan, and Y. Zhuang. Purification and characterization of novel antioxidant peptides from enzymatic hydrolysates of tilapia (*Oreochromis niloticus*) skin gelatin. Peptides, 2012. 38(1): 13–21.
[99] Ruoslahti, E. RGD and other recognition sequences for integrins. Annu Rev Cell Dev Biol, 1996. 12: 697–715.
[100] Schwab, E.H., M. Halbig, K. Glenske, A.S. Wagner, S. Wenisch, and E.A. Cavalcanti-Adam. Distinct effects of RGD-glycoproteins on Integrin-mediated adhesion and osteogenic differentiation of human mesenchymal stem cells. Int J Med Sci, 2013. 10(13): 1846–59.
[101] Sulaiman, S., S.R. Chowdhury, M.B. Fauzi, R.A. Rani, N.H.M. Yahaya, Y. Tabata, Y. Hiraoka, R. Binti Haji Idrus, and N. Min Hwei. 3D Culture of MSCs on a Gelatin Microsphere in a Dynamic Culture System Enhances Chondrogenesis. Int J Mol Sci, 2020. 21(8).
[102] Merten, O.W. Advances in cell culture: anchorage dependence. Philos Trans R Soc Lond B Biol Sci, 2015. 370(1661): 20140040.
[103] Leong, W., T.T. Lau, and D.A. Wang. A temperature-cured dissolvable gelatin microsphere-based cell carrier for chondrocyte delivery in a hydrogel scaffolding system. Acta Biomater, 2013. 9(5): 6459–67.
[104] Gohi, B., X.Y. Liu, H.Y. Zeng, S. Xu, K.M.H. Ake, X.J. Cao, K.M. Zou, and S. Namulondo. Enhanced efficiency in isolation and expansion of hAMSCs via dual enzyme digestion and micro-carrier. Cell Biosci, 2020. 10: 2.
[105] Vardiani, M., M. Ghaffari Novin, M. Koruji, H. Nazarian, E. Goossens, A. Aghaei, A.M. Seifalian, H. Ghasemi Hamidabadi, F. Asgari, and M. Gholipourmalekabadi. Gelatin electrospun mat as a potential co-culture system for *in vitro* production of sperm cells from embryonic stem cells. ACS Biomater Sci Eng, 2020. 6(10): 5823–5832.
[106] Su, K., and C. Wang. Recent advances in the use of gelatin in biomedical research. Biotechnol Lett, 2015. 37(11): 2139–45.
[107] Song, J.H., H.E. Kim, and H.W. Kim. Production of electrospun gelatin nanofiber by water-based co-solvent approach. J Mater Sci Mater Med, 2008. 19(1): 95–102.
[108] George, J., and S.N. Sabapathi. Cellulose nanocrystals: synthesis, functional properties, and applications. Nanotechnol Sci Appl, 2015. 8: 45–54.
[109] Xing, Q., F. Zhao, S. Chen, J. McNamara, M.A. Decoster, and Y.M. Lvov. Porous biocompatible three-dimensional scaffolds of cellulose microfiber/gelatin composites for cell culture. Acta Biomater, 2010. 6(6): 2132–9.
[110] Cheung, R.C., T.B. Ng, J.H. Wong, and W.Y. Chan. Chitosan: An update on potential biomedical and pharmaceutical applications. Mar Drugs, 2015. 13(8): 5156–86.
[111] Mohamed, K.R., H.H. Beherei, and Z.M. El-Rashidy. *In vitro* study of nano-hydroxyapatite/chitosan-gelatin composites for bio-applications. J Adv Res, 2014. 5(2): 201–8.
[112] Miranda, S.C., G.A. Silva, R.C. Hell, M.D. Martins, J.B. Alves, and A.M. Goes. Three-dimensional culture of rat BMMSCs in a porous chitosan-gelatin scaffold: A promising association for bone tissue engineering in oral reconstruction. Arch Oral Biol, 2011. 56(1): 1–15.
[113] Hu, W., Z. Wang, Y. Xiao, and S. Zhang, J. Wang. Advances in crosslinking strategies of biomedical hydrogels. Biomater Sci, 2019. 7(3): 843–855.
[114] Nichol, J.W., S.T. Koshy, H. Bae, C.M. Hwang, S. Yamanlar, and A. Khademhosseini. Cell-laden microengineered gelatin methacrylate hydrogels. Biomaterials, 2010. 31(21): 5536–44.
[115] Ramirez, M.A., E. Pericuesta, R. Fernandez-Gonzalez, B. Pintado, and A. Gutierrez-Adan. Inadvertent presence of pluripotent cells in monolayers derived from differentiated embryoid bodies. Int J Dev Biol, 2007. 51(5): 397–407.

[116] Dang, S.M., M. Kyba, R. Perlingeiro, G.Q. Daley, and P.W. Zandstra. Efficiency of embryoid body formation and hematopoietic development from embryonic stem cells in different culture systems. Biotechnol Bioeng, 2002. 78(4): 442–53.

[117] Janeczek Portalska, K., A. Leferink, N. Groen, H. Fernandes, L. Moroni, C. van Blitterswijk, and J. de Boer. Endothelial differentiation of mesenchymal stromal cells. PLoS One, 2012. 7(10): e46842.

[118] Lin, C.H., J.J. Su, S.Y. Lee, and Y.M. Lin. Stiffness modification of photopolymerizable gelatin-methacrylate hydrogels influences endothelial differentiation of human mesenchymal stem cells. J Tissue Eng Regen Med, 2018. 12(10): 2099–2111.

[119] O'Donnell, B.T., S. Al-Ghadban, C.J. Ives, M.P. L'Ecuyer, T.A. Monjure, M. Romero-Lopez, Z. Li, S.B. Goodman, H. Lin, R.S. Tuan, and B.A. Bunnell. Adipose tissue-derived stem cells retain their adipocyte differentiation potential in three-dimensional hydrogels and bioreactors (dagger), Biomolecules, 2020. 10(7).

[120] Goulao, M., and A.C. Lepore. iPS cell transplantation for traumatic spinal cord injury. Curr Stem Cell Res Ther, 2016. 11(4): 321–8.

[121] Nagoshi, N., and H. Okano. iPSC-derived neural precursor cells: potential for cell transplantation therapy in spinal cord injury. Cell Mol Life Sci, 2018. 75(6): 989–1000.

[122] Fan, L., C. Liu, X. Chen, Y. Zou, Z. Zhou, C. Lin, G. Tan, L. Zhou, C. Ning, and Q. Wang. Directing induced pluripotent stem cell derived neural stem cell fate with a three-dimensional biomimetic hydrogel for spinal cord injury repair. ACS Appl Mater Interfaces, 2018. 10(21): 17742–17755.

[123] Tseng, T.C., L. Tao, F.Y. Hsieh, Y. Wei, I.M. Chiu, and S.H. Hsu. An injectable, self-healing hydrogel to repair the central nervous system. Adv Mater, 2015. 27(23): 3518–24.

[124] Huse, J.T., and E.C. Holland. Targeting brain cancer: advances in the molecular pathology of malignant glioma and medulloblastoma. Nat Rev Cancer, 2010. 10(5): 319–31.

[125] Mukherjee, D., H.A. Zaidi, T. Kosztowski, K.L. Chaichana, H. Brem, D.C. Chang, and A. Quinones-Hinojosa. Disparities in access to neuro-oncologic care in the United States. Arch Surg, 2010. 145(3): 247–53.

[126] Stupp, R., W.P. Mason, M.J. van den Bent, M. Weller, B. Fisher, M.J. Taphoorn, K. Belanger, A.A. Brandes, C. Marosi, U. Bogdahn, J. Curschmann, R.C. Janzer, S.K. Ludwin, T. Gorlia, A. Allgeier, D. Lacombe, J.G. Cairncross, E. Eisenhauer, and R.O. Mirimanoff. Radiotherapy plus concomitant and adjuvant temozolomide for glioblastoma. N Engl J Med, 2005. 352(10): 987–96.

[127] Chaichana, K.L., E.E. Cabrera-Aldana, I. Jusue-Torres, O. Wijesekera, A. Olivi, M. Rahman, and A. Quinones-Hinojosa. When gross total resection of a glioblastoma is possible, how much resection should be achieved? World Neurosurg, 2014. 82(1-2): e257–65.

[128] Chaichana, K.L., I. Jusue-Torres, A.M. Lemos, A. Gokaslan, E.E. Cabrera-Aldana, A. Ashary, A. Olivi, and A. Quinones-Hinojosa. The butterfly effect on glioblastoma: is volumetric extent of resection more effective than biopsy for these tumors? J Neurooncol, 2014. 120(3): 625–34.

[129] Chaichana, K.L., I. Jusue-Torres, R. Navarro-Ramirez, S.M. Raza, M. Pascual-Gallego, A. Ibrahim, M. Hernandez-Hermann, L. Gomez, X. Ye, J.D. Weingart, A. Olivi, J. Blakeley, G.L. Gallia, M. Lim, H. Brem, and A. Quinones-Hinojosa. Establishing percent resection and residual volume thresholds affecting survival and recurrence for patients with newly diagnosed intracranial glioblastoma. Neuro Oncol, 2014. 16(1): 113–22.

[130] McGirt, M.J., K.L. Chaichana, M. Gathinji, F.J. Attenello, K. Than, A. Olivi, J.D. Weingart, H. Brem, and A.R. Quinones-Hinojosa. Independent association of extent of resection with survival in patients with malignant brain astrocytoma. J Neurosurg, 2009. 110(1): 156–62.

[131] Ostrom, Q.T., H. Gittleman, P. Farah, A. Ondracek, Y. Chen, Y. Wolinsky, N.E. Stroup, C. Kruchko, and J.S. Barnholtz-Sloan. CBTRUS statistical report: Primary brain and central nervous system tumors diagnosed in the United States in 2006–2010. Neuro Oncol, 2013. 15 Suppl 2: ii1–56.

[132] Sharma, P., K. Sheets, S. Elankumaran, and A.S. Nain. The mechanistic influence of aligned nanofibers on cell shape, migration and blebbing dynamics of glioma cells. Integr Biol (Camb), 2013. 5(8): 1036–44.

[133] Cuddapah, V.A., S. Robel, S. Watkins, and H. Sontheimer. A neurocentric perspective on glioma invasion. Nat Rev Neurosci, 2014. 15(7): 455–65.

[134] Umesh, V., A.D. Rape, T.A. Ulrich, and S. Kumar. Microenvironmental stiffness enhances glioma cell proliferation by stimulating epidermal growth factor receptor signaling. PLoS One, 2014. 9(7): e101771.

[135] Janmey, P.A., and R.T. Miller. Mechanisms of mechanical signaling in development and disease. J Cell Sci, 2011. 124(Pt 1): 9–18.

[136] Hardin, J.O., T.J. Ober, A.D. Valentine, and J.A. Lewis. Microfluidic printheads for multimaterial 3D printing of viscoelastic inks. Advanced Materials, 2015. 27(21): 3279–84.

[137] Chang, C.C., E.D. Boland, S.K. Williams, and J.B. Hoying. Direct-write bioprinting three-dimensional biohybrid systems for future regenerative therapies, Journal of biomedical materials research. Part B. Applied Biomaterials, 2011. 98(1): 160–70.

[138] Duan, B., L.A. Hockaday, K.H. Kang, and J.T. Butcher. 3D bioprinting of heterogeneous aortic valve conduits with alginate/gelatin hydrogels. Journal of Biomedical Materials Research. Part A, 2013. 101(5): 1255–64.

[139] Schuurman, W., V. Khristov, M.W. Pot, P.R. van Weeren, W.J. Dhert, and J. Malda. Bioprinting of hybrid tissue constructs with tailorable mechanical properties. Biofabrication, 2011. 3(2): 021001.

[140] Xu, T., W. Zhao, J.M. Zhu, M.Z. Albanna, J.J. Yoo, and A. Atala. Complex heterogeneous tissue constructs containing multiple cell types prepared by inkjet printing technology. Biomaterials, 2013. 34(1): 130–9.

[141] Cui, X., K. Breitenkamp, M.G. Finn, M. Lotz, and D.D. D'Lima. Direct human cartilage repair using three-dimensional bioprinting technology. Tissue Engineering. Part A, 2012. 18(11-12): 1304–12.

[142] Cui, X., T. Boland, D.D. D'Lima, and M.K. Lotz. Thermal inkjet printing in tissue engineering and regenerative medicine. Recent Patents on Drug Delivery & Formulation, 2012. 6(2): 149–55.

[143] Christensen, K., C. Xu, W. Chai, Z. Zhang, J. Fu, and Y. Huang. Freeform inkjet printing of cellular structures with bifurcations. Biotechnology and Bioengineering, 2015. 112(5): 1047–55.

[144] Whatley, B.R., X. Li, N. Zhang, and X. Wen. Magnetic-directed patterning of cell spheroids. Journal of Biomedical Materials Research. Part A, 2014. 102(5): 1537–47.

[145] Koch, L., M. Gruene, C. Unger, and B. Chichkov. Laser assisted cell printing. Current Pharmaceutical Biotechnology, 2013. 14(1): 91–7.

[146] Guillotin, B., A. Souquet, S. Catros, M. Duocastella, B. Pippenger, S. Bellance, R. Bareille, M. Remy, L. Bordenave, J. Amedee, and F. Guillemot. Laser assisted bioprinting of engineered tissue with high cell density and microscale organization. Biomaterials, 2010. 31(28): 7250–6.

[147] Michael, S., H. Sorg, C.T. Peck, L. Koch, A. Deiwick, B. Chichkov, P.M. Vogt, and K. Reimers. Tissue engineered skin substitutes created by laser-assisted bioprinting form skin-like structures in the dorsal skin fold chamber in mice. PLoS One, 2013. 8(3): e57741.

[148] Highley, C.B., C.B. Rodell, and J.A. Burdick. Direct 3D printing of shear-thinning hydrogels into self-healing hydrogels. Adv Mater, 2015. 27(34): 5075–9.

[149] Hinton, T.J., Q. Jallerat, R.N. Palchesko, J.H. Park, M.S. Grodzicki, H.J. Shue, M.H. Ramadan, A.R. Hudson, and A.W. Feinberg. Three-dimensional printing of complex biological structures by freeform reversible embedding of suspended hydrogels. Sci Adv, 2015. 1(9): e1500758.

[150] Bhattacharjee, T., S.M. Zehnder, K.G. Rowe, S. Jain, R.M. Nixon, W.G. Sawyer, and T.E. Angelini. Writing in the granular gel medium. Sci Adv, 2015. 1(8): e1500655.

[151] Kolesky, D.B., R.L. Truby, A.S. Gladman, T.A. Busbee, K.A. Homan, and J.A. Lewis. 3D bioprinting of vascularized, heterogeneous cell-laden tissue constructs. Adv Mater, 2014. 26(19): 3124–30.

[152] Muth, J.T., D.M. Vogt, R.L. Truby, Y. Menguc, D.B. Kolesky, R.J. Wood, and J.A. Lewis. Embedded 3D printing of strain sensors within highly stretchable elastomers. Adv Mater, 2014. 26(36): 6307–12.

[153] Li, Y.E., Y.A. Jodat, R. Samanipour, G. Zorzi, K. Zhu, M. Hirano, K. Chang, A. Arnaout, S. Hassan, N. Matharu, A. Khademhosseini, M. Hoorfar, and S.R. Shin. Toward a neurospheroid niche model: optimizing embedded 3D bioprinting for fabrication of neurospheroid brain-like co-culture constructs. Biofabrication, 2020. 13(1).

[154] Luo, L., K. Guo, W. Fan, Y. Lu, L. Chen, Y. Wang, Y. Shao, G. Wu, J. Xu, and L. Lu. Niche astrocytes promote the survival, proliferation and neuronal differentiation of co-transplanted neural stem cells following ischemic stroke in rats. Exp Ther Med, 2017. 13(2): 645–650.

[155] Lee, K.Y., and D.J. Mooney. Alginate: properties and biomedical applications. Prog Polym Sci, 2012. 37(1): 106–126.

[156] Hong, S.H., M. Shin, J. Lee, J.H. Ryu, S. Lee, J.W. Yang, W.D. Kim, and H. Lee. STAPLE: Stable alginate gel prepared by linkage exchange from ionic to covalent bonds. Adv Healthc Mater, 2016. 5(1): 75–9.

[157] Haug, A., O. Smidsrod, B. Larsen, S. Gronowitz, R.A. Hoffman, and A. Westerdahl. The effect of divalent metals on the properties of alginate solutions. II. comparison of different metal ions. Acta Chemica Scandinavica, 1965. 19: 341–351.

[158] Ge Gao, Jun Hee Lee, Jinah Jang, Dong Han Lee, Jeong-Sik Kong, Byoung Soo Kim, Yeong-Jin Choi, Woong Bi Jang, Young Joon Hong, Sang-Mo Kwon, and D.-W. Cho. Tissue engineered bio-blood-vessels constructed using a tissue-specific bioink and 3D coaxial cell printing technique: a novel therapy for ischemic disease. Adv Funct Mater, 2017. 27(33): 1700798.

[159] Ramos, P.E., P. Silva, M.M. Alario, L.M. Pastrana, J.A. Teixeira, M.A. Cerqueira, and A.A. Vicente. Effect of alginate molecular weight and M/G ratio in beads properties foreseeing the protection of probiotics. Food Hydrocolloid, 2018. 77: 8–16.

[160] Fannon, O.M., A. Bithell, B.J. Whalley, and E. Delivopoulos. A fiber alginate co-culture platform for the differentiation of mESC and modeling of the neural tube. Front Neurosci, 2020. 14: 524346.

[161] Schwartz, S.D., C.D. Regillo, B.L. Lam, D. Eliott, P.J. Rosenfeld, N.Z. Gregori, J.P. Hubschman, J.L. Davis, G. Heilwell, M. Spirn, J. Maguire, R. Gay, J. Bateman, R.M. Ostrick, D. Morris, M. Vincent, E. Anglade, L.V. Del Priore, and R. Lanza. Human embryonic stem cell-derived retinal pigment epithelium in patients with age-related macular degeneration and Stargardt's macular dystrophy: follow-up of two open-label phase 1/2 studies. Lancet, 2015. 385(9967): 509–16.

[162] Mellough, C.B., J. Collin, E. Sernagor, N.K. Wride, D.H. Steel, and M. Lako. Lab generated retina: realizing the dream. Vis Neurosci, 2014. 31(4-5): 317–32.

[163] Bidarra, S.J., and C.C. Barrias. 3D culture of mesenchymal stem cells in alginate hydrogels. Methods Mol Biol, 2019. 2002: 165–180.

[164] Pittenger, M.F., D.E. Discher, B.M. Peault, D.G. Phinney, J.M. Hare, and A.I. Caplan. Mesenchymal stem cell perspective: cell biology to clinical progress. NPJ Regen Med, 2019. 4: 22.

[165] Nebel, S., M. Lux, S. Kuth, F. Bider, W. Dietrich, D. Egger, A.R. Boccaccini, and C. Kasper. Alginate core-shell capsules for 3D cultivation of adipose-derived mesenchymal stem cells. Bioengineering (Basel), 2022. 9(2).

[166] Alepee, N., A. Bahinski, M. Daneshian, B. De Wever, E. Fritsche, A. Goldberg, J. Hansmann, T. Hartung, J. Haycock, H. Hogberg, L. Hoelting, J.M. Kelm, S. Kadereit, E. McVey, R. Landsiedel, M. Leist, M. Lubberstedt, F. Noor, C. Pellevoisin, D. Petersohn, U. Pfannenbecker, K. Reisinger, T. Ramirez, B. Rothen-Rutishauser, M. Schafer-Korting, K. Zeilinger, and M.G. Zurich. State-of-the-art of 3D cultures (organs-on-a-chip) in safety testing and pathophysiology, ALTEX, 2014. 31(4): 441–77.

[167] Harper, B.A., S. Barbut, L.T. Lim, and M.F. Marcone. Effect of various gelling cations on the physical properties of "wet" alginate films. J Food Sci, 2014. 79(4): E562–7.

[168] Sreeram, K.J., H. Yamini Shrivastava, and B.U. Nair. Studies on the nature of interaction of iron(III) with alginates, Biochim Biophys Acta, 2004. 1670(2): 121–5.

[169] Hou, J., C. Li, Y. Guan, Y. Zhang, and X.X. Zhu. Enzymatically crosslinked alginate hydrogels with improved adhesion properties. Polymer Chemistry, 2015. 6: 2204–2213.

[170] Shou, K., Y. Huang, B. Qi, X. Hu, Z. Ma, A. Lu, C. Jian, L. Zhang, and A. Yu. Induction of mesenchymal stem cell differentiation in the absence of soluble inducer for cutaneous wound regeneration by a chitin nanofiber-based hydrogel. J Tissue Eng Regen Med, 2018. 12(2): e867–e880.

[171] Chen, Y.H., S.H. Chang, I.J. Wang, and T.H. Young. The mechanism for keratinocyte detaching from pH-responsive chitosan. Biomaterials, 2014. 35(34): 9247–54.

[172] Mishra, S.K., J.M. Ferreira, and S. Kannan. Mechanically stable antimicrobial chitosan-PVA-silver nanocomposite coatings deposited on titanium implants. Carbohydr Polym, 2015. 121: 37–48.

[173] Thein-Han, W.W., and Y. Kitiyanant. Chitosan scaffolds for in vitro buffalo embryonic stem-like cell culture: an approach to tissue engineering. J Biomed Mater Res B Appl Biomater, 2007. 80(1): 92–101.

[174] Davanger, M., and A. Evensen. Role of the pericorneal papillary structure in renewal of corneal epithelium. Nature, 1971. 229(5286): 560–1.

[175] Chien, Y., Y.W. Liao, D.M. Liu, H.L. Lin, S.J. Chen, H.L. Chen, C.H. Peng, C.M. Liang, C.Y. Mou, and S.H. Chiou. Corneal repair by human corneal keratocyte-reprogrammed iPSCs and amphiphatic carboxymethyl-hexanoyl chitosan hydrogel. Biomaterials, 2012. 33(32): 8003–16.

[176] Baraniak, P.R., and T.C. McDevitt. Stem cell paracrine actions and tissue regeneration. Regen Med, 2010. 5(1): 121–43.

[177] Boido, M., M. Ghibaudi, P. Gentile, E. Favaro, R. Fusaro, and C. Tonda-Turo. Chitosan-based hydrogel to support the paracrine activity of mesenchymal stem cells in spinal cord injury treatment. Sci Rep, 2019. 9(1): 6402.

[178] Zuk, P.A., M. Zhu, H. Mizuno, J. Huang, J.W. Futrell, A.J. Katz, P. Benhaim, H.P. Lorenz, and M.H. Hedrick. Multilineage cells from human adipose tissue: implications for cell-based therapies, Tissue Eng, 2001. 7(2): 211–28.

[179] Yang, G., H. Long, X. Ren, K. Ma, Z. Xiao, Y. Wang, and Y. Guo. Regulation of adipose-tissue-derived stromal cell orientation and motility in 2D- and 3D-cultures by direct-current electrical field. Dev Growth Differ, 2017. 59(2): 70–82.

[180] Han, S.S., D.Y. Kang, T. Mujtaba, M.S. Rao, and I. Fischer. Grafted lineage-restricted precursors differentiate exclusively into neurons in the adult spinal cord. Exp Neurol, 2002. 177(2): 360–75.

[181] Wang, G., Q. Ao, K. Gong, H. Zuo, Y. Gong, and X. Zhang. Synergistic effect of neural stem cells and olfactory ensheathing cells on repair of adult rat spinal cord injury. Cell Transplant, 2010. 19(10): 1325–37.

[182] Wang, G., X. Wang, and L. Huang. Feasibility of chitosan-alginate (Chi-Alg) hydrogel used as scaffold for neural tissue engineering: a pilot study *in vitro*. Biotechnology & Biotechnological Equipment, 2017. 31(4): 766–773.

[183] Lim, S.M., D.K. Song, S.H. Oh, D.S. Lee-Yoon, E.H. Bae, and J.H. Lee. *In vitro* and *in vivo* degradation behavior of acetylated chitosan porous beads. J Biomater Sci Polym Ed, 2008. 19(4): 453–66.

[184] Choi, B., S. Kim, B. Lin, B.M. Wu, and M. Lee. Cartilaginous extracellular matrix-modified chitosan hydrogels for cartilage tissue engineering. ACS Appl Mater Interfaces, 2014. 6(22): 20110–21.

[185] Faikrua, A., R. Jeenapongsa, M. Sila-asna, and J. Viyoch. Properties of beta-glycerol phosphate/collagen/chitosan blend scaffolds for application in skin tissue engineering. Scienceasia, 2009. 35(3): 247–254.

[186] Zhu, Z., Y.-M. Wang, J. Yang, and X.-S. Luo. Hyaluronic acid: a versatile biomaterial in tissue engineering. Plastic and Aesthetic Research, 2017. 4: 219–227.

[187] Cowman, M.K., T.A. Schmidt, P. Raghavan, and A. Stecco. Viscoelastic properties of hyaluronan in physiological conditions. F1000Res, 2015. 4: 622.

[188] Pichert, A., D. Schlorke, S. Franz, and J. Arnhold. Functional aspects of the interaction between interleukin-8 and sulfated glycosaminoglycans. Biomatter, 2012. 2(3): 142–8.

[189] Nordsieck, K., L. Baumann, V. Hintze, M.T. Pisabarro, M. Schnabelrauch, A.G. Beck-Sickinger, and S.A. Samsonov. The effect of interleukin-8 truncations on its interactions with glycosaminoglycans. Biopolymers, 2018. 109(10): e23103.

[190] Zhai, P., X. Peng, B. Li, Y. Liu, H. Sun, and X. Li. The application of hyaluronic acid in bone regeneration. Int J Biol Macromol, 2020. 151: 1224–1239.

[191] Turksen, K., and T.C. Troy. Human embryonic stem cells: isolation, maintenance, and differentiation. Methods Mol Biol, 2006. 331: 1–12.

[192] Czyz, J., and A. Wobus. Embryonic stem cell differentiation: the role of extracellular factors. Differentiation, 2001. 68(4-5): 167–74.

[193] Abbott, A. Cell culture: biology's new dimension. Nature, 2003. 424(6951): 870–2.

[194] Gerecht, S., J.A. Burdick, L.S. Ferreira, S.A. Townsend, R. Langer, and G. Vunjak-Novakovic. Hyaluronic acid hydrogel for controlled self-renewal and differentiation of human embryonic stem cells, Proc Natl Acad Sci U S A, 2007. 104(27): 11298–303.

[195] Cho, H., D. Kim, and K. Kim. Engineered co-culture strategies using stem cells for facilitated chondrogenic differentiation and cartilage repair. Biotechnology and Bioprocess Engineering, 2018. 23: 261–270.

[196] Rutgers, M., D.B. Saris, L.A. Vonk, M.H. van Rijen, V. Akrum, D. Langeveld, A. van Boxtel, W.J. Dhert, and L.B. Creemers. Effect of collagen type I or type II on chondrogenesis by cultured human articular chondrocytes. Tissue Eng Part A, 2013. 19(1-2): 59–65.

[197] Brittberg, M., A. Lindahl, A. Nilsson, C. Ohlsson, O. Isaksson, and L. Peterson. Treatment of deep cartilage defects in the knee with autologous chondrocyte transplantation. N Engl J Med, 1994. 331(14): 889–95.

[198] Huang, S., X. Song, T. Li, J. Xiao, Y. Chen, X. Gong, W. Zeng, L. Yang, and C. Chen. Pellet coculture of osteoarthritic chondrocytes and infrapatellar fat pad-derived mesenchymal stem cells with chitosan/hyaluronic acid nanoparticles promotes chondrogenic differentiation. Stem Cell Res Ther, 2017. 8(1): 264.

[199] Deszcz, I., A. Lis-Nawara, P. Grelewski, S. Dragan, and J. Bar. Utility of direct 3D co-culture model for chondrogenic differentiation of mesenchymal stem cells on hyaluronan scaffold (Hyaff-11). Regen Biomater, 2020. 7(6): 543–552.

[200] Wang, C.Z., R. Eswaramoorthy, T.H. Lin, C.H. Chen, Y.C. Fu, C.K. Wang, S.C. Wu, G.J. Wang, J.K. Chang, and M.L. Ho. Enhancement of chondrogenesis of adipose-derived stem cells in HA-PNIPAAm-CL hydrogel for cartilage regeneration in rabbits. Scientific Reports, 2018. 8(1): 10526.

[201] Dong, Y., M. Cui, J. Qu, X. Wang, S.H. Kwon, J. Barrera, N. Elvassore, and G.C. Gurtner. Conformable hyaluronic acid hydrogel delivers adipose-derived stem cells and promotes regeneration of burn injury. Acta Biomaterialia, 2020. 108: 56–66.

[202] Serra, M., C. Brito, C. Correia, and P.M. Alves. Process engineering of human pluripotent stem cells for clinical application. Trends Biotechnol, 2012. 30(6): 350–9.

[203] Ekerdt, B.L., C.M. Fuentes, Y. Lei, M.M. Adil, A. Ramasubramanian, R.A. Segalman, and D.V. Schaffer. Thermoreversible hyaluronic acid-PNIPAAm hydrogel systems for 3D stem cell culture. Adv Healthc Mater, 2018. 7(12): e1800225.

[204] Wu, S., R. Xu, B. Duan, and P. Jiang. Three-dimensional hyaluronic acid hydrogel-based models for *in vitro* human iPSC-derived NPC culture and differentiation. J Mater Chem, 2017. B 5(21): 3870–3878.

[205] Prange, M.T., and S.S. Margulies. Regional, directional, and age-dependent properties of the brain undergoing large deformation. J Biomech Eng, 2002. 124(2): 244–52.

[206] Wobma, H., and G. Vunjak-Novakovic. Tissue engineering and regenerative medicine 2015: A Year in Review. Tissue Eng Part B Rev, 2016. 22(2): 101–13.

[207] Lantoine, J., T. Grevesse, A. Villers, G. Delhaye, C. Mestdagh, M. Versaevel, D. Mohammed, C. Bruyere, L. Alaimo, S.P. Lacour, L. Ris, and S. Gabriele. Matrix stiffness modulates formation and activity of neuronal networks of controlled architectures. Biomaterials, 2016. 89: 14–24.

[208] Saha, K., A.J. Keung, E.F. Irwin, Y. Li, L. Little, D.V. Schaffer, and K.E. Healy. Substrate modulus directs neural stem cell behavior. Biophys J, 2008. 95(9): 4426–38.

[209] Kook, Y.M., Y. Jeong, K. Lee, and W.G. Koh. Design of biomimetic cellular scaffolds for co-culture system and their application. J Tissue Eng, 2017. 8: 2041731417724640.

[210] Hong, Y., J. Chen, H. Fang, G. Li, S. Yan, K. Zhang, C. Wang, and J. Yin. All-in-one hydrogel realizing adipose-derived stem cell spheroid production and *in vivo* injection via "Gel-Sol" transition for angiogenesis in hind limb ischemia. ACS applied materials & interfaces, 2020. 12(10): 11375–11387.

[211] Dong, Y., Y. Qin, M. Dubaa, J. Killion, Y. Gao, T. Zhao, D. Zhou, D. Duscher, L. Geever, G.C. Gurtner, and W. Wang. A rapid crosslinking injectable hydrogel for stem cell delivery, from multifunctional hyperbranched polymers via RAFT homopolymerization of PEGDA. Polymer Chemistry, 2015. 6(34): 6182–6192.

[212] Xu, Q., S. A, Y. Gao, L. Guo, J. Creagh-Flynn, D. Zhou, U. Greiser, Y. Dong, F. Wang, H. Tai, W. Liu, W. Wang, and W. Wang. A hybrid injectable hydrogel from hyperbranched PEG macromer as a stem cell delivery and retention platform for diabetic wound healing. Acta Biomaterialia, 2018. 75: 63–74.

[213] Diniz, I.M., C. Chen, X. Xu, S. Ansari, H.H. Zadeh, M.M. Marques, S. Shi, and A. Moshaverinia. Pluronic F-127 hydrogel as a promising scaffold for encapsulation of dental-derived mesenchymal stem cells. J Mater Sci Mater Med, 2015. 26(3): 153.

[214] Ruel-Gariepy, E., and J.C. Leroux. *In situ*-forming hydrogels—review of temperature-sensitive systems. Eur J Pharm Biopharm, 2004. 58(2): 409–26.

[215] Silva, A.K.A., S. Perretta, G. Perrod, L. Pidial, V. Lindner, F. Carn, S. Lemieux, D. Alloyeau, I. Boucenna, P. Menasché, B. Dallemagne, F. Gazeau, C. Wilhelm, C. Cellier, O. Clément, and G. Rahmi. Thermoresponsive gel embedded with adipose stem-cell-derived extracellular vesicles promotes esophageal fistula healing in a thermo-actuated delivery strategy. ACS nano, 2018. 12(10): 9800–9814.

[216] Rahman, N., K.A. Purpura, R.G. Wylie, P.W. Zandstra, and M.S. Shoichet. The use of vascular endothelial growth factor functionalized agarose to guide pluripotent stem cell aggregates toward blood progenitor cells. Biomaterials, 2010. 31(32): 8262–70.

[217] Forget, A., A. Blaeser, F. Miessmer, M. Kopf, D.F.D. Campos, N.H. Voelcker, A. Blencowe, H. Fischer, and V.P. Shastri. Mechanically tunable bioink for 3D bioprinting of human cells. Adv Healthc Mater, 2017. 6(20).

[218] Oliveira, J.T., L. Martins, R. Picciochi, P.B. Malafaya, R.A. Sousa, N.M. Neves, J.F. Mano, and R.L. Reis. Gellan gum: a new biomaterial for cartilage tissue engineering applications. Journal of Biomedical Materials Research. Part A, 2010. 93(3): 852–63.

[219] Quinn, F.X., T. Hatakeyama, H. Yoshida, M. Takahashi, and H. Hatakeyama. The conformational properties of gellan gum hydrogels. Polymer Gels and Networks, 1993. 1(2): 93–114.

[220] Oliveira, J.T., L.S. Gardel, T. Rada, L. Martins, M.E. Gomes, and R.L. Reis. Injectable gellan gum hydrogels with autologous cells for the treatment of rabbit articular cartilage defects. Journal of Orthopaedic Research : Official Publication of the Orthopaedic Research Society, 2010. 28(9): 1193–9.

[221] Nair, S.S., S. Majeed, S. Sankar, and M. MJ. Formulation of some antioxidant herbal creams. Hygeia, 2009. 1: 44–45.

[222] Bidra, A.S., J.J. Tarrand, D.B. Roberts, K.V. Rolston, and M.S. Chambers. Antimicrobial efficacy of oral topical agents on microorganisms associated with radiated head and neck cancer patients: an *in vitro* study. Quintessence international (Berlin, Germany : 1985), 2011. 42(4): 307–15.

[223] Habeeb, F., G. Stables, F. Bradbury, S. Nong, P. Cameron, R. Plevin, and V.A. Ferro. The inner gel component of Aloe vera suppresses bacterial-induced pro-inflammatory cytokines from human immune cells. Methods (San Diego, Calif.), 2007. 42(4): 388–93.

[224] Sholehvar, F., D. Mehrabani, P. Yaghmaei, and A. Vahdati. The effect of Aloe vera gel on viability of dental pulp stem cells. Dent Traumatol, 2016. 32(5): 390–6.

[225] Jittapiromsak, N., D. Sahawat, W. Banlunara, P. Sangvanich, and P. Thunyakitpisal. Acemannan, an extracted product from Aloe vera, stimulates dental pulp cell proliferation, differentiation, mineralization, and dentin formation. Tissue Engineering. Part A, 2010. 16(6): 1997–2006.

[226] Oryan, A., E. Alemzadeh, A.A. Mohammadi, and A. Moshiri. Healing potential of injectable Aloe vera hydrogel loaded by adipose-derived stem cell in skin tissue-engineering in a rat burn wound model, Cell and Tissue Research, 2019. 377(2): 215–227.

[[227] Guilhem Godeau, Julie Bernard, Cathy Staedel and Philippe Barthélémy. Glycosyl-nucleoside-lipid based supramolecular assembly as a nanostructured material with nucleic acid delivery capabilities. Chem. Commun., 2009. 14: 5127–5129.

[228] Coviello, T., P. Matricardi, C. Marianecci, and F. Alhaique. Polysaccharide hydrogels for modified release formulations. J Control Release, 2007. 119(1): 5–24.

[229] Ball, D.H., B.J. Wiley, and E.T. Reese. Effect of substitution at C-6 on the susceptibility of pullulan to pullulanases. Enzymatic degradation of modified pullulans. Can J Microbiol, 1992. 38(4): 324–7.

[230] Thebaud, N.B., D. Pierron, R. Bareille, C. Le Visage, D. Letourneur, and L. Bordenave. Human endothelial progenitor cell attachment to polysaccharide-based hydrogels: a pre-requisite for vascular tissue engineering. J Mater Sci Mater Med, 2007. 18(2): 339–45.

[231] Leathers, T.D. Biotechnological production and applications of pullulan. Appl Microbiol Biotechnol, 2003. 62(5-6): 468–73.

[232] Wong, V.W., K.C. Rustad, M.G. Galvez, E. Neofytou, J.P. Glotzbach, M. Januszyk, M.R. Major, M. Sorkin, M.T. Longaker, J. Rajadas, and G.C. Gurtner. Engineered pullulan-collagen composite dermal hydrogels improve early cutaneous wound healing. Tissue Engineering. Part A, 2011. 17(5-6): 631–44.

[233] Autissier, A., D. Letourneur, and C. Le Visage. Pullulan-based hydrogel for smooth muscle cell culture. Journal of Biomedical Materials Research. Part A, 2007. 82(2): 336–42.

[234] da Silva, L.P., R.L. Reis, V.M. Correlo, and A.P. Marques. Hydrogel-based strategies to advance therapies for chronic skin wounds. Annual Review of Biomedical Engineering, 2019. 21: 145–169.

[235] Kosaraju, R., R.C. Rennert, Z.N. Maan, D. Duscher, J. Barrera, A.J. Whittam, M. Januszyk, J. Rajadas, M. Rodrigues, and G.C. Gurtner. Adipose-derived stem cell-seeded hydrogels increase endogenous progenitor cell recruitment and neovascularization in wounds. Tissue Engineering. Part A, 2016. 22(3-4): 295–305.

7

In vitro Biosystems—Stem Cell-based Organ on Chips

I-Chi Lee,[1,*] *Nai-Chen Cheng*,[2] *Chia-Ning Shen*,[3] *Min-Huey Chen*,[4] *Wen-Yen Huang*,[5] *Sung-Jan Lin*[5] and *Yi-Chen Ethan Li*[6,*]

Currently, there is an urgent need to develop effective preclinical testing models that accommodate better precision in drug discovery and reduce costly failures in clinical trials. Generally, conventional *in vitro* cell-based drug discovery is implemented under two dimensional (2D) culture conditions. However, a 2D culture system may not achieve the biomimetic environment since real tissues actually includes multiple types of cells, several kinds of surrounding extracellular matrices (ECM), and cell-cell interaction. To evaluate the drug effects in 2D culture systems may lead to the loss of some key information and results in distortion [1]. To address the abovementioned limitations, the development of 3D cell/tissue culture models and microfabrication technique have enabled the creation of a new paradigm, organ on chips (OoC). OoC utilizes the microfluidic technology and 3D cell culture principle to reproduce organ- or tissue-level functionality at a small scale that may provide novel testing approaches with more accurate testing results than traditional methods for generating reliable predictions and safety information of drugs. Because primary human cells are not capable of differentiation and are difficult to culture *in vitro* and

1 Department of Biomedical Engineering and Environmental Sciences, National Tsing Hua University, Taiwan.
2 Department of Surgery, National Taiwan University Hospital, Taiwan.
3 Genomics Research Center, Academia Sinica, Taiwan.
4 Graduate Institute of Clinical Dentistry, School of Dentistry, National Taiwan University, Taiwan.
5 Department of Biomedical Engineering, College of Medicine and College of Engineering, National Taiwan University, Taiwan.
6 Department of Chemical Engineering, Feng Chia University, Taiwan.
* Corresponding authors: iclee@mx.nthu.edu.tw; yicli@fcu.edu.tw

the functionality of cell lines differs from real tissues, implying that stem cells are the most suitable candidates on the development of OoCs. Therefore, stem cell-based OoC have become a promising tool for personalized medicine and drug development, providing an alternative to bridge conventional animal models and cell assays in the recent decade. Generally, the field of OoC engineering is focused on dynamically controlling the microenvironments of the living cells to produce biomimetic and functional 3D tissue/organ-like construction with crucial physiological functions for understanding the drug effects, improving preclinical assessment and efficacy testing on the liver, heart, lung, intestine, kidney, brain and bone [2]. Microfluidic technology enables OoCs to regulate cellular morphology, shear stress, mechanical forces, and induction factors in a biomimetic microenvironment. Stem cell-based OoCs integrated microfabrication, stem cell technology, biomaterials, and microfluidic techniques to drive the establishment of *in vitro* 3D models and often additionally integrated transepithelial/transendothelial electrical resistance (TEER) and biosensing system to monitor biomarkers, chemokines, and critical physiological properties in real time [2]. Furthermore, multiorgan OoC models have also been demonstrated to explore organ–organ interactions and to determine the organ-specific responses to drug candidates.

In this chapter, we discuss the value of this new approach to scientists in basic and applied research. The innovations and development of different types of tissue/organ-like construction techniques such as *in vitro* spheroid and organoid are introduced. Combination of 3D tissue/organ—like construction, microfluidic technique, and biosensors to achieve OoC in *in vitro* organ environment modeling and a disease model are also included. We also describe the technical challenges that must be overcome to develop OoCs into robust, predictive models of human physiology and disease, address the opportunities for drug discovery, and discuss the challenges in moving those techniques into mainstream drug discovery. We believe OoCs model will be powerful techniques for pathophysiological applications and drug screening.

In vitro stem cell-based spheroid and organoid culture

In vitro cell-based assays supply simple and cost-effective tools to avoid the disadvantages of large-scale of animal life losses, but conventional 2D cell-based assays cannot provide accurate information due to its limitations. Morphological differences correlate with functional differences. Relative to the conventional 2D culture, 3D culture-based models can mimic the real microenvironment *in vivo* and reduce the reliance on animal models while increasing the efficiency of drug screening [3]. The 3D culture methods facilitate greater cell–cell contacts and interactions of cells with the surrounding ECM. By allowing cells to adapt to their native morphology, significant differences between the cellular phenotype and biological response of cells cultured in monolayer and 3D cell culture have been observed [4–6]. It is becoming increasingly accepted that 3D culture methods provide a cellular environment more consistent with that *in vivo* [5].

Spheroids and organoids are 3D cultures without the use of scaffold materials, in which cells can grow as an adherent population in a spherical shape that creates its own ECM [7]. Advances in these models have revolutionized the tools of *in vitro* technology by establishing 3D systems to recapitulate the cellular heterogeneity, structure, and functions of the primary or stem cell sources [8–11]. Although the terms spheroid and organoid are sometimes used interchangeably, they actually have different definitions. Spheroids are the more general definition for 3D aggregates of cells. Organoids are a sub-category of them and described as more complex clusters, representing mini-organs/tissues or simplified versions which consist of tissue- relevant cell types derived from progenitor cells grown in 3D *in vitro*. Organoids exhibit self-renewal and self-organizational capabilities providing them with prominent potential to model animal organ development and diseases in a dish [12–13]. These features constitute a substantial benefit of organoids over simplified spheroids and are advantageous for drug discovery and regenerative medicine [14].

Real tissue, organ, and tumors are composed of heterogeneous cell types and complex cell-cell and cell-ECM interactions. Therefore, mixed heterotypic spheroids composed of two or more different cell sources could help to study heterologous interactions and mimic real tissue niches. 3D construction multicellular spheroids are self-assembled cell aggregations which provide tissue mimic construction and biomimetic microenvironments with several physiological parameters [15]. The core principle are based on anchorage-independent methodology or biomaterials that promote cell–cell interactions which dominate over cell-substrate interactions [15]. Also, colony/spheroid formation has been widely used as the 3D model in the application of stem cells niche mimicking, tumor progression, metastasis investigation, and oncology drug testing [16–19]. Colony/spheroid models' critical physiologic parameters present *in vivo*, include complex multicellular architecture with spatial structures and ECM deposition and packed 3D structures that include tight junctions, barriers to mass transport of drugs, nutrients and other factors, which are close to real tissues or tumor environments [20]. Fabrication of spheroids have been demonstrated to improve stem cell function; however, the hypoxic inner-center and limited penetration of nutrients and signaling cues to the interior of the spheroid are potential challenges. The incorporation of different materials such as hydrogels [21–22], gelatin [23–24], and chitosan [25] in spheroids resulted in relatively relaxed assembly of composite spheroids with enhanced transport of nutrient and biological gas. Because of the low surface area of cells within spheroids and the heterogeneous distribution of polymers throughout the spheroid, these polymers cannot increase the cell activity to extracellular matrix interactions required to support differentiation. Hence, Lee et al. developed spheroids incorporating polydopamine coated single-segmented fibers from an electrospun nanofiber sheet. The spheroids incorporating polydopamine coated single-segmented fibers showed enhanced viability regardless of sizes and increased their functionality by regulating the size of spheroids which may be used for various tissue reconstruction and therapeutic applications [26]. Besides, in order to gather uniform sized spheroids, microfabrication techniques were generally used to prepare spheroids with good size controllability, to be cultured for long periods and permit morphologies investigation conveniently; have space for spheroids to grow; and form spheroids with a high throughput. Unlike the macroscale approaches, the purpose of these micro-manufacturing technologies is to produce uniform tumor spheroids across cellular-

and tissue-length scales. These manufacture methods offer the exciting prospect of standardized spheroid mass production to progress drug testing and high-throughput screening. Especially, uniform spheroids closed the 3D tumor-like construct and pathophysiological gradients but also provide the advantages of easy operation and assays [27].

Adult stem cells such as mesenchymal stem cells (MSCs) and adipose stem cell (ASCs) have been considered as potent cell source for tissue regeneration and organ on chip. 3D construction provides cell-to-ECM interaction and cell-to-cell interaction, which control the localization of endogenous growth factors and affect the differentiation [28]. With a cell aggregation approach, cell-to-cell connections are not disrupted and cells are harvested as a contiguous cell sheet, thus preserving cell-to-cell junction proteins and the native ECM secreted by the cells [29]. ASCs loss of the stemness for monolayer cell culture and the stemness marker expression were highly enhanced during aggregation culture [30]. Zhang et al. developed a biomaterial-free spheroid forming model using microgravity bioreactor, and significantly upregulated expression levels of E-cadherin and pluripotent markers were noted compared with 2D culture [31]. Human ASC-derived neurospheroids displayed self-renewal properties to form secondary neurospheroids, which could be induced to become neurons and glial cells [32]. Yoon et al. suggested that in spheroids, ASCs could be triggered to undergo chondrogenesis via hypoxia-related cascades and increased cell-cell interactions [33]. Tu et al. also created cartilage tissues from scaffold-free ASC-derived spheroids [34]. Moreover, ASC-derived spheroids showed higher capability of osteogenic differentiation [35–36]. Bernard et al. have also been using the PEG microwell system to collect 3D pancreatic cell spheroids and it is demonstrated such 3D spheroids were found to secrete higher insulin levels in response to a glucose change in comparison with single cells [37].

Besides normal stem cells, cancer stem cells (CSCs) spheroid are also generally used on the determination of tumor drug sensitivity, and it is revealed that chemoresistance, metastasis, and recurrence were successfully represented in a 3D spheroid model by including the role of CSCs in comparison with 2D cell-based models [38–39]. In addition, these models also allow for the analysis of different growth constraints such as the hypoxia environment, chemo- and radiation effects, and angiogenesis [40]. Co-culture models mimicking tumor and stromal cell interactions have been recently shown to present powerful tools to delineate cancer resistance mechanisms. Previous literature has shown that hybrid spheroid models offer the most promising approach for studying heterologous interactions in tumorigenesis. Several types of hybrid tumor spheroids have been reported in the literature, and the most common cell types that form these spheroids are tumor cells mixed with immune cells, fibroblasts, and endothelial cells [41]. It is revealed that CSCs upregulated the key macrophage biomarkers significantly compared with other ovarian cancer cells in a co-culture model of ovarian CSCs and macrophages spheroids by hanging-drop-made [42].

3D organoid culture systems permit a more complex structure to mimic organ development in comparison with the methods describe above. The organoid is an ideal *in vitro* analysis system which comprises human cells in a construct complex enough to demonstrate physiologic-like composition, morphology, and heterogeneity. Such a construct captures some of the complexity of a human organ in a dish but also maintains the benefits for convenient analysis and investigation *in vitro* [14]. Organoids have been created using tissue adult stem cells or pluripotent stem cells from humans for a variety of organs, including the brain, retina, intestine, kidney, liver, lung, and stomach. Regardless of the organ type, three key components are required for the formation of organoids: (1) precisely timed addition of small molecules or growth factors that activate or inhibit signalling pathways involved in organ development, (2) formulation of medium with proper nutrient replenishment to sustain organoid growth, (3) enabling of expansion via 3D culturing system including embedding into an extracellular matrix such as Matrigel, generation of embryoid bodies, or organoid formation using air-liquid interface.

Organoid models include 3D cell culture systems that reproduce the complex spatial morphology of a differentiated epithelium and create a more comprehensive model of the living tissue. In addition, organoids also include the following characteristics: self-organization, multicellularity, and functionality. The stem cell-based organoid system develops the right culture conditions to induce stem cells to self-renew and grow into organ- or tissue-like structures. Generally, in order to generate the correct tissue type, stem cells must be taken through a series of carefully choreographed steps using different media cocktails and a biomimetic microenvironment. During 2009, Sato and colleagues succeeded in establishing the minimal requirements for sustainable growth of mouse crypt-villus structures without mesenchyme [43]. Then, organoids are established from surgically resected intestinal tissue and endoscopic biopsies of patients suffering from adenomas and adenocarcinomas [44]. In the past decade, organoid-forming methods have been used to formed several other organoid culture of multiple mouse and human epithelia, including the colon, liver, pancreas, prostate, stomach, fallopian tube, salivary glands, lung, and breast [45]. Therefore, organoids have opened new avenues for the development of novel, more physiological cancer models on new treatment methods and drug screening. Furthermore, white adipose tissue (WAT) is a complex organ composed of differentiated adipocytes, adipose stromal cells, as well as endothelial vascular cells and infiltrating leukocytes. Daquinag et al. used a 3D levitation tissue culture system based on magnetic nanoparticle assembly to create organoids with vascular-like structures containing luminal endothelial and perivascular stromal cell layers, mimicking the development and growth of adipose tissue [46]. Notably, adipose tissue organoids often lack resident immune cell populations and therefore require the addition of immune cells isolated from other organs. Taylor et al. developed an adipose tissue organoid which contains and maintains the resident immune cell populations of the stromal vascular fraction [47]. Additional studies have applied the ASC-derived organoids for therapeutic purposes, as ASC-derived organoids display changes in transcriptional profile and their biological potency. Smolar et al. generated hybrid organoids performed with autologous pre-differentiated smooth muscle-like ASCs and smooth muscle cells, providing an evidence that pre-differentiated smooth muscle-like ASCs can partially form smooth muscle tissue for detrusor muscle regenerative purposes [48]. Todorov et al. embedded stromal cells from human adipose tissue within a gel along with the devitalized cartilage matrix, leading to a composite construct with enhanced capacity to form de novo bone tissue [49].

In essence, the generation of iPSC-derived organoids attempts to replicate the developmental processes and organ regeneration *in vitro*. Since it is impossible to precisely provide the biochemical and physical cue present during *in vivo* organ development, organoids rely mostly on self-organization triggered by defined culture conditions. The following section will describe published methodologies and techniques utilized to generate iPSC-derived organoids while highlighting advantages of each method with the aim of providing an overview of current and future development of organoids.

Generation of iPSC-derived organoids via ECM-dependent methods

Organoid formation using ECM was first utilized by the Clevers group to generate intestinal organoid from adult $Lgr5^+$ stem cells present at the bottom of small intestinal crypts. These $Lgr5^+$ stem cells can self-organize into crypt-villus organoids without the presence of the mesenchymal niche when cultured in laminin-rich Matrigel along with key niche signals including Wnt, Noggin, R-spondin and EGF [50]. Following this discovery, a similar strategy was employed to generate iPSC-derived human intestinal organoids. Indeed, mesenchyme-free intestinal organoids was generated from human iPSCs by first differentiating into $CXCR4^+$/c-Kit^+ definitive endoderm, then inhibiting the BMP/TGFβ signaling pathways to promote the formation of anterior foregut endoderm. Lineage specification was further achieved through Wnt activation and enrichment of $NKX2.1^-$ or $CD47^{lo}$ fraction into 3D Matrigel droplets. Culturing of Matrigel droplets in a defined serum-free medium containing CHIR, KGF along with Dexamethasone, cAMP and IBMX resulted in single cell outgrowth into self-organizing human intestinal organoids (HIOs). These HIOs were shown to be able to model monogenic disorders such as Familial Adenomatous Polyposis (FAP) and Cystic Fibrosis (CF) using patient-derived iPSCs. CF-patient derived HIOs exhibited significantly reduced organoid size, lumen area, and forskolin-induced swelling compared to control organoids. The correction of the disease-causing allele via gene editing in a rescued CF-patient derived HIOs to levels comparable to control organoids [51].

Apart from HIOs, the generation of human hepatic organoids (HHOs) can also be approached the same way. First, the 2D monolayer induction of definitive endoderm was achieved followed by enrichment and seeding of $EpCAM^+$ hepatic progenitors into Matrigel. Second, addition of defined liver organoid expansion medium enabled the formation of 3D organoids that can be passaged and expanded for more than 30 passages without significant reduction in differentiation capacity. Switching to differentiation medium resulted in the maturation of HHOs as demonstrated by the expression of hepatocyte markers as well as mature hepatocyte functions including CYP3A4 activity, low-density lipoprotein (LDL) uptake, and glycogen storage. Like HIOs, HHOs can be utilized to model monogenic diseases such as classical citrullinemia type 1 (CTLN1). Additionally, it has also been shown that dissociated hepatic organoids can be transplanted into the liver of humanized FRG mice ($Fah^{-/-}Rag^{-/-}Il2rg^{-/-}$ mice) and repopulate the damaged liver. By incorporating the human fetal liver mesenchymal cells or immune cells such as Kupffer cells into hepatic organoids, it is also possible to model alcohol-induced liver injuries or steatohepatitis, enabling rapid screening of new drug candidates. Together, these studies demonstrate the potential for the use of an extracellular matrix such as Matrigel to generated human iPSC-derived organoids [52].

Despite the successful generation of iPSC-derived organoids using Matrigel as a scaffold for organoid formation, there are several limitations that still warrant further development [53]. As Matrigel is derived from mouse sarcoma, the reliance of many protocols on it meant these organoids are not suitable for direct use in human transplantation. Moreover, the undefined nature of Matrigel with over 18,000 unique proteins makes it extremely difficult to tailor and may therefore not achieve optimal differentiation niche for different organoids. This is further compounded by lot-to-lot variations which can lead to poor reproducibility and hampers its potential use in the clinical setting. Due to these limitations, there are currently active development of Matrigel-free organoid culture methods utilizing decellularized extracellular matrix, synthetic hydrogels, or recombinant proteins matrices.

To mimic the structure and composition of the organ of interest, decellularized organs can be utilized to support the growth of organoids. Decellularization for each organ type can be different depending on tissue composition, cellularity, and lipid content. Therefore, the method in which each organ is decellularized also differs. To obtain ECM from organs, various techniques has been conducted using a combination of mechanical, chemical, or enzymatic dissociation. For example, in human pancreas, the high lipid content meant the inclusion of a homogenization step is crucial for the removal of lipid as well as increase the cytocompatibility of the resulting ECM with hESC-derived pancreatic cells [54]. Whilst for liver organoids, acellular liver-specific ECM discs can be generated by perfusing liver with a decellularization detergent followed by freezing and cryo-sectioning of acellular liver lobes into 300 μm-thick sections. By implanting human fetal liver progenitor cells into these acellular liver dishes, self-assembled liver organoids can form with the hepatocytes and bile duct structure present. Decellularized ECM represent one of the most efficient strategies to mimic the composition of the organ of interest. Despite its advantages, the quantity and quality of decellularized ECM can be limiting due to the availability and condition of the donor. Even obtaining organs from seemingly healthy donor may result in differing ECM stiffness due to individual variations that are difficult to monitor or control. Therefore, to further standardize ECM for iPSC-derived organoids, many organoid cultures began to utilize synthetic hydrogels.

Synthetic hydrogels are composed of synthetic polymers such as PEG and PLGA with specific biochemical cues incorporated to allow for cell attachment and organoid formation, thus enabling control of ECM mechanical and chemical properties. In terms of biochemical cues, fibronectin, laminin, or collagen are often incorporated into synthetic hydrogels to permit organoid formation and differentiation. For example, the use of 3D-hexagonally arrayed inverted colloidal crystal scaffold that is composed of PEG functionalized with collagen I, allowed for the establishment of liver organoids when seeded with iPSC-derived hepatic progenitors [55]. These liver organoids demonstrated enhanced functionality and drug metabolism capacity compared to 2D cultured hepatic progenitors. Stiffness of the synthetic hydrogen is also an important point to consider when fabricating the ECM. High PEG density, and thus increased matric stiffness, significantly reduced iPSC-derived intestinal organoid viability compared to a softer and degradable matrix [56]. One potential disadvantage of synthetic hydrogel ECM such as PEG is the potential to trigger immune response when implanted *in vivo* [57]. Therefore, for use in transplantation, additional care must be given to the selection of polymer that serves as scaffold for the growth of organoids. Alternatively, peptide or recombinant protein matrices maybe used to replace synthetic polymers. By combining recombinantly synthesized hyaluronan and elastin-like protein via a bioconjugation reaction, a PEG-free hydrogel network can be generated to support the formation and differentiation of patient-derived intestinal organoids [58]. Furthermore, this protein matrix can be tuned for matrix stiffness, stress relaxation rate, and integrin-ligand concentration by chemical modification of functional groups such as benzaldehyde and hydrazine. Thus, it can also be applied for intestinal organoids and epithelial organoids from hepatocytes. The advantage of hyaluronan and elastin-like protein are that they are well tolerated *in vivo*, potentially serving as a more suitable platform for generating iPSC-derived organoids for transplantation purposes [59].

Generation of iPSC-derived organoids via embryoid bodies

Apart from utilizing ECM-dependent methods, it is also possible to generate iPSC-derived organoids in suspension culture via the formation of embryoid bodies (EBs) [60]. EBs describe multicellular aggregates that spontaneously form when iPSCs are cultured under suspension conditions. These EBs partly mimic early embryonic developmental stages with potential to develop into three germ layers. Thus, compared to 2D cultures, EBs serves as a superior 3D niche for lineage specific iPSC differentiation or organoid formation. Formation of EBs can be achieved via self-aggregating or forced aggregating methods [61]. A self-aggregating method typically involves the use of low-attachment dishes and the seeding of iPSCs at appropriate density to enable spontaneous formation of EBs at various sizes. However, in some cases the size of EBs is crucial in determining the efficiency of differentiation. For example, EBs of 100 μm in diameter were demonstrated to be optimal for cardiomyocyte differentiation whilst EBs with 300 μm was shown to be more suitable for melanocyte differentiation. To obtain uniform size EBs hanging drop, U-bottom- or microwell-plates can be used. With the hanging drop method, complete medium droplets containing an appropriate number of cells are seeded on to either the lid of petri dishes or using hanging drop culture plates. Upon overnight culture, uniform size EBs should begin to form. These EBs can be transferred to low attachment dishes for further culture and expansion. Similarly, with either 96 well U-bottom or microwell plates, accurate determination of cell count is required prior to seeding the same cell number into each well to ensure uniform EB formation and consistent differentiation. Changing to differentiation medium for individual wells or transferring EBs to a low attachment dish for bulk culture can be performed to direct differentiation to the desired lineage.

By using EB as an intermediate, the formation of an iPSC-derived brain organoid has been achieved. When dissociated iPSC was reaggregated using a 96-well U-bottomed plate to form EBs and cultured in the presence of a differentiation medium, polarized cortical neuroepithelial rosettes that recapitulate early corticogenesis could be generated [62]. These pattered structures exhibit distinct functional regions of cortical neurons such as the outer subventricular zone that is present in human neocortex [63]. This approach was further elaborated upon to generate an iPSC-derived cerebral organoid [64]. Without using patterning growth factors, it was demonstrated that optimizing growth conditions alone can trigger intrinsic cues for cerebral development. First, EB was utilized to generate neuroectoderm which was then embedded into a Matrigel droplet to provide the necessary scaffold for neuroepithelium formation. By transferring the neuroepithelium containing droplet into a spinning bioreactor to enhance nutrient exchange, cerebral organoids were formed containing defined brain regions such as prefrontal cortex, occipital lobe, cortical lobes, hippocampus, and ventral forebrain. More recently, iPSC-derived brain region-specific organoids was used to model a Zika virus infection [65]. A miniaturized spinning bioreactor was used to reduce the cost of generating multiple iPSC-derived cerebral organoids under different conditions. With the pre-patterning of EBs towards the forebrain organoid, it was shown that African or Asian Zika virus strains preferred the infection of neural progenitors, leading to increased apoptosis and microcephaly-like reduction in neuronal volume. Together, the examples in generating iPSC-derived cerebral organoids demonstrated the advantages of using EBs as self-organizing niches with intrinsic cues to mimic *in vivo* developmental stages.

Generation of iPSC-derived organoids via air-liquid interface and organ-on-a-chip model

In addition to ECM- and EB- dependent methods, other methods of generating iPSC-derived organoids are also being attempted to recapture the multicellular composition of organs; these include the use of air-liquid interface or organ-on-a-chip strategies. Immunotherapy can be key to developing effective treatment against infectious diseases and cancer. Engineered T cells, such as chimeric antigen receptor (CAR) T cells, hold great potential in the field of immunotherapy. Therefore, developing the protocol to generate T cells from iPSCs can allow the mass production of donor specific or universal T cells from a self-renewing source. Many have attempted the use of planar culture to generate functionally mature iPSC-derived T cells with limited success [66]. These protocols lacked the spatiotemporal interactions with thymic epithelial, mesenchymal, and hematopoietic cells that exist within the thymus. Therefore, the establishment of a 3D artificial thymic organoid (ATO) culture is a significant advancement in the generation of iPSC-derived T cells [67–68]. The first mesoderm induction from iPSC was performed using planar culture and enriched for a human embryonic mesodermal progenitor (hEMP). Second, hematopoietic induction was performed by aggregating hEMPs with a stromal cell line expressing Notch ligand, DLL4, crucial for T-lineage commitment. These embryonic mesodermal organoids (EMOs) were then plated onto an air-liquid interface with a porous membrane to induce hemato-endothelial commitment. In the last phase, T cell differentiation and the establishment of ATOs were induced by the addition of cytokines required for T cell maturation. iPSC-derived T cells that matured within the ATO demonstrated diverse T cell receptor repertoire and if engineered to express antigen-specific TCR, can produce potent antigen-specific cytotoxicity.

Although iPSC-derived organoids have proven beneficial in modeling cellularity and structure of a specific organ, organs within the human body do not act in isolation. Thus, a strategy to generate an interconnected *in vitro* organ system is necessary to better recapitulate the disease or drugs that can affect multiple organs. In this respect, a micro-physiological model that combines advancement in iPSC culture, organoid formation, and development of microfluidics may be key to truly mimicking human physiology. An organ-on-a-chip strategy that allows for microfluidic connection between organ chips containing iPSC-derived organoids of different organ type can provide a platform for more accurate disease modeling and drug screening. Indeed, it was demonstrated that by developing a two-organ-chip containing iPSC-derived pancreas and liver organoids, it was possible to mimic functional coupling of the pancreas and liver [69]. This level of connectivity not only enhanced the maturation of an iPSC-derived cerebral organoid, and using a patient-derived iPSC carrying Parkinson's disease-causing mutation confirmed that the short-chain fatty acid associated with the gut-microbiome increased the pathology associated with the disease.

Although spheroids and organoids can reflect parts of the complex 3D organization of organs, the lack of vascularization in cell aggregation still limits their capacity to recapitulate the entire *in vivo* situation [70]. The vascularization is necessary for transcellular transport, absorption, or secretion. This also extends to another major drawback that many systems actually lack the multiscale architecture and tissue-tissue interfaces [13], such as interactions between different cell types [71]. Additionally, organoids are highly variable in size and shape, and it is difficult to maintain cells in consistent positions in these structures for an extended analysis.

iPSC-based disease modelling and drug discovery

Establishing the pathological pathways that underpin human diseases is critical for developing new therapeutic options. Animal models have shown to be excellent tools for modeling human diseases, allowing pathogenic processes to be identified *in vivo* at various developmental stages and in specific cell types. Furthermore, in mice, it is feasible to develop both *in vitro* and *in vivo* iPSC-based disease models at the same time. The strength and limitations of the *in vitro* human iPSC-based models could be better understood by comparing the phenotypes seen with similar *in vitro* and *in vivo* mice models. However, significant species differences may preclude whole human disease symptoms from being replicated in animals like mice, which are the most widely utilized animal models. The lack of disease recapitulation between mouse and human cells is most likely owing to fundamental species differences. As a result, human disease modeling platforms are urgently needed to supplement studies that use animal models for biomedical research.

Modeling human genetic diseases has been done using both human ESCs and iPSCs [72]. ESCs were used in the previous models, but since the introduction of human iPSC technology, human iPSCs have become the favored option due to their availability and lack of potential ethical difficulties related with human ESCs. Human iPSCs are very similar to human embryonic stem cells (ESCs). Both types of cells carry human pluripotent factors and ESC surface marks, as well as the ability to differentiate into three germ layers during development. In iPSCs residual epigenetic memory of somatic cells may remain, affecting the differentiation potential of these cells. Although the persistence of parental cell epigenetic memory has been found in iPSCs, similar phenotypes have been documented in disease modeling utilizing human ESCs and iPSCs in the majority of cases indicating that disease modeling utilizing patient-derived iPSCs is effective [73–74].

Many drug screenings are focused on targets thought to be related to disease mechanisms. Furthermore, due to the poor success rates of compounds discovered by target-based screening, phenotypic screening has gained popularity [75]. The discovery of iPSCs has facilitated the revival of phenotypic screening for a variety of reasons, including the scalability of iPSC production, which makes assay development easier. Furthermore, because iPSCs are pluripotent, they can be developed into a variety of disease-relevant cell types, including those that are otherwise difficult to access, such as neurons [76]. In a culture dish, patient-derived iPSC models can mimic disease phenotypes and pathologies. Molecular and cellular traits may be present in cells developed from patient-derived iPSCs. If the gene responsible for the disease characteristics is identified, a gene-editing strategy can establish whether the phenotype chosen as a readout for a drug screen is indeed relevant to the condition, and can be further validated in patient samples and/or animal models [77]. iPSCs can be used for target-based screening in addition to phenotypic screening. Many drug screens have been performed using human iPSC models, and potential drug candidates have been found using either phenotypic or target-based screening. The neural crest precursors for autonomic neurons were sorted and purified from iPSCs derived from patients with familial dysautonomia, a monogenic early-onset disease characterized by degeneration of neurons in the sensory and autonomic nervous systems, in the first report of a large-scale drug screen using an iPSC-based disease model [78]. The use of human iPSC models to find disease-relevant targets could be an important component of future drug development. After initial screening utilizing a HEK293 cell line, Naryshkin et al. discovered that a patient-derived iPSC model of SMA could be used to evaluate human- and disease-specific drug responses. These drugs were subsequently tested in patient-specific fibroblasts and motor neurons produced from patient-derived iPSCs, which served as a disease-relevant and patient-specific cellular model. Finally, the *in vivo* activity of the hit drug was tested in a mouse model [79]. Taking advantage of the patient-specific and disease-relevant characteristics of motor neurons created from patient iPSCs, this drug development methodology used a patient-derived iPSC model as one of the validation processes. Over 1,000 compounds have been evaluated using iPSC-based drug screening for a variety of disorders, and numerous clinical candidates have been found.

Drug repositioning is another application of disease-specific iPSCs, in which existing drugs that have already been approved for specific disorders are examined for new use in other diseases. Patient-derived iPSCs did not develop efficiently into cartilage tissue in a human iPSC model obtained from individuals with achondroplasia with mutations in fibroblast growth factor receptor 3 (FGFR3). A screen for compounds that rescued chondrogenically differentiated iPSCs from the defective cartilage phenotype was conducted using this model, and numerous statins, which are approved cardiovascular medicines, were discovered. In a mouse model of FGFR3-linked disease, statins were found to promote the formation of shorter limbs. These findings suggest that statins could be repositioned as potential achondroplasia treatments [80].

Another example of drug repositioning is the anti-epileptic drug ezogabine, which was shown to be effective in an iPSC model of the motor neuron disease ALS and is now being tested on humans [81]. The scientists used an iPSC model obtained from patients with ALS who had mutations in the superoxide dismutase 1 (SOD1) gene as well as individuals with ALS who had mutations in other ALS-related genes such C9ORF72 and FUS. It has also been shown that iPSC motor neurons generated from ALS patients exhibit a hyperexcitable condition at first, followed by a reduction in excitability. This data shows that ezogabine treatment for ALS should begin as soon as possible. The generalization of treatment responsiveness across ALS types was made possible by observing similar drug responses in different patient groups. Drug development with iPSCs produced from multiple genetic variants of a disease is advantageous since it allows for an assessment of the drug response in a large patient population. Analysis of a drug's effect in several mouse models at the same time, on the other hand, is difficult. Disease repositioning will be aided by accumulating information from disease iPSC research in combination with patients' personalized clinical experience, in which diseases are defined not by clinical but by cellular characteristics. If the cellular phenotypes of *in vitro* iPSC models of clinically distinct diseases are identical or comparable, then a treatment that works in one condition may also work in others. For example, hyperexcitable neuronal cells were discovered in an iPSC model of bipolar illness. A patient with ALS203 had iPSC-derived motor neurons with similar hyperexcitability. As a result, the same therapeutic drug could work for several clinically dissimilar but cellularly comparable disorders. Accumulating data on cellular phenotypes in iPSC models from a cross-sectional range of diseases could lead to novel disease stratifications and knowledge, as well as new cross-sectional therapeutic methods [82].

Microfluidic device (pattern, mechanical strains and dynamic environmental factors)

Microfluidic technology enables the manipulation and control of fluidic volumes at a micro-scale on a device. Unlike the conventional macroscale cell-screening methods, microfluidic platforms provide several unique advantages based on the independent functional units' potential, a smaller reagent used, with high sensitivity and high throughput [83]. Consequently, it miniaturizes basic conventional biological or chemical laboratory operations in a chip, such as sample preparation, cell isolation, and bioassay. The microfluidic system has been quickly and well designed by the biological and medical research communities as a powerful tool for reconstructing microenvironments of cells. In addition, microfluidic-based cell cultures provide a dynamic system which could continuously supply cells with fresh media containing oxygen and nutrition at a controlled flow rate [84]. Poly-dimethylsiloxane (PDMS) soft lithography techniques also benefited the development of cost-efficient microfluidics platform, with less cells and high throughput determination. Stem cell-based OoC with micro-manufacturing devices generally composed with microfluidic channels, designed chambers, pumps, valves, and biosensors supply cell patterning, shear stress, fluid flow, or mechanical stimulation.

3T3-L1 preadipocytes, murine primary preadipocytes, and human ASCs have been cultured and differentiated in various microfluidic devices to study adipokine secretions, glycerol secretions, free fatty acid secretions, as well as mammalian target of rapamycin (mTOR) regulations [85–88]. To protect adipocytes from direct fluidic shear stress, the dual-layer microfluidic devices was developed, in which porous membranes separate cell culture chambers from the fluid flow [89]. In most microfluidic systems, preadipocytes were either differentiated on glass coverslips and later sealed into the microfluidic device or seeded to surface-coated microfluidic chambers before differentiation, making the systems essentially 2D models.

Recently, efforts have been made to incorporate 3D adipose tissue into microfluidics. For instance, Godwin et al. integrated a reservoir into a passively operated microfluidic device to culture primary murine adipocytes suspended and entrapped in collagen hydrogel [90]. Yang et al. created a microfluidic device that has five connected cell culture chambers in which the 3D tissue resides, flanked by two side channels that deliver the medium to the cell culture chambers [85–88]. This design has enabled ASCs to differentiate within the microfluidic system to form a dense lipid-loaded mass with the expression of adipose tissue genetic markers. The engineered adipose tissue showed a decreased adiponectin secretion and increased free fatty acid secretion with increasing shear stress. Adipogenesis markers were downregulated with increasing shear stress. This is the first report on 3D adipose tissue differentiation and formation in a microfluidic device, showing the adipose tissue response under physiological level interstitial shear stress stimulation.

Furthermore, microfluidics allows spatio-temporal control culture system and enable the biomimetic scale-down of differentiation processes. The germ layer specification and phenotypic differentiation of iPSCs can be controlled and regulated in a microfluidic platform. A multistage microfluidic-based process with extrinsic signal modulation and optimal frequency of medium delivery was designed for controlling hiPSC expansion and differentiation into functional tissue-specific cells. They demonstrated that human cardiomyocytes and hepatocytes derived on chips, displayed functional phenotypes and responses to dynamics of induction with temporally defined factors treatments [91].

Mechanical strains have also been incorporated into microfluidics to induce biophysical cues on stem cells and different type of stretching platform includes unidirectional, bi-directional, and three-dimensional strain have been developed [92]. Generally, it is revealed that most microfluidics devices used to generate a cyclic mechanical strain of relatively small amplitude at a frequency comprised between 0.2 and 2 Hz, and 10% linear strain are most widely used [92]. A programmable uniaxial cyclic stretch with a compliant microfluidic platform was applied on the induction of ESCs differentiated into cardiomyocytes [93]. In addition, the impact of the mechanical strain on the Lung-on-a -hip has increased importantly. Platforms equipped with a porous membrane are widely used on the Lung-on-a-chip to supply transport across the cultured cell layers and to reproduce of the air-liquid interface [94]. A lung-on-a-chip array that mimics the lung parenchyma with its thin alveolar barrier has also been developed by co-culturing epithelial and endothelial cells that form tight monolayers on each side of a porous and stretchable membrane. The microsystem also provides the 3D cyclic strain with the porous membrane to modulate breathing movements [95].

3D bioprinting on stem cell based OoCs

3D bioprinting technology is a promising tool for constructing complex structures that mimic *in vivo* architecture and cellular interactions. This technology can be used to fabricate biomimetic *in vitro* organ-like constructs and to develop disease models. Dias et al. used laser direct-write to control the size of the EB formed by mouse ESCs and the local cell density in printed colonies, potentially leading to better stem cell maintenance and directed differentiation [96]. In addition, hiPSC-cardiomyocytes were bioprinted into gelatin methacrylic (GelMA) microfibrous scaffolds with endothelial cells to fabricate an endothelialized myocardium by an extruder 3D bioprinting and generate an aligned myocardium capable of spontaneous and synchronous contraction [97]. Multicellular spheroids comprising human iPSC-derived cardiomyocytes, endothelial cells, and cardiac fibroblasts have also been printed to form 3D cardiac tissues [98]. iPSC have also been embedded into alginate hydrogels for hepatocyte differentiation and to form a mini 3D liver by using a drop-based printing platform [99]. Also, a microscale hepatic construct consisting of physiologically relevant hexagonal units of liver cells and supporting cells have been developed by using iPSC-derived hepatic progenitor cells and 3D bioprinting technology [100]. Specially fabricated nanothin and highly porous membranes were fabricated to bioprint highly viable, homogenous cell sheets, which were derived from co-culture of MSCs and cardiomyoblast cells. The results demonstrated that this method facilitated the formation of direct gap junctions between MSCs and cardiomyoblast cells [101]. Furthermore, multiple stem cells have also been used to fabricate neural tissue, including neural stem cells (NSCs), ESCs, iPSCs, ADSCs, and adult MSCs by 3D bioprinting and these studies demonstrated better functionality in 3D construction in comparison with 2D culture [102]. Gu et al. used polysaccharide-based bioink comprising alginate, carboxymethyl-chitosan, and agarose and NSCs to construct neural tissue and these differentiated neurons form synaptic contacts, establish networks, and increased the functionality [103]. Ma et al. applied 3D digital bioprinting technology to create a 3D hydrogel-based culture system that embedded iPSC-derived hepatic progenitor cells with human umbilical vein endothelial cells and ADSCs in a microscale hexagonal architecture. It is revealed that the liver-like microenvironment enhanced high levels of liver-specific gene expression and metabolic product secretion [100]. MSC are generally used for 3D bioprinting skin tissue. Microextrusion bioprinting technology was applied to print amniotic fluid-derived stem cells (AFSCs) and MSCs suspended in fibrin-collagen hydrogel. In addition, MSCs, keratinocytes, and fibroblasts were used to fabricate a skin-like tissue by laser bioprinting with functional layers printing [104].

Furthermore, one of the main challenges in constructing biomimetic tissue models for long-term culture is the lack of functional vasculature. Vasculature plays an important role in transporting nutrients and oxygen to the cells and removing waste. Kolesky et al. have used 3D cell-laden bioprinting to fabricate a vascularized tissues that exceed 1 cm in thickness and they demonstrated this structure can be perfused on chip for more than six weeks of culture period [105]. In their study, multiple inks composed of human hMSCs and human neonatal dermal fibroblasts (hNDFs) within a customized extracellular matrix were used and lined with human umbilical vein endothelial cells (HUVECs). *In vitro* cardiac tissue-like constructs also require the nutrient transport and waste disposal offered by blood vessel structures. Maiullari et al. have fabricated bioprinted cardiac tissue products containing vessel-like networks by encapsulating iPSCs-derived cardiomyocytes (iPSC-CMs) and HUVECs in hydrogel containing alginate and PEG-Fibrinogen [106]. Furthermore, Skylar-Scott et al. developed a biomanufacturing method that relies on sacrificial writing into functional tissue composed of a living organ building blocks (OBB) matrix consisting of iPSCs derived organoids to construct viable cardiac tissue with high cell density, maturation, and desired functionality [107]. They pattern a sacrificial ink within the matrix via embedded 3D printing, which upon removal yields perfusable channels in the form of single or branching conduits.

Development of stem cell- based OoCs

A microfluidic OoCs system provides continuously engineered microfluidic tissue culture platforms which integrate multicellular architectures in a micrometer-sized chamber [108]. The goal is not building a whole living organ but synthesizing minimal functional units that recapitulate tissue- and organ-level functions, combining the advantages of *in vivo* (animal models) and *in vitro* (cell culture) approaches. The new systems allow small-scale cell culture with a constant flow that can represent *in vivo* physiological mass transport conditions [90]. Typically, the microfluidic OoCs consist of designed culture chambers which are connected to each other via semi-permeable barriers or micro-channels. The culture chambers contain cells, culture/induction medium, and materials such as hydrogel or 3D bioprinting construction. Continuous flow with controllable flow rate is perfused into the culture chambers and the molecular factors may also deliver into the chambers dynamically. OoC platforms equipped with a porous membrane can sometimes be actively deformed and stretched.

Lung-on-a-chip: The first OoC model was Lung-on-a-chip. This biomimetic microsystem was able to reconstitute the expansion and contraction movements of the alveolus, reproduce inflammatory reactions, and reveal enhanced nanoparticle uptake in epithelial and endothelial cells, which was similar to that observed in a whole mouse lung [109]. Other OoC models were evolved rapidly, such as Liver-on-a-chip [110], Gut-on-a-chip [111] and Kidney-on-a-chip [112].

Brain on chip: NSCs and iPSCs-based Brain-on-chip have also been established in several studies and demonstrated a great potential for exploring the mechanisms underlying the neurodegenerative diseases and toxicity to neonatal neural development. A microphysiological system combine with 3D cell aggregation model from human iPSCs for four weeks of neural induction was presented to generate a Brain-on-a-chip model [113]. Besides, our lab has also constructed a Brain-on-a-chip model by patterning NSCs spheroids, guiding NSCs differentiation, and promoting network formation [114]. Park et al. further developed an NSCs spheroids-based Brain-on-a- chip with a constant flow fluid by osmotic micropump that more closely mimics the interstitial flow of the *in vivo* brain microenvironment [115]. Moreno et al. used a phase-guided 3D microfluidic cell culture bioreactor to derive human neuroepithelial cells from hiPSCs and further induced them into dopaminergic neurons, which may provide an efficient route to personalized drug discovery for Parkinson's disease [116]. A system to generate myelinating oligodendrocytes from neurons derived from mouse ESCs in microfluidic devices has also been established [117]. This study open avenues for modeling some aspects of demyelinating diseases like multiple sclerosis.

Liver-on-a-chip: Preclinical and clinical studies for liver drug screening and chemical-safety testing are important. However, conventional *in vitro* models are generally failed to mimic human liver physiology [2]. Therefore, it is urgent to fabricate better prediction models with the properties of high throughput, low cost, and being easy to operate for testing of drug absorption, distribution, metabolism, excretion, and toxicity. A simple Liver-on-a-Chip platform with mature and functional hepatocyte-like cells derived from hPSCs was established in a microfluidic device with a 3D culture environment. The results showed significant improvements of hepatic functions such as drug uptake/excretion capabilities [118]. Another device consists of an integrated network of hexagonal tissue-culture chambers, which contains a central outlet and culture with hiPSC-derived hepatocytes was designed to mimic the central vein of a liver lobule [119]. The 3D tissue-like structure and bile-canaliculi network formation were observed in this study. Furthermore, a study combining stem cell biology with microengineering technology to fabricated liver organoids derived from hiPSCs in a 3D perfusable chip system have been established. The results demonstrated the liver organoids not only expressed higher expression of endodermal genes but also showed a marked enhancement of hepatic-specific functions, including albumin and urea production and metabolic capabilities. Most importantly, they presented the liver organoids exhibited hepatotoxic response after exposure to acetaminophen in a dose- and time-dependent manner [120]. Furthermore, another liver organoid system was established to characterize the pathological features of a common metabolic and progressive disease in liver organoids by exposure to free fatty acids in perfused 3D cultures during a prolonged period [121]. Upon FFA induction, liver organoids represented the typical biochemical characteristics of NAFLD progression included lipid droplet formation, triglyceride accumulation, and upregulation of lipid metabolism-associated genes.

Gut-on-a-chip: Gut-microbiota interaction have gained attention recently, and the gut is also an important part of the immune and endocrine systems. Gut-on-a-chip can help scientists to understand the basic functions of the gut and how they are influenced by environmental conditions, drugs, and other cells. However, endocrine and immunological functions in Gut-on-a-chip models are still poorly represented and stem cell-based Gut-on-a-chip are even less. Two types of mini-gut models were developed and used generally. One of them is formed from the cells derived from iPSCs and the other one is cells generated from stem cells present in an intestinal crypt niche. A mini-gut model derived from primary single stem cells or isolated intestinal crypts was initially developed by Hans Clevers et al. [122]. It is revealed that these cells grew faster than cells derived from iPSC and a mature mini-gut consisting of > 40 crypt domains surrounding a central lumen were obtained.

Barrier-on-a-chip: Biomimetic barrier construction and measuring barrier function for disruption of tissue barriers are also important for OoC disease models [123]. Several OoC models in which epithelial cells (ECs) are co-cultured with other cells have been developed and mimic barriers such as the blood-brain barrier (BBB), blood retinal barrier (BRB), and the pulmonary air-liquid interface (ALI). Trans-endothelial electrical resistance (TEER) is a typical method to determine the barrier function by directly measuring diffusion or migration of tracers and cells. OoC supply advance platform with dynamic physicochemical microenvironment combine several factor controls, which could modulate blood flow, interstitial flow, tissue shape, mechanical strain, dynamic signaling, cell-cell interaction, and cell-ECM interaction [123]. A recent study have design an *in vitro* microfluidic BBB chip lined by human brain microvascular endothelium derived from hiPSCs interfaced with primary human brain astrocytes and pericytes that recapitulates the high level of tight junction proteins and functional efflux pumps in an hypoxia microenvironment [124].

Multi-organ-on-a-chip: Further advancements focus on the construction of complex systems, which simulate multi-organ interactions by linking many organs- and tissues-on-a-chip [125]. However, most of the studies have been carried out using cell lines or primary cells to model various organ or biological systems. Precise control of stem cell differentiation in the microfluidic microenvironment makes OoC developments more promising. hiPSC and hESC-derived EC OoC model have been published on several previous literature. Moreno et al. have used neuroepithelial cells derived from human iPSC and these cells were successfully differentiated into functional dopaminergic neurons and demonstrated electrophysiological characterization in a 3D microfluidic cell culture system [126].

Disease-on-achip: Besides organ models, the promising OoC technology has evolved as a powerful tool for translational biomedical applications and inherent understanding of disease mechanisms. Wang et al. have also designed an iPSC-derived cardiomyocytes model of Barth syndrome (BTHS), a kind of mitochondrial cardiomyopathies, through Heart-on-chip techniques to elucidate the pathophysiology of BTHS and the results demonstrated that the cells assembled into sparse and irregular sarcomeres, which contracted weakly similar to *in vivo* situations [127]. Park et al. also tested the toxic effects of amyloid-β as AD-on-a-chip and demonstrated that a dynamic system caused a significantly more destruction of neural networks in comparison with the amyloid-β treatment under static conditions [115]. Obesity is one of the most common causes of insulin resistance. WAT is an insulin sensitive organ, as well as a critical storage site for excess dietary energy, so it is a potential tissue for pharmacotherapies of associated diseases such as obesity and type 2 diabetes. In 2017, Loskill et al. created a WAT-on-a-chip system enabling the control of nanoliter fluid volumes and flows and computationally predictable character, the system has the potential to be a powerful tool for the study of adipose tissue [128]. Tanataweethum et al. also established a microfluidic OoC model of insulin-resistant adipocytes and confirmed defects of disrupted insulin signaling through reduction of lipid accumulation from fatty acid uptake and elevation of glycerol secretion [129]. Furthermore, a novel integration of a label-free, multiplexed biosensor into a complex biomimetic obese adipose tissue microenvironment allowed quantitative characterization of the dynamic cytokine secretion behaviors in a spatiotemporal manner [130]. Liu et al. also employed a silicon-based microfluidic chip which comprises the co-culture of adipocytes and immune cells in an enclosed system that provides an *in vivo*-like environment, including continuous nutrient supply and cell–cell interaction between the two cell types [131].

Biosensors of OoCs

In a previous review article, we had discussed the development of biosensors in OoCs and the application on disease model which has published on Biosensors and reproduced in the following paragraph [132].

Biosensor-based systems contribute convenient and rapid tools to detect various signals from [an] OoC. Compared to standard ELISA analytical technology, biosensors provide an alternative for real-time and on-site monitor micro biophysiological signals via [a] combination of biological, chemical, and physical technologies. Key developments in the cell/tissue-based biosensors, biomolecular sensing strategies and the expansion of several biochip approaches such as organs-on-chips, paper based-biochip, and flexible biosensors are available. Cell polarity, cell proliferation, stem cells differentiation, stimulation response and metabolism detection were included. Biosensors applied on OoC and their derivative disease models such as brain, heart, lung and liver system have been widely discussed in the past decade.

Evaluation of electrophysiological properties of neural-related cells and cardiomyocytes is important for tissue engineering and microelectrode arrays (MEAs) is a candidate for measuring electrophysiological properties of cells. Chowdhury et al. considered that the contact electrogram is related with cellular action potential and should change in conjunction with each other during arrhythmogenesis. They developed a novel technique combining MEA recordings with optical mapping to simultaneously record the contact electrograms and action potentials [133]. Also, biomechanical measurements on cardiomyocytes for investigating the pathophysiological electro-mechanical coupling were done by using the stretcher device previously [134]. Further, Caluori et al. combined the MEA technology and atomic force microscope system for recording the beating force of the cardiomyocytes cluster and the triggering electric events [135].

Neurological research is another field of investigation where cell-based biosensors have proven to be significant and MEA technology is also the primary mode of determining neuronal circuits, physiology, and abnormalities [136]· MEA provides the advantages of non-invasive monitoring the electrophysiological activity of neurons for a long-term culture, multi-site recording, and high-throughput screening [137]. The collective electrophysiological behavior of the neuronal network in terms of burst activity on MEA was applied on pharmaceutical agent testing. In addition, in order to analyze spreading depolarization, Lourenco et al. developed an innovative multimodal approach for metabolic, electric, and hemodynamic measurements with neuronal activity [138].

Based on the MEA technique, our lab has developed a novel interdigitated microelectrode arrays, which combined electric stimulation on the guidance of neurite outgrowth ITO-PEM microfluidic system and the neuronal network development on different patterns by an impedimetric monitoring system [139]. The biochip was constructed with a PDMS culture chamber layer, an SU-8 structural layer, an indium tin oxide (ITO)-glass detector layer, and the structural layer was treated with oxygen plasma and polyelectrolyte multilayer films (PEMs) to guide neural stem cell differentiation [139]. The electrical connections of two neurospheroids were determined by measuring the impedance across two electrodes. Because cell membranes are electrically insulated, impedance measurements have been used to analyze cellular responses [140–142]. Therefore, a threshold of 40 kΩ was defined to determine the electrical connections of the neurospheroids. Neural communication and regeneration were investigated using not only immunocytochemistry but also electrical stimulations and recordings. Impedance measurements were conducted across two neurospheroids to provide quantitative evidence and validate the connections of the neural network. The development of an NSC-based OoC provided an alternative to monitor stem cells proliferation, differentiation, neural network formation, electrophysiological properties, and stimulation response which can be applied on drug discovery and tissue engineering.

Furthermore, ultra-flexible and micro-electrocorticography arrays fabricated by glassy carbon electrodes have also been used to stimulate and record brain activity with low background noise [143]. Other applications where live cells-based devices have been applied on monitoring of endothelial barrier functions, live cell secretion, and molecules releasing [144]. Recently, Li et al. developed a reversible electrode for rapidly diagnosis Alzheimer's disease by using magnetic graphene nanomaterials [145]. They conjugated the antibody of Alzheimer's disease biomarker, Amyloid-beta peptide 1–42 (Aβ42), on a magnetic nitrogen-doped graphene (MNG). Then, an Alzheimer's disease biosensor could be rapidly constructed by dropping the magnetic MNG immunocarriers on an Au electrode surface, which has a tapping permeant magnet at the underside of electrode. Afterward, the used MNG biosensors could be removed and the Au electrode could be reproduced by switching off the tapped permeant magnet. Moreover, another biosensor with rapid detection of PD has been reported by Yang and co-workers [146]. They generated a self-assembled monolayer (SAM) by grafting a DNA aptamer and an SH-spacer on the surface of the substrate. After incubation with a PD biomarker (α-synuclein), the DNA aptamer provides a specific binding to capture the α-synuclein protein, and the changed optical signals caused from the binding of α-synuclein could be easily recognized by using a microscope.

Noticeably, bioelectrical activity is also an important myocardial function which can report the health the of heart tissues. In general, the bioelectrical activity could be generated from the cardiomyocytes, which induced the change of action potential of the cellular membrane; through the change of action potential, the heart could be induced a synchronized pumping behavior via this organized electrical propagation [147]. An earlier study has been reported that the change of oxygen level in tissues significantly interrupts the regular organized duration of action potential [148]. Therefore, a continuous electrocardiogram (ECG) monitoring technique provides a clinical standard method to detect the cardiac rhythm signal for diagnosing cardiac-related diseases. To easily monitor cardiac rhythm signals, Lee et al. developed a small wearable flexible cardiac sensor, which integrates an electrode, a near-field-communication chip, and a battery in polyurethane substrates [149]. The small cardiac biosensor offers a real-time visible signal that people could directly observe the change of cardiac rhythm signals on their smart phones. Furthermore, the Heart-on-a-chip system containing biosensors also contributes to a living heart cell system to mimic heart-related diseases. Liu and co-workers fabricated a Pt nanopillar array on an Au electrode [150]. Then, a PDMS microfluid channel was stacked on the Au electrode and cardiac cells were perfused and seeded on that to create a hypoxia-like heart disease environment such as myocardial infraction. Significantly, the biosensor displays narrow action potential signals consistent with a possible mechanism that oxygen-deficits enhance the activity of ATP-K channel and the repolarization of cellular membrane. This Heart-on-a-chip type biosensor provides an *in vitro* disease model for investigating and understanding the effect of hypoxia on the electrophysiological behaviors of heart.

Similar to Brain- and Heart-on-a-chip systems, other types tissue-like models such as liver and lung are also designed with embedded biosensors to detect the function of tissue-like living constructs. For example, mitochondrial dysfunction involves in the development of chemical or pharmaceutical toxicity. Balvi et al. used HepG2 cell-based liver organoids as a tissue model and cultured the organoids in a microfluidic device [151]. By real-time monitoring the metabolic function of liver organoids, it endows the microdevice as a biosensor with a feature to track the dynamic of mitochondrial dysfunction. Via sensing the oxidative phosphorylation of glycolysis or glutaminolysis, the liver organoid-based system permits the evaluation of the safety and effect of drug concentration on mitochondrial damage.

Conventional systems suffer from not fully recapitulating the complexity of the microenvironment as well as an inclusion of physical forces that have an impact on the development and differentiation of cells. Stem cell-based OoCs integrate stem cell co-culture systems, physical stimulations, dynamic factors, barrier models, and sensor monitoring to mimic the physiological tissue environment more realistically, and act as a great tool to overcome the challenges of a conventional *in vitro* systems. However, we still have some challenges vis-a-vis improving the OoCs. It is revealed that local gas concentration control and exact barrier function measurement are still the problems of OoCs design. Because most OoCs are currently fabricated from PDMS, the high gas permeability of this material makes it challenging to control gas pressures locally in different organ model, especially in a multiorgan cascade model. In addition, the barrier function of many tissues are affected by local oxygen concentrations [123].

References

[1] Matsusaki, M., C.P. Case, and M. Akashi. Three-dimensional cell culture technique and pathophysiology. Advanced Drug Delivery Reviews, 2014. 74: 95–103, DOI: 10.1016/j.addr.2014.01.003.

[2] Geraili, A., P. Jafari, M.S. Hassani, B.H. Araghi, M.H. Mohammadi, A.M. Ghafari, S.H. Tamrin, H.P. Modarres, A.R. Kolahchi, S. Ahadian, and A. Sanati-Nezhad. Controlling differentiation of stem cells for developing personalized organ-on-chip platforms. Advanced Healthcare Materials, 2018. 7(2): 1700426, DOI: https://doi.org/10.1002/adhm.201700426.

[3] Prantil-Baun, R., R. Novak, D. Das, M.R. Somayaji, A. Przekwas, and D.E. Ingber. Physiologically based pharmacokinetic and pharmacodynamic analysis enabled by microfluidically linked organs-on-chips. Annu Rev Pharmacol Toxicol, 2018. 58: 37–64, DOI: 10.1146/annurev-pharmtox-010716-104748.

[4] Bartosh, T.J., Z. Wang, A.A. Rosales, S.D. Dimitrijevich, and R.S. Roque. 3D-model of adult cardiac stem cells promotes cardiac differentiation and resistance to oxidative stress. J Cell Biochem, 2008. 105(2): 612–23, DOI: 10.1002/jcb.21862.

[5] Frith, J.E., B. Thomson, and P.G. Genever. Dynamic three-dimensional culture methods enhance mesenchymal stem cell properties and increase therapeutic potential. Tissue Eng Part C Methods, 2010. 16(4): 735–49, DOI: 10.1089/ten.TEC.2009.0432.

[6] Lin, S.J., S.H. Jee, W.C. Hsiao, H.S. Yu, T.F. Tsai, J.S. Chen, C.J. Hsu, and T.H. Young. Enhanced cell survival of melanocyte spheroids in serum starvation condition. Biomaterials, 2006. 27(8): 1462–9, DOI: S0142-9612(05)00799-4 [pii] 10.1016/j.biomaterials.2005.08.031.

[7] Fennema, E., N. Rivron, J. Rouwkema, C. van Blitterswijk, and J. de Boer. Spheroid culture as a tool for creating 3D complex tissues. Trends Biotechnol, 2013. 31(2): 108–15, DOI: 10.1016/j.tibtech.2012.12.003.

[8] van der Vaart, J., and H. Clevers. Airway organoids as models of human disease. J Intern Med, 2021. 289(5): 604–613, DOI: doi.org/10.1111/joim.13075.

[9] Sato, T., and H. Clevers. Growing self-organizing mini-guts from a single intestinal stem cell: mechanism and applications. Science, 2013. 340(6137): 1190–4, DOI: 10.1126/science.1234852.

[10] Huch, M., and B.K. Koo. Modeling mouse and human development using organoid cultures. Development, 2015. 142(18): 3113–25, DOI: 10.1242/dev.118570.

[11] Lancaster, M.A., and J.A. Knoblich. Organogenesis in a dish: modeling development and disease using organoid technologies. Science, 2014. 345(6194): 1247125, DOI: 10.1126/science.1247125.

[12] Clevers, H. Modeling development and disease with organoids. Cell, 2016. 165(7): 1586–1597, DOI: 10.1016/j.cell.2016.05.082.

[13] Fatehullah, A., S.H. Tan, and N. Barker. Organoids as an *in vitro* model of human development and disease. Nat Cell Biol, 2016. 18(3): 246–254, DOI: 10.1038/ncb3312.

[14] Ranga, A., N. Gjorevski, and M.P. Lutolf. Drug discovery through stem cell-based organoid models. Adv Drug Deliv Rev, 2014. 69–70: 19–28, DOI: 10.1016/j.addr.2014.02.006.

[15] Friedrich, J., C. Seidel, R. Ebner, and L.A. Kunz-Schughart. Spheroid-based drug screen: considerations and practical approach. Nature Protocols, 2009. 4: 309, DOI: 10.1038/nprot.2008.226.

[16] Debnath, J., and J.S. Brugge. Modelling glandular epithelial cancers in three-dimensional cultures. Nature reviews. Cancer, 2005. 5(9): 675–88, DOI: 10.1038/nrc1695.

[17] Lee, G.Y., P.A. Kenny, E.H. Lee, and M.J. Bissell. Three-dimensional culture models of normal and malignant breast epithelial cells. Nature Methods, 2007. 4(4): 359–65, DOI: 10.1038/nmeth1015.

[18] Dolznig, H., A. Walzl, N. Kramer, M. Rosner, P. Garin-Chesa, and M. Hengstschläger. Organotypic spheroid cultures to study tumor–stroma interaction during cancer development. Drug Discovery Today: Disease Models, 2011. 8 (2-3): 113–119, DOI: http://dx.doi.org/10.1016/j.ddmod.2011.06.003.

[19] Sharma, R., S. Greenhough, C.N. Medine, and D.C. Hay. Three-dimensional culture of human embryonic stem cell derived hepatic endoderm and its role in bioartificial liver construction. Journal of Biomedicine & Biotechnology, 2010. 236147, DOI: 10.1155/2010/236147.

[20] Mehta, G., A.Y. Hsiao, M. Ingram, G.D. Luker, and S. Takayama. Opportunities and challenges for use of tumor spheroids as models to test drug delivery and efficacy. Journal of Controlled Release, 2012. 164(2): 192–204, DOI: https://doi.org/10.1016/j.jconrel.2012.04.045.

[21] Kim, G., Y. Jung, K. Cho, H.J. Lee, and W.G. Koh. Thermoresponsive poly(N-isopropylacrylamide) hydrogel substrates micropatterned with poly(ethylene glycol) hydrogel for adipose mesenchymal stem cell spheroid formation and retrieval. Mater Sci Eng C Mater Biol Appl, 2020. 115: 111128, DOI: 10.1016/j.msec.2020.111128.

[22] Tsai, C.C., Y.J. Hong, R.J. Lee, N.C. Cheng, and J. Yu. Enhancement of human adipose-derived stem cell spheroid differentiation in an in situ enzyme-crosslinked gelatin hydrogel. J Mater Chem B, 2019. 7(7): 1064–1075, DOI: 10.1039/c8tb02835d.

[23] Lou, Y.R., L. Kanninen, B. Kaehr, J.L. Townson, J. Niklander, R. Harjumaki, C. Jeffrey Brinker, and M. Yliperttula. Silica bioreplication preserves three-dimensional spheroid structures of human pluripotent stem cells and HepG2 cells. Sci Rep, 2015. 5: 13635, DOI: 10.1038/srep13635.

[24] Wei, J., J. Lu, Y. Liu, S. Yan, and X. Li. Spheroid culture of primary hepatocytes with short fibers as a predictable in vitro model for drug screening. J Mater Chem B, 2016. 4 (44): 7155–7167, DOI: 10.1039/C6TB02014C.

[25] Cheng, N.C., S. Wang, and T.H. Young. The influence of spheroid formation of human adipose-derived stem cells on chitosan films on stemness and differentiation capabilities. Biomaterials, 2012. 33(6): 1748–58, DOI: 10.1016/j.biomaterials.2011.11.049.

[26] Lee, J., S. Lee, S.M. Kim, and H. Shin. Size-controlled human adipose-derived stem cell spheroids hybridized with single-segmented nanofibers and their effect on viability and stem cell differentiation. Biomater Res, 2021. 25(1): 14: DOI: 10.1186/s40824-021-00215-9.

[27] Hirschhaeuser, F., H. Menne, C. Dittfeld, J. West, W. Mueller-Klieser, and L.A. Kunz-Schughart. Multicellular tumor spheroids: An underestimated tool is catching up again. Journal of Biotechnology, 2010. 148(1): 3–15, DOI: http://dx.doi.org/10.1016/j.jbiotec.2010.01.012.

[28] Sart, S., A.C. Tsai, Y. Li, and T. Ma. Three-dimensional aggregates of mesenchymal stem cells: cellular mechanisms, biological properties, and applications. Tissue Eng Part B Rev, 2014. 20(5): 365–80, DOI: 10.1089/ten.TEB.2013.0537.

[29] Cheng, N.C., Y.B. Tang, C.W. Liang, and H.F. Chien. Myxoid solitary fibrous tumour of the axilla. Journal of Plastic Reconstructive and Aesthetic Surgery, 2006. 59(1): 86–89, DOI: 10.1016/j.bjps.2005.04.053.

[30] Cheng, N.C., S. Wang, and T.H. Young. The influence of spheroid formation of human adipose-derived stem cells on chitosan films on sternness and differentiation capabilities. Biomaterials, 2012. 33(6): 1748–1758, DOI: 10.1016/j.biomaterials.2011.11.049.

[31] Zhang, S., P. Liu, L. Chen, Y. Wang, Z. Wang, and B. Zhang. The effects of spheroid formation of adipose-derived stem cells in a microgravity bioreactor on stemness properties and therapeutic potential. Biomaterials, 2015. 41: 15–25, DOI: 10.1016/j.biomaterials.2014.11.019.

[32] Peng, C., L. Lu, Y. Li, and J. Hu. Neurospheres induced from human adipose-derived stem cells as a new source of neural progenitor cells. Cell Transplant, 2019. 28(1_suppl): 66S–75S, DOI: 10.1177/0963689719888619.

[33] Yoon, H.H., S.H. Bhang, J.Y. Shin, J. Shin, and B.S. Kim. Enhanced cartilage formation via three-dimensional cell engineering of human adipose-derived stem cells. Tissue Engineering Part A, 2012. 18(19-20): 1949–56, DOI: 10.1089/ten.TEA.2011.0647.

[34] Tu, V.T.-K., H.T.-N. Le, X.H.-V. To, P.D.-N. Nguyen, P.D. Huynh, T.M. Le, and N.B. Vu. Method for in production of cartilage from scaffold-free spheroids composed of human adipose-derived stem cells. Biomed Res Ther, 2020. 7(4): 3697–3708, DOI: 10.15419/bmrat.v7i4.597.

[35] Shen, F.H., B.C. Werner, H. Liang, H. Shang, N. Yang, X. Li, A.L. Shimer, G. Balian, and A.J. Katz. Implications of adipose-derived stromal cells in a 3D culture system for osteogenic differentiation: an *in vitro* and *in vivo* investigation. Spine Journal, 2013. 13(1): 32–43, DOI: 10.1016/j.spinee.2013.01.002.

[36] Tae, J.-y., S.-I. Lee, Y. Ko, and J. Park. The use of adipose-derived stem cells for the fabrication of three-dimensional spheroids for the osteogenic differentiation. Biomedical Research-tokyo, 2017. 28: 7098–7103.

[37] Bernard, A.B., C.C. Lin, and K.S. Anseth. A microwell cell culture platform for the aggregation of pancreatic β-cells. Tissue Eng Part C Methods, 2012. 18(8): 583–92, DOI: 10.1089/ten.TEC.2011.0504.

[38] Breslin, S., and L. O'Driscoll. Three-dimensional cell culture: the missing link in drug discovery. Drug Discovery Today, 2013. 18(5): 240–249, DOI: https://doi.org/10.1016/j.drudis.2012.10.003.

[39] Dogan, E., A. Kisim, G. Bati-Ayaz, G.J. Kubicek, D. Pesen-Okvur, and A.K. Miri. Cancer stem cells in tumor modeling: challenges and future directions. Advanced NanoBiomed Research, 2021. 1(11): 2100017, DOI: https://doi.org/10.1002/anbr.202100017.

[40] Kelm, J.M., and M. Fussenegger. Microscale tissue engineering using gravity-enforced cell assembly. Trends in Biotechnology, 2004. 22(4): 195–202, DOI: https://doi.org/10.1016/j.tibtech.2004.02.002.

[41] Thoma, C.R., M. Zimmermann, I. Agarkova, J.M. Kelm, and W. Krek. 3D cell culture systems modeling tumor growth determinants in cancer target discovery. Advanced Drug Delivery Reviews, 2014. 69-70: 29–41, DOI: https://doi.org/10.1016/j.addr.2014.03.001.

[42] Raghavan, S., P. Mehta, Y. Xie, Y.L. Lei, and G. Mehta. Ovarian cancer stem cells and macrophages reciprocally interact through the WNT pathway to promote pro-tumoral and malignant phenotypes in 3D engineered microenvironments. Journal for ImmunoTherapy of Cancer, 2019. 7(1): 190, DOI: 10.1186/s40425-019-0666-1.

[43] Sato, T., R.G. Vries, H.J. Snippert, M. van de Wetering, N. Barker, D.E. Stange, J.H. van Es, A. Abo, P. Kujala, P.J. Peters, and H. Clevers. Single Lgr5 stem cells build crypt–villus structures *in vitro* without a mesenchymal niche. Nature, 2009. 459: 262, DOI: 10.1038/nature07935. https://www.nature.com/articles/nature07935#supplementary-information.

[44] Sato, T., D.E. Stange, M. Ferrante, R.G.J. Vries, J.H. van Es, S. van den Brink, W.J. van Houdt, A. Pronk, J. van Gorp, P.D. Siersema, and H. Clevers. Long-term expansion of epithelial organoids from human colon, adenoma, adenocarcinoma, and Barrett's Epithelium. Gastroenterology, 2011. 141(5): 1762–1772, DOI: https://doi.org/10.1053/j.gastro.2011.07.050.

[45] Drost, J., and H. Clevers. Organoids in cancer research. Nature Reviews Cancer, 2018. 18(7): 407–418, DOI: 10.1038/s41568-018-0007-6.

[46] Daquinag, A.C., G.R. Souza and M.G. Kolonin. Adipose tissue engineering in three-dimensional levitation tissue culture system based on magnetic nanoparticles. Tissue Eng Part C Methods, 2013. 19(5): 336–44, DOI: 10.1089/ten.TEC.2012.0198.

[47] Taylor, J., J. Sellin, L. Kuerschner, L. Krähl, Y. Majlesain, I. Förster, C. Thiele, H. Weighardt, and E. Weber. Generation of immune cell containing adipose organoids for *in vitro* analysis of immune metabolism. Sci Rep, 2020. 10(1): 21104, DOI: 10.1038/s41598-020-78015-9.

[48] Smolar, J., M. Horst, S. Salemi, and D. Eberli. Predifferentiated smooth muscle-like adipose-derived stem cells for bladder engineering. Tissue Eng Part A, 2020. 26(17-18): 979–992, DOI: 10.1089/ten.tea.2019.0216.

[49] Todorov, A., M. Kreutz, A. Haumer, C. Scotti, A. Barbero, P.E. Bourgine, A. Scherberich, C. Jaquiery, and I. Martin. Fat-derived stromal vascular fraction cells enhance the bone-forming capacity of devitalized engineered hypertrophic cartilage matrix. Stem Cells Transl Med, 2016. 5(12): 1684–1694, DOI: 10.5966/sctm.2016-0006.

[50] Basak, O., J. Beumer, K. Wiebrands, H. Seno, A. van Oudenaarden, and H. Clevers. Induced quiescence of Lgr5+ stem cells in intestinal organoids enables differentiation of hormone-producing enteroendocrine cells. Cell Stem Cell, 2017. 20(2): 177–190.e4, DOI: https://doi.org/10.1016/j.stem.2016.11.001.

[51] Mithal, A., A. Capilla, D. Heinze, A. Berical, C. Villacorta-Martin, M. Vedaie, A. Jacob, K. Abo, A. Szymaniak, M. Peasley, A. Stuffer, J. Mahoney, D.N. Kotton, F. Hawkins, and G. Mostoslavsky. Generation of mesenchyme free intestinal organoids from human induced pluripotent stem cells. Nature Communications, 2020. 11(1): 215, DOI: 10.1038/s41467-019-13916-6.

[52] Panwar, A., P. Das, and L.P. Tan. 3D Hepatic Organoid-based advancements in LIVER tissue engineering. Bioengineering, 2021. 8(11): 185.

[53] Kozlowski, M.T., C.J. Crook, and H.T. Ku. Towards organoid culture without Matrigel. Communications Biology, 2021. 4(1): 1387, DOI: 10.1038/s42003-021-02910-8.

[54] Sackett, S.D., D.M. Tremmel, F. Ma, A.K. Feeney, R.M. Maguire, M.E. Brown, Y. Zhou, X. Li, C. O'Brien, L. Li, W.J. Burlingham, and J.S. Odorico. Extracellular matrix scaffold and hydrogel derived from decellularized and delipidized human pancreas. Scientific Reports, 2018. 8(1): 10452, DOI: 10.1038/s41598-018-28857-1.
[55] Ng, S.S., K. Saeb-Parsy, S.J.I. Blackford, J.M. Segal, M.P. Serra, M. Horcas-Lopez, D.Y. No; S. Mastoridis, W. Jassem, C.W. Frank, N.J. Cho, H. Nakauchi, J.S. Glenn, and S.T. Rashid. Human iPS derived progenitors bioengineered into liver organoids using an inverted colloidal crystal poly (ethylene glycol) scaffold. Biomaterials, 2018. 182: 299–311, DOI: https://doi.org/10.1016/j.biomaterials.2018.07.043.
[56] Cruz-Acuña, R., M. Quirós, A.E. Farkas, P.H. Dedhia, S. Huang, D. Siuda, V. García-Hernández, A.J. Miller, J.R. Spence, A. Nusrat, and A.J. García. Synthetic hydrogels for human intestinal organoid generation and colonic wound repair. Nature Cell Biology, 2017. 19(11): 1326–1335, DOI: 10.1038/ncb3632.
[57] Fernandez-Yague, M.A., L.A. Hymel, C.E. Olingy, C. McClain, M.E. Ogle, J.R. García, D. Minshew, S. Vyshnya, H.S. Lim, P. Qiu, A.J. García, and E.A. Botchwey. Analyzing immune response to engineered hydrogels by hierarchical clustering of inflammatory cell subsets. Science Advances, 2022. 8(8): eabd8056, DOI: doi:10.1126/sciadv.abd8056.
[58] Hunt, D.R., K.C. Klett, S. Mascharak, H. Wang, D. Gong, J. Lou, X. Li, P.C. Cai, R.A. Suhar, J.Y. Co, B.L. LeSavage, A.A. Foster, Y. Guan, M.R. Amieva, G. Peltz, Y. Xia, C.J. Kuo, and S.C. Heilshorn. Engineered matrices enable the culture of human patient-derived intestinal organoids. Advanced Science, 2021. 8(10): 2004705, DOI: https://doi.org/10.1002/advs.202004705.
[59] Zamboni, F., S. Vieira, R.L. Reis, J. Miguel Oliveira, and M.N. Collins. The potential of hyaluronic acid in immunoprotection and immunomodulation: Chemistry, processing and function. Progress in Materials Science, 2018. 97: 97–122, DOI: https://doi.org/10.1016/j.pmatsci.2018.04.003.
[60] Guo, N.N., L.P. Liu, Y.W. Zheng, and Y.M. Li. Inducing human induced pluripotent stem cell differentiation through embryoid bodies: A practical and stable approach. World J Stem Cells, 2020. 12(1): 25–34, DOI: 10.4252/wjsc.v12.i1.25.
[61] Lin, Y., and G. Chen. Embryoid body formation from human pluripotent stem cells in chemically defined E8 media. Harvard Stem Cell Institute, Cambridge (MA): 2008.
[62] Eiraku, M., K. Watanabe, M. Matsuo-Takasaki, M. Kawada, S. Yonemura, M. Matsumura, T. Wataya, A. Nishiyama, K. Muguruma, and Y. Sasai. Self-organized formation of polarized cortical tissues from ESCs and its active manipulation by extrinsic signals. Cell Stem Cell, 2008. 3(5): 519–532, DOI: https://doi.org/10.1016/j.stem.2008.09.002.
[63] Kadoshima, T., H. Sakaguchi, T. Nakano, M. Soen, S. Ando, M. Eiraku, and Y. Sasai. Self-organization of axial polarity, inside-out layer pattern, and species-specific progenitor dynamics in human ES cell–derived neocortex. Proceedings of the National Academy of Sciences, 2013. 110(50): 20284–20289, DOI: doi:10.1073/pnas.1315710110.
[64] Lancaster, M.A., M. Renner, C.-A. Martin, D. Wenzel, L.S. Bicknell, M.E. Hurles, T. Homfray, J.M. Penninger, A.P. Jackson, and J.A. Knoblich. Cerebral organoids model human brain development and microcephaly. Nature, 2013. 501(7467): 373–379, DOI: 10.1038/nature12517.
[65] Qian, X., Nguyen, Ha N., Song, Mingxi M., C. Hadiono, Ogden, Sarah C., C. Hammack, B. Yao, Hamersky, Gregory R., F. Jacob, C. Zhong, K.-j. Yoon; W. Jeang, L. Lin, Y. Li, J. Thakor, Berg, Daniel A., C. Zhang, E. Kang, M. Chickering, D. Nauen, C.-Y. Ho; Z. Wen, Christian, Kimberly M., P.-Y. Shi, Maher, J. Brady, H. Wu, P. Jin, H. Tang, H. Song, and G.-l. Ming. Brain-region-specific organoids using mini-bioreactors for modeling ZIKV exposure. Cell, 2016. 165(5): 1238–1254, DOI: https://doi.org/10.1016/j.cell.2016.04.032.
[66] Montel-Hagen, A., and G.M. Crooks. From pluripotent stem cells to T cells. Experimental Hematology, 2019. 71: 24–31, DOI: https://doi.org/10.1016/j.exphem.2018.12.001.
[67] Montel-Hagen, A., C.S. Seet, S. Li, B. Chick, Y. Zhu, P. Chang, S. Tsai, V. Sun, S. Lopez, H.-C. Chen, C. He, C.J. Chin, D. Casero, and G.M. Crooks. Organoid-induced differentiation of conventional T cells from human pluripotent stem cells. Cell Stem Cell, 2019. 24(3): 376–389.e8, DOI: https://doi.org/10.1016/j.stem.2018.12.011.
[68] Seet, C.S., C. He, M.T. Bethune, S. Li, B. Chick, E.H. Gschweng, Y. Zhu, K. Kim, D.B. Kohn, D. Baltimore, G.M. Crooks, and A. Montel-Hagen. Generation of mature T cells from human

hematopoietic stem and progenitor cells in artificial thymic organoids. Nat Methods, 2017. 14(5): 521–530, DOI: 10.1038/nmeth.4237.

[69] Bauer, S., C. Wennberg Huldt, K.P. Kanebratt, I. Durieux, D. Gunne, S. Andersson, L. Ewart, W.G. Haynes, I. Maschmeyer, A. Winter, C. Ämmälä, U. Marx, and, T.B. Andersson. Functional coupling of human pancreatic islets and liver spheroids on-a-chip: Towards a novel human ex vivo type 2 diabetes model. Scientific Reports, 2017. 7(1): 14620, DOI: 10.1038/s41598-017-14815-w.

[70] Velma T.E. Aho, Madelyn C. Houser, Pedro A.B. Pereira, Jianjun Chang, Knut Rudi, Lars Paulin, Vicki Hertzberg, Petri Auvinen, Malú G. Tansey and Filip Scheperjans. Relationships of gut microbiota, short-chain fatty acids, inflammation, and the gut barrier in Parkinson's disease. Molecular Neurodegeneration 2021. 16: 6.

[71] Pasca, S.P. The rise of three-dimensional human brain cultures. Nature, 2018. 553(7689): 437–445, DOI: 10.1038/nature25032.

[72] Ben-David, U., O. Kopper, and N. Benvenisty. Expanding the boundaries of embryonic stem cells. Cell Stem Cell, 2012. 10(6): 666–677, DOI: 10.1016/j.stem.2012.05.003.

[73] Bar-Nur, O., H.A. Russ, S. Efrat, and N. Benvenisty. Epigenetic memory and preferential lineage-specific differentiation in induced pluripotent stem cells derived from human pancreatic islet beta cells. Cell Stem Cell, 2011. 9(1): 17–23, DOI: 10.1016/j.stem.2011.06.007.

[74] Kim, K., R. Zhao, A. Doi, K. Ng, J. Unternaehrer, P. Cahan, H. Hongguang, Y.-H. Loh, M.J. Aryee, M.W. Lensch, H. Li, J.J. Collins, A.P. Feinberg, and G.Q. Daley. Donor cell type can influence the epigenome and differentiation potential of human induced pluripotent stem cells. Nature Biotechnology, 2011. 29(12): 1117–1119, DOI: 10.1038/nbt.2052.

[75] Vincent, F., P. Loria, M. Pregel, R. Stanton, L. Kitching, K. Nocka, R. Doyonnas, C. Steppan, A. Gilbert, T. Schroeter, and M.-C. Peakman. Developing predictive assays: The phenotypic screening “rule of 3&#x201d. Science Translational Medicine, 2015. 7(293): 293ps15-293ps15, DOI: doi:10.1126/scitranslmed.aab1201.

[76] Inoue, H., N. Nagata, H. Kurokawa, and S. Yamanaka. iPS cells: a game changer for future medicine. The EMBO Journal, 2014. 33(5): 409–417, DOI: https://doi.org/10.1002/embj.201387098.

[77] Matsa, E., Burridge, Paul W., K.-H. Yu, Ahrens, John H., V. Termglinchan, H. Wu, C. Liu, P. Shukla, N. Sayed, Churko Jared M., N. Shao, Woo, Nicole A., Chao, Alexander S., Gold, Joseph D., I. Karakikes, Snyder, Michael P., and Wu, Joseph C. Transcriptome profiling of patient-specific human iPSC-cardiomyocytes predicts individual drug safety and efficacy responses *in vitro*. Cell Stem Cell, 2016. 19(3): 311–325, DOI: https://doi.org/10.1016/j.stem.2016.07.006.

[78] Lee, G., C.N. Ramirez, H. Kim, N. Zeltner, B. Liu, C. Radu, B. Bhinder, Y.J. Kim, I Y. Choi, B. Mukherjee-Clavin, H. Djaballah, and L. Studer. Large-scale screening using familial dysautonomia induced pluripotent stem cells identifies compounds that rescue IKBKAP expression. Nat Biotechnol, 2012. 30(12): 1244–8, DOI: 10.1038/nbt.2435.

[79] Naryshkin, N.A., M. Weetall, A. Dakka, J. Narasimhan, X. Zhao, Z. Feng, K.K.Y. Ling, G.M. Karp, H. Qi, M.G. Woll, G. Chen, N. Zhang, V. Gabbeta, P. Vazirani, A. Bhattacharyya, B. Furia, N. Risher, J. Sheedy, R. Kong, J. Ma, A. Turpoff, C.-S. Lee, X. Zhang, Y.-C. Moon, P. Trifillis, E.M. Welch, J.M. Colacino, J. Babiak, N.G. Almstead, S.W. Peltz, L.A. Eng, K.S. Chen, J.L. Mull, M.S. Lynes, L.L. Rubin, P. Fontoura, L. Santarelli, D. Haehnke, K.D. McCarthy, R. Schmucki, M. Ebeling, M. Sivaramakrishnan, C.-P. Ko, S.V. Paushkin, H. Ratni, I. Gerlach, A. Ghosh, and F. Metzger. <i>SMN2</i> splicing modifiers improve motor function and longevity in mice with spinal muscular atrophy. Science, 2014. 345(6197): 688–693, DOI: doi:10.1126/science.1250127.

[80] Yamashita, A., M. Morioka, H. Kishi, T. Kimura, Y. Yahara, M. Okada, K Fujita, H. Sawai, S. Ikegawa, and N. Tsumaki. Statin treatment rescues FGFR3 skeletal dysplasia phenotypes. Nature, 2014. 513(7519): 507–511, DOI: 10.1038/nature13775.

[81] McNeish, J., Gardner, Jason P., Wainger, Brian J., Woolf, Clifford J., and K. Eggan. From dish to bedside: lessons learned while translating findings from a stem cell model of disease to a clinical trial. Cell Stem Cell, 2015. 17(1): 8–10, DOI: https://doi.org/10.1016/j.stem.2015.06.013.

[82] Wainger, B.J., E. Kiskinis, C. Mellin, O. Wiskow, S.S. Han, J. Sandoe, N.P. Perez, L.A. Williams, S. Lee, G. Boulting, J.D. Berry, R.H., Jr. Brown, M.E. Cudkowicz, B.P. Bean, K. Eggan, and C.J. Woolf. Intrinsic membrane hyperexcitability of amyotrophic lateral sclerosis patient-derived motor neurons. Cell Rep, 2014. 7(1): 1–11, DOI: 10.1016/j.celrep.2014.03.019.

[83] Hong, J., J.B. Edel, and A.J. deMello. Micro- and nanofluidic systems for high-throughput biological screening. Drug Discovery Today, 2009. 14(3-4): 134–46, DOI: 10.1016/j.drudis.2008.10.001.
[84] Zhao, L., Z. Wang, S. Fan, Q. Meng, B. Li, S. Shao, and Q. Wang. Chemotherapy resistance research of lung cancer based on micro-fluidic chip system with flow medium. Biomedical Microdevices, 2010. 12(2): 325–332, DOI: 10.1007/s10544-009-9388-3.
[85] Lai, N., J.K. Sims, N.L. Jeon, and K. Lee. Adipocyte induction of preadipocyte differentiation in a gradient chamber. tissue engineering part C: Methods, 2012. 18(12): 958–967, DOI: 10.1089/ten.tec.2012.0168.
[86] Clark, A.M., K.M. Sousa, C. Jennings, O.A. MacDougald, and R.T. Kennedy. Continuous-flow enzyme assay on a microfluidic chip for monitoring glycerol secretion from cultured adipocytes. Analytical Chemistry, 2009. 81(6): 2350–2356, DOI: 10.1021/ac8026965.
[87] Dugan, C.E., J.P. Grinias, S.D. Parlee, M. El-Azzouny, C.R. Evans, and R.T. Kennedy. Monitoring cell secretions on microfluidic chips using solid-phase extraction with mass spectrometry. Analytical and Bioanalytical Chemistry, 2017. 409(1): 169–178, DOI: 10.1007/s00216-016-9983-0.
[88] Wu, X., N. Schneider, A. Platen, I. Mitra, M. Blazek, R. Zengerle, R. Schule, M. Meier. *In situ* characterization of the mTORC1 during adipogenesis of human adult stem cells on chip. Proc Natl Acad Sci U S A, 2016. 113(29): E4143–50, DOI: 10.1073/pnas.1601207113.
[89] Tanataweethum, N., A. Zelaya, F. Yang, R.N. Cohen, E.M. Brey, and A. Bhushan. Establishment and characterization of a primary murine adipose tissue-chip. Biotechnol Bioeng, 2018. 115(8): 1979–1987, DOI: https://doi.org/10.1002/bit.26711.
[90] Godwin, L.A., J.C. Brooks, L.D. Hoepfner, D. Wanders, R.L. Judd, and C.J. Easley. A microfluidic interface for the culture and sampling of adiponectin from primary adipocytes. Analyst, 2015. 140(4): 1019–1025, DOI: 10.1039/c4an01725k.
[91] Giobbe, G.G., F. Michielin, C. Luni, S. Giulitti, S. Martewicz, S. Dupont, A. and Floreani, N. Elvassore. Functional differentiation of human pluripotent stem cells on a chip. Nature Methods, 2015. 12(7): 637–640, DOI: 10.1038/nmeth.3411.
[92] Guenat, O.T., and F. Berthiaume. Incorporating mechanical strain in organs-on-a-chip: Lung and skin. Biomicrofluidics, 2018. 12(4): 042207, DOI: 10.1063/1.5024895.
[93] Wan, C.R., S. Chung, and R.D. Kamm. Differentiation of embryonic stem cells into cardiomyocytes in a compliant microfluidic system. Ann Biomed Eng, 2011. 39(6): 1840–7, DOI: 10.1007/s10439-011-0275-8.
[94] Huh, D., B.D. Matthews, A. Mammoto, M. Montoya-Zavala, H.Y. Hsin, and D.E. Ingber. Reconstituting organ-level lung functions on a chip. Science, 2010. 328(5986): 1662–1668, DOI: 10.1126/science.1188302.
[95] Stucki, A.O., J.D. Stucki, S.R.R. Hall, M. Felder, Y. Mermoud, R.A. Schmid, T. Geiser, and O.T. Guenat. A lung-on-a-chip array with an integrated bio-inspired respiration mechanism. Lab on a Chip, 2015. 15(5): 1302–1310, DOI: 10.1039/C4LC01252F.
[96] Dias, A.D., A.M. Unser, Y. Xie, D.B. Chrisey, and D.T. Corr. Generating size-controlled embryoid bodies using laser direct-write. Biofabrication, 2014. 6(2): 025007, DOI: 10.1088/1758-5082/6/2/025007.
[97] Zhang, Y.S., A. Arneri, S. Bersini, S.-R. Shin, K. Zhu, Z. Goli-Malekabadi, J. Aleman, C. Colosi, F. Busignani, V. Dell'Erba, C. Bishop, T. Shupe, D. Demarchi, M. Moretti, M. Rasponi, M.R. Dokmeci, A. Atala, and A. Khademhosseini. Bioprinting 3D microfibrous scaffolds for engineering endothelialized myocardium and heart-on-a-chip. Biomaterials, 2016. 110: 45–59, DOI: https://doi.org/10.1016/j.biomaterials.2016.09.003.
[98] Ong, C.S., T. Fukunishi, A. Nashed, A. Blazeski, H. Zhang, S. Hardy, D. DiSilvestre, L. Vricella, J. Conte, L. Tung, G. Tomaselli, and N. Hibino. Creation of cardiac tissue exhibiting mechanical integration of spheroids using 3D bioprinting. J Vis Exp, 2017. (125), DOI: 10.3791/55438.
[99] Faulkner-Jones, A., C. Fyfe, D.-J. Cornelissen, J. Gardner, J. King, A. Courtney, and W. Shu. Bioprinting of human pluripotent stem cells and their directed differentiation into hepatocyte-like cells for the generation of mini-livers in 3D. Biofabrication, 2015. 7(4): 044102, DOI: 10.1088/1758-5090/7/4/044102.
[100] Ma, X., X. Qu, W. Zhu, Y.-S. Li, S. Yuan, H. Zhang, J. Liu, P. Wang, C.S.E. Lai, F. Zanella, G.-S. Feng, F. Sheikh, S. Chien, and S. Chen. Deterministically patterned biomimetic human

iPSC-derived hepatic model via rapid 3D bioprinting. Proceedings of the National Academy of Sciences, 2016. 113(8): 2206–2211, DOI: 10.1073/pnas.1524510113.

[101] Ryu, S., J. Yoo, Y. Jang, J. Han, S.J. Yu, J. Park, S.Y. Jung, K.H. Ahn, S.G. Im, K. Char, and B.-S. Kim. Nanothin coculture membranes with tunable pore architecture and thermoresponsive functionality for transfer-printable stem cell-derived cardiac sheets. ACS Nano, 2015. 9(10): 10186–10202, DOI: 10.1021/acsnano.5b03823.

[102] Hsieh, F.-Y., and S.-h. Hsu. 3D bioprinting: A new insight into the therapeutic strategy of neural tissue regeneration. Organogenesis, 2015. 11(4): 153–158, DOI: 10.1080/15476278.2015.1123360.

[103] Gu, Q., E. Tomaskovic-Crook, R. Lozano, Y. Chen, R.M. Kapsa, Q. Zhou, G.G. Wallace, and J.M. Crook. Functional 3D neural mini-tissues from printed gel-based bioink and human neural stem cells. Advanced Healthcare Materials, 2016. 5(12): 1429–1438, DOI: https://doi.org/10.1002/adhm.201600095.

[104] Ong, C.S., P. Yesantharao, C.Y. Huang, G. Mattson, J. Boktor, T. Fukunishi, H. Zhang, and N. Hibino. 3D bioprinting using stem cells. Pediatric Research, 2018. 83(1): 223–231, DOI: 10.1038/pr.2017.252.

[105] Kolesky, D.B., K.A. Homan, M.A. Skylar-Scott, and J.A. Lewis. Three-dimensional bioprinting of thick vascularized tissues. Proceedings of the National Academy of Sciences, 2016. 113(12): 3179–3184, DOI: doi:10.1073/pnas.1521342113.

[106] Maiullari, F., M. Costantini, M. Milan, V. Pace, M. Chirivì, S. Maiullari, A. Rainer, D. Baci, H.E. Marei, D. Seliktar, C. Gargioli, C. Bearzi, and R. Rizzi. A multi-cellular 3D bioprinting approach for vascularized heart tissue engineering based on HUVECs and iPSC-derived cardiomyocytes. Sci Rep, 2018. 8(1): 13532, DOI: 10.1038/s41598-018-31848-x.

[107] Skylar-Scott, M.A., S.G.M. Uzel, L.L. Nam, J.H. Ahrens, R.L. Truby, S. Damaraju, and J.A. Lewis. Biomanufacturing of organ-specific tissues with high cellular density and embedded vascular channels. Sci Adv, 2019. 5(9): eaaw2459, DOI: 10.1126/sciadv.aaw2459.

[108] Bhatia, S.N., and D.E. Microfluidic organs-on-chips. Nat Biotechnol, 2014. 32(8): 760–772, DOI: 10.1038/nbt.2989.

[109] Huh, D., B.D. Matthews, A. Mammoto, M. Montoya-Zavala, H.Y. Hsin, and D.E. Ingber. Reconstituting organ-level lung functions on a chip. Science, 2010. 328(5986): 1662–8, DOI: 10.1126/science.1188302.

[110] Powers, M.J., K. Domansky, M.R. Kaazempur-Mofrad, A. Kalezi, A. Capitano, A. Upadhyaya, P. Kurzawski, K.E. Wack, D.B. Stolz, R. Kamm, and L.G. Griffith. A microfabricated array bioreactor for perfused 3D liver culture. Biotechnol Bioeng, 2002. 78(3): 257–69, DOI: 10.1002/bit.10143.

[111] Kimura, H., T. Yamamoto, H. Sakai, Y. Sakai, and T. Fujii. An integrated microfluidic system for long-term perfusion culture and on-line monitoring of intestinal tissue models. Lab Chip, 2008. 8(5): 741–6, DOI: 10.1039/b717091b.

[112] Jang, K.J., and K.Y. Suh. A multi-layer microfluidic device for efficient culture and analysis of renal tubular cells. Lab Chip 2010, 10(1): 36–42, DOI: 10.1039/b907515a.

[113] Pamies, D., T. Hartung, and H.T. Hogberg. Biological and medical applications of a brain-on-a-chip. Exp Biol Med (Maywood), 2014. 239(9): 1096–1107, DOI: 10.1177/1535370214537738.

[114] Liu, Y.C., I.C. Lee, and K.F. Lei. Toward the development of an artificial brain on a micropatterned and material-regulated biochip by guiding and promoting the differentiation and neurite outgrowth of neural stem/progenitor cells. ACS Appl Mater Interfaces, 2018. 10(6): 5269–5277, DOI: 10.1021/acsami.7b17863.

[115] Park, J., B.K. Lee, G.S. Jeong, J.K. Hyun, C.J. Lee, and S.H. Lee. Three-dimensional brain-on-a-chip with an interstitial level of flow and its application as an *in vitro* model of Alzheimer's disease. Lab Chip, 2015. 15(1): 141–50, DOI: 10.1039/c4lc00962b.

[116] Moreno, E.L., S. Hachi, K. Hemmer, S.J. Trietsch, A.S. Baumuratov, T. Hankemeier, P. Vulto, J.C. Schwamborn, and R.M. Fleming. Differentiation of neuroepithelial stem cells into functional dopaminergic neurons in 3D microfluidic cell culture. Lab Chip, 2015. 15(11): 2419–28, DOI: 10.1039/c5lc00180c.

[117] Kerman, B.E., H.J. Kim, K. Padmanabhan, A. Mei, S. Georges, M.S. Joens, J.A. Fitzpatrick, R. Jappelli, K.J. Chandross, P. August, and F.H. Gage. *In vitro* myelin formation using embryonic stem cells. Development, 2015. 142(12): 2213–25, DOI: 10.1242/dev.116517.

[118] Kamei, K.-i., M. Yoshioka, S. Terada, Y. Tokunaga, and Y. Chen. Three-dimensional cultured liver-on-a-Chip with mature hepatocyte-like cells derived from human pluripotent stem cells. Biomedical Microdevices, 2019. 21(3): 73, DOI: 10.1007/s10544-019-0423-8.

[119] Banaeiyan, A.A., J. Theobald, J. Paukštyte, S. Wölfl, C.B. Adiels, and M. Goksör. Design and fabrication of a scalable liver-lobule-on-a-chip microphysiological platform. Biofabrication, 2017. 9(1): 015014, DOI: 10.1088/1758-5090/9/1/015014.

[120] Wang, Y., H. Wang, P. Deng, W. Chen, Y. Guo, T. Tao, and J. Qin. *In situ* differentiation and generation of functional liver organoids from human iPSCs in a 3D perfusable chip system. Lab on a Chip, 2018. 18(23): 3606–3616, DOI: 10.1039/C8LC00869H.

[121] Wang, Y., H. Wang, P. Deng, T. Tao, H. Liu, S. Wu, W. Chen, and J. Qin. Modeling human nonalcoholic fatty liver disease (NAFLD) with an Organoids-on-a-Chip System. ACS Biomaterials Science & Engineering, 2020. 6(10): 5734–5743, DOI: 10.1021/acsbiomaterials.0c00682.

[122] Sato, T., R.G. Vries, H.J. Snippert, M. van de Wetering, N. Barker, D.E. Stange, J.H. van Es, A. Abo, P. Kujala, P.J. Peters, and H. Clevers. Single Lgr5 stem cells build crypt-villus structures *in vitro* without a mesenchymal niche. Nature, 2009. 459(7244): 262–265, DOI: 10.1038/nature07935.

[123] Arık, Y.B., M.W.v.d. Helm, M. Odijk, L.I. Segerink, R. Passier, A.v.d. Berg, and A.D.v.d. Meer. Barriers-on-chips: Measurement of barrier function of tissues in organs-on-chips. Biomicrofluidics, 2018. 12(4): 042218, DOI: 10.1063/1.5023041.

[124] Park, T.-E., N. Mustafaoglu, A. Herland, R. Hasselkus, R. Mannix, E.A. FitzGerald, R. Prantil-Baun, A. Watters, O. Henry, M. Benz, H. Sanchez, H.J. McCrea, L.C. Goumnerova, H.W. Song, S.P. Palecek, E. Shusta, and D.E. Ingber. Hypoxia-enhanced Blood-Brain Barrier Chip recapitulates human barrier function and shuttling of drugs and antibodies. Nature Communications, 2019. 10(1): 2621, DOI: 10.1038/s41467-019-10588-0.

[125] Lee, S.H., J.H. Sung. Organ-on-a-chip technology for reproducing multiorgan physiology. Adv Healthc Mater, 2018. 7(2), DOI: 10.1002/adhm.201700419.

[126] Moreno, E.L., S. Hachi, K. Hemmer, S.J. Trietsch, A.S. Baumuratov, T. Hankemeier, P. Vulto, J.C. Schwamborn, and R.M.T. Fleming. Differentiation of neuroepithelial stem cells into functional dopaminergic neurons in 3D microfluidic cell culture. Lab on a Chip, 2015. 15(11): 2419–2428, DOI: 10.1039/C5LC00180C.

[127] Wang, G., M.L. McCain, L. Yang, A. He, F.S. Pasqualini, A. Agarwal, H. Yuan, D. Jiang, D. Zhang, L. Zangi, J. Geva, A.E. Roberts, Q. Ma, J. Ding, J. Chen, D.-Z. Wang, K. Li, J. Wang, R.J.A. Wanders, W. Kulik, F.M. Vaz, M.A. Laflamme, C.E. Murry, K.R. Chien, R.I. Kelley, G.M. Church, K.K. Parker, and W.T. Pu. Modeling the mitochondrial cardiomyopathy of Barth syndrome with induced pluripotent stem cell and heart-on-chip technologies. Nature Medicine, 2014. 20(6): 616–623, DOI: 10.1038/nm.3545.

[128] Loskill, P., T. Sezhian, K.M. Tharp, F.T. Lee-Montiel, S. Jeeawoody, W.M. Reese, P.H. Zushin, A. Stahl, and K.E. Healy. WAT-on-a-chip: a physiologically relevant microfluidic system incorporating white adipose tissue. Lab Chip, 2017. 17(9): 1645–1654, DOI: 10.1039/c6lc01590e.

[129] Tanataweethum, N., F. Zhong, A. Trang, C. Lee, R.N. Cohen, and A. Bhushan. Towards an insulin resistant adipose model on a chip. Cell Mol Bioeng, 2021. 14(1): 89–99, DOI: 10.1007/s12195-020-00636-x.

[130] Zhu, J., J. He, M. Verano, A.T. Brimmo, A. Glia, M.A. Qasaimeh, P. Chen, J.O. Aleman, and W. Chen. An integrated adipose-tissue-on-chip nanoplasmonic biosensing platform for investigating obesity-associated inflammation. Lab Chip, 2018. 18(23): 3550–3560, DOI: 10.1039/c8lc00605a.

[131] Liu, Y., P. Kongsuphol, S.Y. Chiam, Q.X. Zhang, S.B.N. Gourikutty, S. Saha, S.K. Biswas, and Q. Ramadan. Adipose-on-a-chip: a dynamic microphysiological in vitro model of the human adipose for immune-metabolic analysis in type II diabetes. Lab Chip, 2019. 19(2): 241–253, DOI: 10.1039/c8lc00481a.

[132] Li, Y.-C.E., and I.-C. Lee. The current trends of biosensors in tissue engineering. Biosensors, 2020. 10(8): 88.

[133] Chowdhury, R.A., K.N. Tzortzis, E. Dupont, S. Selvadurai, F. Perbellini, C.D. Cantwell, F.S. Ng, A.R. Simon, C.M. Terracciano, and N.S. Peters. Concurrent micro- to macro-cardiac electrophysiology in myocyte cultures and human heart slices. Sci Rep, 2018. 8(1): 6947–6947, DOI: 10.1038/s41598-018-25170-9.

[134] Knöll, R., M. Hoshijima, H.M. Hoffman, V. Person, I. Lorenzen-Schmidt, M.-L. Bang, T. Hayashi, N. Shiga, H. Yasukawa, W. Schaper, W. McKenna, M. Yokoyama, N.J. Schork, J.H. Omens, A.D. McCulloch, A. Kimura, C.C. Gregorio, W. Poller, J. Schaper, H.P. Schultheiss, and K.R. Chien. The cardiac mechanical stretch sensor machinery involves a Z disc complex that is defective in a subset of human dilated cardiomyopathy. Cell, 2002. 111(7): 943–955, DOI: https://doi.org/10.1016/S0092-8674(02)01226-6.

[135] Caluori, G., J. Pribyl, M. Pesl, S. Jelinkova, V. Rotrekl, P. Skladal, and R. Raiteri. Non-invasive electromechanical cell-based biosensors for improved investigation of 3D cardiac models. Biosensors and Bioelectronics, 2019. 124-125: 129–135, DOI: https://doi.org/10.1016/j.bios.2018.10.021.

[136] Seymour, J.P., F. Wu, K.D. Wise, and E. Yoon. State-of-the-art MEMS and microsystem tools for brain research. Microsystems & Nanoengineering, 2017. 3(1): 16066, DOI: 10.1038/micronano.2016.66.

[137] Chiappalone, M., A. Vato, M. Tedesco, M. Marcoli, F. Davide, and S. Martinoia. Networks of neurons coupled to microelectrode arrays: a neuronal sensory system for pharmacological applications. Biosensors and Bioelectronics, 2003. 18(5): 627–634, DOI: https://doi.org/10.1016/S0956-5663(03)00041-1.

[138] Lourenço, C.F., A. Ledo, G.A. Gerhardt, J. Laranjinha, and R.M. Barbosa. Neurometabolic and electrophysiological changes during cortical spreading depolarization: multimodal approach based on a lactate-glucose dual microbiosensor arrays. Sci Rep, 2017. 7(1): 6764, DOI: 10.1038/s41598-017-07119-6.

[139] Liu, Y.-C., I.C. Lee, and K.F. Lei. Toward the development of an artificial brain on a micropatterned and material-regulated biochip by guiding and promoting the differentiation and neurite outgrowth of neural stem/progenitor cells. ACS Applied Materials & Interfaces, 2018. 10(6): 5269–5277, DOI: 10.1021/acsami.7b17863.

[140] Liu, L., X. Xiao, K.F. Lei, and C.-H. Huang. Quantitative impedimetric monitoring of cell migration under the stimulation of cytokine or anti-cancer drug in a microfluidic chip. Biomicrofluidics, 2015. 9(3): 034109, DOI: 10.1063/1.4922488.

[141] Lei, K.F., H.-P. Tseng, C.-Y. Lee, and N.-M. Tsang. Quantitative study of cell invasion process under extracellular stimulation of cytokine in a microfluidic device. Scientific Reports, 2016. 6: 25557, DOI: 10.1038/srep25557.

[142] Lei, K.F., B.-Y. Lin, and N.-M. Tsang. Real-time and label-free impedimetric analysis of the formation and drug testing of tumor spheroids formed via the liquid overlay technique. RSC Advances, 2017. 7(23): 13939–13946, DOI: 10.1039/C7RA00209B.

[143] Vomero, M., E. Castagnola, F. Ciarpella, E. Maggiolini, N. Goshi, E. Zucchini, S. Carli, L. Fadiga, S. Kassegne, and D. Ricci. Highly stable glassy carbon interfaces for long-term neural stimulation and low-noise recording of brain activity. Sci Rep, 2017. 7(1): 40332, DOI: 10.1038/srep40332.

[144] Li, X., M. Soler, C.I. Özdemir, A. Belushkin, F. Yesilköy, and H. Altug. Plasmonic nanohole array biosensor for label-free and real-time analysis of live cell secretion. Lab on a Chip, 2017. 17(13): 2208–2217, DOI: 10.1039/C7LC00277G.

[145] Li, S.S., C.W. Lin, K.C. Wei, C.Y. Huang, P.H. Hsu, H.L. Liu, Y.J. Lu, S.C. Lin, H.W. Yang, and C.C. Ma. Non-invasive screening for early Alzheimer's disease diagnosis by a sensitively immunomagnetic biosensor. Sci Rep, 2016. 6: 25155, DOI: 10.1038/srep25155.

[146] Yang, X., H. Li, X. Zhao, W. Liao, C.X. Zhang, and Z. Yang. A novel, label-free liquid crystal biosensor for Parkinson's disease related alpha-synuclein. Chem Commun (Camb), 2020. 56(40): 5441–5444, DOI: 10.1039/d0cc01025a.

[147] Duranteau, J., N.S. Chandel, A. Kulisz, Z. Shao, and P.T. Schumacker. Intracellular signaling by reactive oxygen species during hypoxia in cardiomyocytes. J Biol Chem,1998. 273(19): 11619–24. DOI: 10.1074/jbc.273.19.11619.

[148] Dutta, S., A. Minchole, T.A. Quinn, and B. Rodriguez. Electrophysiological properties of computational human ventricular cell action potential models under acute ischemic conditions. Prog Biophys Mol Biol, 2017. 129: 40–52, DOI: 10.1016/j.pbiomolbio.2017.02.007.

[149] Lee, S.P., G. Ha, D.E. Wright, Y. Ma, E. Sen-Gupta, N.R. Haubrich, P.C. Branche, W. Li, G.L. Huppert, M. Johnson, H.B. Mutlu, K. Li, N. Sheth, J.A. Wright Jr., Y. Huang, M. Mansour, J.A. Rogers, and R. Ghaffari. Highly flexible, wearable, and disposable cardiac biosensors for remote and ambulatory monitoring. NPJ Digit Med, 2018. 1: 2, DOI: 10.1038/s41746-017-0009-x.

[150] Liu, H., O.A. Bolonduro, N. Hu, J. Ju, A.A. Rao, B.M. Duffy, Z. Huang, L.D. Black, and B.P. Timko. Heart-on-a-chip model with integrated extra- and intracellular bioelectronics for monitoring cardiac electrophysiology under acute hypoxia. Nano Lett, 2020. 20(4): 2585–2593, DOI: 10.1021/acs.nanolett.0c00076.

[151] Bavli, D., S. Prill, E. Ezra, G. Levy, M. Cohen, M. Vinken, J. Vanfleteren, M. Jaeger, and Y. Nahmias. Real-time monitoring of metabolic function in liver-on-chip microdevices tracks the dynamics of mitochondrial dysfunction. Proc Natl Acad Sci U S A, 2016. 113(16): E2231-40, DOI: 10.1073/pnas.1522556113.

8

The Nanotechnology for Stem Cells

Yi-Chen Ethan Li,[1,*] *I-Chi Lee*,[2,*] *Nai-Chen Cheng*,[3] *Wen-Yen Huang*,[4] *Sung-Jan Lin*,[4] *Chia-Ning Shen*[5] and *Min-Huey Chen*[6,*]

Nanotechnology is an interdisciplinary science that integrates physics, chemistry, engineering, biology, life sciences, and medicine [1]. Through the exchange of knowledge and technology between various disciplines, nanotechnology has successfully bridged the gap of the traditional research between various disciplines and then established multidisciplinary research technologies and knowledge platforms. In general, nanotechnology refers to the creation of novel physical, chemical, and biological structures by controlling the size and shape of objects in the nanoscale (10^{-9} m) range utilizing design, characterization, and manufacturing, equipment, and systems expertise.

In recent years, nanotechnology applications have been extended from basic disciplines to the field of biomedicine. In order to meet the rapid increase in the demand for nanotechnology in various areas, the development of miniaturization technology and materials has shifted from the micron size to the nanoscale level [2]. Compared with traditional biomaterials, nanobiotechnology and nanomaterials possess many unique physical, chemical, and biological properties [3]. Therefore, it is widely used in the fields of multidisciplinary research and industry.

[1] Department of Chemical Engineering, Feng Chia University, Taiwan.
[2] Department of Biomedical Engineering and Environmental Sciences, National Tsing Hua University, Taiwan.
[3] Department of Surgery, National Taiwan University Hospital, Taiwan.
[4] Department of Biomedical Engineering, College of Medicine and College of Engineering, National Taiwan University, Taiwan.
[5] Genomics Research Center, Academia Sinica, Taiwan.
[6] Graduate Institute of Clinical Dentistry, School of Dentistry, National Taiwan University, Taiwan.
* Corresponding authors: iclee@mx.nthu.edu.tw; yicli@fcu.edu.tw

In recent years, biomedicine has integrated the knowledge and techniques of various areas and evolved into a new study field with high development potential. Stem cell biology, especially, is the subject of active research by scientists. Nanotechnology brings a promising new strategy to diagnosis and cell therapy applications. Therefore, stem cell research has greatly increased attention and development in the past two decades due to the rapid development of nanotechnology [4].

Nanotechnology applications play a new and important role in biology mainly due to the unique properties of nanomaterials, such as specificity, biocompatibility, water solubility, and low cytotoxicity. Therefore, nanomaterials can be widely used in *in vivo* experiments and reduce unexpected damage or injury to cells. Therefore, using the properties of nanomaterials to assist in the development of *in vitro* stem cell systems and their application in medical diagnosis or treatment has gradually developed into a rather forward-looking field of biomedical research [5].

In general, nanomaterials are the foundation of nanoscience, and the structure and size of nanomaterials generally refer to powders, granules, fibers, films, or bulks with grain sizes between 1–100 nm in structure. Due to the large surface area and special quantum effects of nanomaterials, their optical, thermal, electrical, magnetic, mechanical, and even chemical properties are very different from the same materials at the micron size. After traditional materials are nano-processed, many different effects from micro/macro materials will occur, such as surface effect, that is to say, the change of chemical activity, optical, thermal properties, etc. [6]. In addition, the quantum size effects are defined in optics and magnetism, meaning that the band gap of the valence band and the energy band between the metal and the semiconductor is widened, and the characteristic of insulation is exhibited [7].

Different nanomaterials will exhibit versatile properties and effects under various external field effects (such as electric or magnetic fields). Through the development and advancement of nanobiotechnology, the applications of nanomaterials have been extended to semiconductor technology, electronic materials, optical materials, and biomedical materials. For example, gold nanoparticles can be used in DNA diagnosis, and quantum dots prepared from cadmium selenium (CdSe) will have stronger fluorescence intensity, narrower emission spectrum, and longer characteristics such as half-life [8]; therefore, it can be applied to the development of molecular image analysis. In addition, iron oxide nanoparticles have been used in the diagnosis of magnetic resonance imaging (MRI) in recent years because of their superparamagnetic properties [9–11]. Furthermore, due to the vigorous development of nanotechnology, there have also been major breakthroughs in traditional medicine, especially in developing innovative biomedical nanomaterials. For example, liposomes and dendrimers provide novel mechanisms of drug delivery and control release, which can be applied to assist in diagnosis and improve treatment methods [12, 13]. Therefore, many diseases are cured by surgery only, which can be treated with nanomaterials in the future. So, clinicians can use less invasive treatment methods to reduce patients' dependence on surgical treatments and reduce the demand for postoperative patient rehabilitation. Currently, nanomaterials have been widely used in different medical research fields, among which the more common applications include: contrast agents, molecule images, molecule diagnosis, biosensors, bio-targeting, drug control release, and others [14]. Among them, the most representative ones in medical applications are optical molecular, cellular imaging diagnosis, and MRI contrast agents.

First, for detecting optical molecules and cell/stem cell images, in recent years, nanotechnology can create new fluorescent labeling nanomaterials by modifying the surface of semiconductors with quantum dots and biorecognition molecules [15]. Compared with conventional organic dyes, these quantum dots and biorecognition molecules have better water solubility and biocompatibility, and the wavelength of luminescence can be adjusted by regulating the size of the nanometer [16]. Therefore, using a single excitation light source can obtain different luminescent colors by exciting quantum dots of various sizes. Importantly, the advantage of quantum dots lies in their longer luminescence lifetime and negligible photobleaching; therefore, controlling time-gating can solve the problem of images with autofluorescence in living animal samples or living stem cell samples [17]. In addition, since quantum dots have narrow emission bands, their luminescence spectrum can be adjusted by modulating the composition and physical dimensions of nanomaterials [18]. Based on the abovementioned optical properties, quantum dots have gradually replaced traditional organic dyes in the technical field of stem cell labeling or living molecular imaging.

Many technologies using quantum dot bioanalysis methods have been gradually developed. For example, antibody-conjugated quantum dots are a new reagent for conjugating quantum dots to antibodies [19]. If injected into mice by intravenous injection, images of live stem cells can be captured directly. Through the conjugation of antibodies and antigens, the quantum dots can be highly specifically adsorbed on the cell surface of specific stem cells to achieve the function of cell targeting. Using quantum dots can improve the sensitivity of stem cell images and track stem cells or cancer cells in the *in vivo* and *in vitro* culture systems [20].

In addition, another nanomaterial often used for biomedical imaging observation is gold nanoparticles. Using surface plasmon resonance, called SPR technology, gold nanoparticles can be used for colorimetric contrast applications. The factors that determine the SPR properties of these gold nanoparticles are particle shape, size, dielectric properties, solvent, ligan, aggregate morphology, surface functionalization, and refractive index of the surrounding liquid [21]. Gold nanoparticles have better stability and biocompatibility than other reagents and do not produce the photo decomposition usually occurs in common fluorescent dyes. A gold nanoparticle can also adjust its SPR frequency to a specific spectrum area according to the requirement of experiments, promoting applications in nanobioassays.

In addition to gold nanoparticles, nanoparticles with magnetite (Fe_3O_4), maghemite (Fe_2O_3), or other ferrite cores are often used as superparamagnetic contrast agents currently. This type of nanoparticle is called a superparamagnetic iron oxide particle (SPIO). SPIO can be used to make a highly-sensitive MRI contrast agent because MRI can detect SPIO at the micromolar level [22]. Furthermore, a layer of hydrophilic polymers can also be coated on the surface of the nanoparticles to improve the water solubility of SPIO. This method can alter the distribution and pharmacokinetic properties of SPIO *in vivo* [23]. Moreover, according to the needs of tests, the ligands can be modified on the surface of SPIO to design the functionalized SPIO with selective binding characteristics. In addition to detecting specific molecules, the ligand modification of SPIO can also reduce the image background and toxicity caused by non-selective binding. In addition to the nanoparticles mentioned above for use in MRI, many other nanomaterials are becoming more popular, such as nanoparticles, nanofibers, and nanowires [24–26]. These nanomaterials can be used in different biomedical applications because of their various properties. First, nanoparticles can be prepared by using polymers, dendrimers, microemulsions, and others. The advantage of the method is that the structures of nanoparticles can contain a core and shell [27–30]. So if it is used as a drug release carrier, the gene or drug carried can be encapsulated in the core of nanoparticles to isolate the factors in the microenvironment such as pH value, ionic strength, chemical oxidation/reduction, and catalytic enzymes. Such a strategy can avoid the destruction of genes or drugs by the factors in the microenvironment before they are delivered to the target and can also avoid contact with other non-targeted cells or tissues and reduce side effects. For example, liposomes are hollow microspheres formed from lipids and can be suspended in water [31]. The surface of liposomes is mainly fabricated via lipid bilayers composed of phospholipids. The structure of lipid bilayers is very similar to the cell membrane, so it has the property of being able to fuse with the cell membrane or enter cells via endocytosis. In addition, the surface features of liposomes can also be designed so that liposomes can be used for molecular recognition and permeability control *in vivo*. Generally speaking, the surface of nanoparticles can be modified by binding ligands or antibodies, and then specific genes or drugs can be delivered to targeted cells through specific binding with receptors on targeted cells [32]. Subsequently, a controlled release function is carried out to produce the drug effects on targeted cells.

Nanofibers are one of the new nanomaterials developed using commercial nanotechnology [33]. The main manufacturing methods include electrospinning, self-assembly, and phase separation. The most commonly used method is the electrospinning method [34]. Overall, nanofibers have the advantages of high biocompatibility and biodegradability so that they can be decomposed, absorbed, or excreted through metabolic mechanisms in the human body. Furthermore, nanofibers have a high surface area as well as porosity properties, which endow nanofibers with a wide range of uses in many fields. For example, nanofibers are often used as scaffolds for culturing skeletal muscles, skin, blood vessels, nerves, and stem cells and then used in tissue engineering [35], artificial dressings, or as a surface-coating layer on implants. In addition, nanofibers are also used as drugs, proteins, DNA, or cell carriers [36].

Nanowires are often used in nanosensor applications because biomolecules with molecular recognition capabilities can be used to modify the surface characteristic of nanowires, thereby detecting and modifying molecules that can specifically interact with nanowires [37]. Nanowires have extremely small sizes and high sensitivity. Therefore, the concentration and quantity of sensing targets can be reduced to the minimum when using nanowires for sensing applications. In addition, nanowires can also be used for early disease and diagnosis applications. It can also be integrated to fabricate lab-on-a-chip systems, in which nanowires can be used for disease diagnosis and drug delivery [38]. Therefore, the extremely small size and high sensitivity of nanofibers can be beneficial to be integrated into instruments or other systems and can also be used for high-sensitivity detection for different purposes, so it has promising for broad applications.

After understanding the characteristics of nanomaterials, we will further discuss the applications of nanomaterials in stem cell biomedicine. Recently, the most important application of nanomaterials is for disease diagnosis via advanced imaging scanning equipment, including MRI and positron emission tomography (PET), which are often used to diagnose diseases in medical institutions. MRI is currently the most advanced way of scanning the human body [39]. Through MRI diagnosis, high-resolution images can be obtained without using any radioactive substance. The inspection process is also very safe and comfortable for the patient and does not harm the human body [40]. PET is a non-invasive nuclear medicine imaging technique [41]. First, positron radiopharmaceuticals can be inhaled, swallowed, or injected into the human body. Then, PET is used to detect the distribution of the radiopharmaceuticals used in the body that can carry out the functional examination of organs and tissues of the whole body. The above two imaging techniques are similar and can be used to rectify diseases. In addition, nanomaterials provide a novel approach to image capture. Currently, the most commonly used nanomaterials for imaging are the SPIO mentioned above [40]. SPIO used in imaging is mainly a micro-contrast agent combined with MRI to diagnose inflammatory and degenerative diseases such as focal ischemic lesions, atherosclerosis, multiple sclerosis, kidney diseases, osteoarthritis, etc. [42]. Furthermore, the highly sensitive biosensor made of nanomaterials such as nanowires can not only be used for early disease diagnosis but also can provide more accurate and precise diagnosis results. Moreover, nanomaterials can also be used to detect specific disease cells by modifying fluorescent molecules and specific antibodies on the surface of nanomaterials and then through specific binding between antibodies and antigens [43]. After binding nanomaterials with fluorescent molecules and antibodies to antigens or receptors on cells, the distribution and content of specific cells in the body, such as stem cells or cancer stem cells, can be effectively detected. Then, the labeled stem cells or cancer stem cells can be used to study the related processes of cell development and disease severity and progression that can facilitate future treatment strategies. In addition to diagnosis, nanomaterials, such as SPIO, can also be modified with the antibody on the surface, specifically bind with the surface antigen of stem cells, and separate and collect stem cells under the action of a magnetic field. SPIO has shown a fast and high recovery rate of stem cells in stem cell separation technology compared with the cell sorting method from conventional cell culture or flow cytometry instruments [44]. In addition to the functions mentioned earlier of cell tracking and molecular targeting to achieve targeted therapy, we will next introduce the biomedical application of nanomaterials in the treatment and repair of diseases. Drug delivery and controlling release is also one of the important applications of nanomaterials in therapy, especially cancer treatment. In recent years, this method of molecular labeling for targeted therapy has gradually developed into an important field for disease therapy. The usual systemic drug delivery in the past often fails to meet the needs of effective treatment. To more accurately deliver the drugs, stem cells, or genes to inflamed, damaged, or damaged tissues for repair, biomaterials developed via nanotechnology bring a promising tool for targeted therapy and drug-controlled release [45].

The current therapeutic method for cancer diseases is systemic chemotherapy because this therapeutic method is usually not specific [46]. In addition to killing cancer cells, systemic chemotherapy also causes normal cells to die together during the treatment process, causing many cancer patients to die not because of cancer progression but because of the side effects of chemotherapy [47–49]. To improve the quality of life of patients with cancer and the effects of treatments, scientists are constantly looking for new therapeutic methods. Through in-depth studies, it can be found that most cancer tissues contain a small number of cancer stem cells with stem cell characteristics [47]. Recently, many studies have pointed out that cancer stem cells may be the few cancer cells with self-renewal in cancer tissues. So, the occurrence and recurrence of various types of cancer may also be caused by the unlimited proliferation and differentiation of these very few cancer stem cells. Stem cells are characterized by their ability to undergo asymmetric division to form new stem cells or differentiate into other cells. On the other hand, normal stem cells are controlled by biological mechanisms in their dividing behavior, thus repairing injured or apoptotic cells through limited proliferation. Compared with normal stem cells, cancer stem cells are characterized by their abnormal proliferation to produce cancer cells in large numbers [50]. Many studies have found cancer stem cells in cancer cells in different body parts, including the blood, breast, central nervous system, skin, pancreas, brain, neck, rectum, and prostate [51, 52]. Moreover, current research also points out that if only differentiated cancer cells are removed, the remaining cancer stem cells will continue to proliferate [53]. So, if the cancer stem cells remain in the body, the cancer treatment is not efficient, which leads to a 100% possibility of recurrence and metastasis. However, the biggest challenge facing traditional chemotherapy methods is that they cannot remove or eliminate 100% of cancer stem cells. Therefore, the targeted therapeutic strategy using nanomaterials to track and observe cancer stem cells has very good application potential in cancer treatment and diagnosis. For example, many cancer stem cells have a specific CD133 antigen [54], so after modifying the CD133 antibody and fluorescent molecule on the nanomaterials, the location and distribution of the cancer stem cells can be identified and calibrated through the specific binding of the antibody of antigen. Then the anticancer drugs encapsulated in the core of nanoparticles can be directly delivered from nanoparticles to the location of the cancer stem cells for cancer treatment at a specific site. The advantage of this method is not only to improve the therapeutic effect but also to avoid contact between the drug and normal cells, which can reduce the death rate of normal cells and the occurrence of drug side effects.

As mentioned above, nanoparticles can be used as carriers for drug delivery [55]. With the development of science and technology, clinical drugs for treatment needs must be more precise. However, precision control is less likely to be achieved if the drug is administered directly systemically, such as orally or by injection. After injection or being taken orally, drugs will be transported to various organs in the body through blood circulation. So, in addition to the targeted tissues or cells, most drugs may also be diluted to other parts, resulting in the lack of drug concentration. In addition to insufficient drug concentration, other tissues also have adverse reactions or side effects. To improve this shortcoming, the most common method at present is to encapsulate drugs in nanoparticles. The performance of nanoparticles is connected with the relevant antibody to increase the specificity for the surface antigens or receptors of the therapeutic cells. As mentioned in the above paragraph, this method allows nanoparticles to be guided to bind with targeted cells during the delivery process. Subsequently, through changes in the surrounding environment, the drug is released to the target cell to ensure that the drug is released to cells without losing its original activity before reaching a specific tissue to improve the effectiveness of the treatment.

Besides nanoparticles, nanofibers can also be used in drug delivery and controlled release applications [56]. The advantage of nanofibers is that combing nanofibers with cells or drugs forms scaffolds to allow cells to attach, grow, and provide a mechanism for controlled drug release. In terms of drug delivery, nanofibers can coat drugs to form a hydrogel and extend the time of drugs in contact with cells in the tissue after being injected into the body to improve the absorption efficiency of the drug. Similar to nanoparticles, the surface of nanofibers can also be modified to optimize their water solubility, immunocompatibility, and cellular specificity of nanofibers after modification. Therefore, the problems of traditional drug delivery methods, such as bioavailability, can also be improved, thereby improving the accuracy of target drugs and reducing adverse reactions. As far as drug carriers are concerned, nanofibers can carry a wide variety of drugs, including macromolecules, small molecules, amino acids, proteins, DNA, and siRNA [57]. In tissue engineering applications, nanofibers can be combined with stem cells or growth factors such as vascular endothelial growth factor (VEGF), platelet-derived growth factors (PDGF), etc., to form composite artificial scaffolds and then directly applied to damaged tissues [58, 59]. This method can effectively help repair and regenerate muscles, skin, nerves, blood vessels, and other tissues. Furthermore, since various growth factors differ in their ability to have bidding affinity with nanofibers, the time for growth factors uptake by cells is also different after release from nanofibers. Therefore, when using nanofibers to carry growth factors or drugs, it is necessary to design various controlled release mechanisms according to different purposes to achieve the best release time and rate. For example, previous studies have used self-assembling peptide nanofibers carrying PDGF to repair damaged myocardial tissues [60]. Compared with the group directly injected with PDGF, self-assembling peptide nanofibers had higher significant difference effects. The complex scaffolds formed by nanofibers and PDGF can effectively retain PDGF on the damaged myocardium without being removed by blood circulation, which can prolong the time for PDGF to repair the myocardium to improve the therapeutic effect. In addition to repairing myocardium, we mentioned the application of many biomaterials combined with stem cells in Chapters 2, 4, and 5. In this chapter, although nanomaterials are mostly used for drug control release, there are also applications of nanomaterials as artificial biomimetic materials. The development of nano-inorganic/organic biomedical composites is derived from the idea of natural tissues. The teeth and bones in the human body are nano-composite materials composed of nano hydroxyapatite and macromolecules. Teeth and bones have good mechanical properties that inspire scientists to prepare nanobiomedical composite materials to mimic natural hard tissues for tissue engineering applications. For example, in treating oral frontal bone defects, nano hydroxyapatite/collagen composites can be used to promote the formation of new bone after implantation in the narrow cavity.

Generally speaking, there are three main mechanisms for the controlled drug release of nanofibers [61]. The first one is that nanofibers decompose themselves with time and then release the growth factors coated in the inner layer. This method is mainly used for drugs encapsulated in nanofibers. Due to without covalent bonds, so after reaching the destination, the drug is released from nanofibers with the difference in concentration gradient. This method has the shortest release time. The second method is to release the carried drugs or growth factors from nanofiber hydrogels via the cleavage of chemical bonds. In this method, the typical way is to graft drugs or growth factors on nanofibers scaffolds through ionic or chemical bondings first. With the self-degradation of nanofibers, the bonded drugs/growth factors can be released. Noticeably, the release rate of drugs or growth factors can be affected by the stimulation of surrounding microenvironments. Compared with the first method, the release mechanism of this method takes longer and is difficult to control. The third method uses enzymes to trigger the release of drugs or growth factors. Generally, using this way for release, the chemical bonds of nanofibers need to be broken by the action of proteolytic enzymes, so the drugs/growth factors are only released in the presence of enzymes. According to different needs, the release mechanisms can be applied individually, meaning that each release mechanism can be integrated into one nanofiber scaffold, which can be designed as a scaffold with multifunctional releases for cell or tissue repair.

Finally, we will mention another application that combines nanomaterials with stem cells for stem cell therapy. The current limitation of stem cell therapy is to have a low retention rate and survival rate of stem cells *in vivo* [62]. In many studies, stem cells were tried to inject stem cells into the body for therapy directly, and the results indicated that most of the stem cells died immediately, and the blood circulation would rapidly remove the stem cells from the injured tissues. As a result, the stem cells stayed in the tissues for only a short time, which reduced the therapeutic efficiency [63–65]. Therefore, if combined with nanomaterials and stem cells as cell-laden scaffolds are directly injected into damaged tissues such as damaged heart muscle caused by myocardial infarction, the retention rate and survival of stem cells can be effectively improved for promoting the effects of stem cell therapy.

The main purpose of applying nanotechnology in stem cell biology is to achieve early diagnosis of diseases, improve the sensitivity of diagnostic results, precisely control the release mechanism for drug delivery, and use targeted therapy to improve therapeutic effects. Applying nanomaterials to the applications mentioned above is a long, laborious, and costly process, and involves continuously developing and producing nanomaterials with high specificity, precision, and sensitivity as diagnostic tools for pairing and treatment techniques. The understanding of stem cells in various fields has been deepening. So, scientists can explore the interaction mechanism between biomedical materials and stem cells at the nanoscale with the assistance of innovative nanotechnology and further develop a novel nanomaterials-based culture system for culturing and regulating stem cells. Therefore, developing a platform related to stem cell research combined with nanotechnology has a high potential for developing innovative clinical diagnosis and treatment methods in the future.

References

[1] Ray, S.S., and J. Bandyopadhyay. Nanotechnology-enabled biomedical engineering: Current trends, future scopes, and perspectives. Nanotechnol Rev, 2021. 10(1): 728–743.

[2] Mir, M., S. Ishtiaq, S. Rabia, M. Khatoon, A. Zeb, G.M. Khan, A. Ur Rehman, and F. Ud Din. Nanotechnology: from *in vivo* imaging system to controlled drug delivery. Nanoscale Res Lett, 2017. 12(1): 500.

[3] Melchor-Martinez, E.M., N.E. Torres Castillo, R. Macias-Garbett, S.L. Lucero-Saucedo, R. Parra-Saldivar, and J.E. Sosa-Hernandez,. Modern world applications for nano-bio materials: Tissue engineering and COVID-19. Front Bioeng Biotechnol, 2021. 9: 597958.

[4] Dong, Y., X. Wu, X. Chen, P. Zhou, F. Xu, and W. Liang. Nanotechnology shaping stem cell therapy: Recent advances, application, challenges, and future outlook. Biomed Pharmacother, 2021. 137: 111236.

[5] Asil, S.M., J. Ahlawat, G.G. Barroso, and M. Narayan. Application of nanotechnology in stem-cell-based therapy of neurodegenerative diseases. Appl Sci-Basel, 2020. 10(14).

[6] Zakrzewski, W., M. Dobrzynski, A. Zawadzka-Knefel, A. Lubojanski, W. Dobrzynski, M. Janecki, K. Kurek, M. Szymonowicz, R.J. Wiglusz, and Z. Rybak. Nanomaterials application in endodontics. Materials (Basel), 2021. 14(18).

[7] Chaves, A., J.G. Azadani, H. Alsalman, D.R. da Costa, R. Frisenda, A.J. Chaves, S.H. Song, Y.D. Kim, D.W. He, J.D. Zhou, A. Castellanos-Gomez, F.M. Peeters, Z. Liu, C.L. Hinkle, S.H. Oh, P.D. Ye, S.J. Koester, Y.H. Lee, P. Avouris, X.R. Wang, and T. Low. Bandgap engineering of two-dimensional semiconductor materials. Npj 2d Mater Appl, 2020. 4(1).

[8] Huang, D.M., H. Liu, B. Zhang, K. Jiao, and X. Fu. Highly sensitive electrochemical detection of sequence-specific DNA of 35S promoter of cauliflower mosaic virus gene using CdSe quantum dots and gold nanoparticles. Microchim Acta, 2009. 165(1-2): 243–248.

[9] Wabler, M., W. Zhu, M. Hedayati, A. Attaluri, H. Zhou, J. Mihalic, A. Geyh, T.L. DeWeese, R. Ivkov, and D. Artemov. Magnetic resonance imaging contrast of iron oxide nanoparticles developed for hyperthermia is dominated by iron content. Int J Hyperthermia, 2014. 30(3): 192–200.

[10] Alphandery, E. Iron oxide nanoparticles as multimodal imaging tools. RSC Adv, 2019. 9(69): 40577–40587.

[11] Geppert, M., and M. Himly. Iron oxide nanoparticles in bioimaging - an immune perspective. Front Immunol, 2021. 12: 688927.

[12] Mody, N., R.K. Tekade, N.K. Mehra, P. Chopdey, and N.K. Jain. Dendrimer, liposomes, carbon nanotubes and PLGA nanoparticles: one platform assessment of drug delivery potential, AAPS PharmSciTech, 2014. 15(2): 388–99.

[13] Franco, M.S., E.R. Gomes, M.C. Roque, and M.C. Oliveira. Triggered drug release from liposomes: exploiting the outer and inner tumor environment. Front Oncol, 2021. 11: 623760.

[14] Han, X., K. Xu, O. Taratula, and K. Farsad. Applications of nanoparticles in biomedical imaging. Nanoscale, 2019. 11(3): 799–819.

[15] Peserico, A., C. Di Berardino, V. Russo, G. Capacchietti, O. Di Giacinto, A. Canciello, C. Camerano Spelta Rapini, and B. Barboni. Nanotechnology-assisted cell tracking. Nanomaterials (Basel), 2022. 12(9).

[16] Kumar, Y.R., K. Deshmukh, K.K. Sadasivuni, and S.K.K. Pasha. Graphene quantum dot based materials for sensing, bio-imaging and energy storage applications: A review. RSC Adv, 2020. 10(40): 23861–23898.

[17] Wagner, A.M., J.M. Knipe, G. Orive, and N.A. Peppas., Quantum dots in biomedical applications. Acta Biomater, 2019. 94: 44–63.

[18] Wang, Z., X.Z. Dong, S.Y. Zhou, Z. Xie, and Z. Zalevsky. Ultra-narrow-bandwidth graphene quantum dots for superresolved spectral and spatial sensing. Npg Asia Mater, 2021. 13(1).

[19] Gorshkov, K., K. Susumu, J. Chen, M. Xu, M. Pradhan, W. Zhu, X. Hu, J.C. Breger, M. Wolak, and E. Oh. Quantum dot-conjugated SARS-CoV-2 spike pseudo-virions enable tracking of angiotensin converting enzyme 2 binding and endocytosis. ACS Nano, 2020. 14(9): 12234–12247.

[20] Lin, S., X. Xie, M.R. Patel, Y.H. Yang, Z. Li, F. Cao, O. Gheysens, Y. Zhang, S.S. Gambhir, J.H. Rao, and J.C. Wu. Quantum dot imaging for embryonic stem cells. BMC Biotechnol, 2007. 7: 67.

[21] Chang, C.C., C.P. Chen, T.H. Wu, C.H. Yang, C.W. Lin, and C.Y. Chen. Gold nanoparticle-based colorimetric strategies for chemical and biological sensing applications. Nanomaterials (Basel), 2019. 9(6).
[22] Wei, H., O.T. Bruns, M.G. Kaul, E.C. Hansen, M. Barch, A. Wisniowska, O. Chen, Y. Chen, N. Li, S. Okada, J.M. Cordero, M. Heine, C.T. Farrar, D.M. Montana, G. Adam, H. Ittrich, A. Jasanoff, P. Nielsen, and M.G. Bawendi. Exceedingly small iron oxide nanoparticles as positive MRI contrast agents. Proc Natl Acad Sci U S A, 2017. 114(9): 2325–2330.
[23] Jiang, M., Q. Liu, Y. Zhang, H. Wang, J. Zhang, M. Chen, Z. Yue, Z. Wang, X. Wei, S. Shi, M. Wang, Y. Hou, Z. Wang, F. Sheng, N. Tian, and Y. Wang. Construction of magnetic drug delivery system and its potential application in tumor theranostics. Biomed Pharmacother, 2022. 154: 113545.
[24] Richards, D.J., Y. Tan, R. Coyle, Y. Li, R. Xu, N. Yeung, A. Parker, D.R. Menick, B. Tian, and Y. Mei. Nanowires and electrical stimulation synergistically improve functions of hiPSC cardiac spheroids. Nano Lett, 2016. 16(7): 4670–8.
[25] Yu, D., J. Wang, K.J. Qian, J. Yu, and H.Y. Zhu. Effects of nanofibers on mesenchymal stem cells: environmental factors affecting cell adhesion and osteogenic differentiation and their mechanisms, J Zhejiang Univ Sci B, 2020. 21(11): 871–884.
[26] Sun, Y., Y. Lu, L. Yin, and Z. Liu. The roles of nanoparticles in stem cell-based therapy for cardiovascular disease. Front Bioeng Biotechnol, 2020. 8: 947.
[27] Chenthamara, D., S. Subramaniam, S.G. Ramakrishnan, S. Krishnaswamy, M.M. Essa, F.H. Lin, and M.W. Qoronfleh. Therapeutic efficacy of nanoparticles and routes of administration. Biomater Res, 2019. 23: 20.
[28] Yousefi, M., A. Narmani, and S.M. Jafari. Dendrimers as efficient nanocarriers for the protection and delivery of bioactive phytochemicals. Adv Colloid Interface Sci, 2020. 278: 102125.
[29] Mahdavi, Z., H. Rezvani, and M. Keshavarz Moraveji. Core-shell nanoparticles used in drug delivery-microfluidics: a review. RSC Adv, 2020. 10(31): 18280–18295.
[30] Lin, Y.S., S.H. Wu, C.T. Tseng, Y. Hung, C. Chang, and C.Y. Mou. Synthesis of hollow silica nanospheres with a microemulsion as the template. Chem Commun (Camb), 2009. (24): 3542–4.
[31] Barea, M.J., M.J. Jenkins, Y.S. Lee, P. Johnson, and R.H. Bridson. Encapsulation of liposomes within pH responsive microspheres for oral colonic drug delivery. Int J Biomater, 2012. 458712.
[32] Richards, D.A., A. Maruani, and V. Chudasama. Antibody fragments as nanoparticle targeting ligands: a step in the right direction. Chem Sci, 2017. 8(1): 63–77.
[33] Kenry, C.T. Lim, Nanofiber technology: current status and emerging developments. Prog Polym Sci, 2017. 70: 1–17.
[34] Shahriar, S.M.S., J. Mondal, M.N. Hasan, V. Revuri, D.Y. Lee, and Y.K. Lee. Electrospinning nanofibers for therapeutics delivery. Nanomaterials (Basel), 2019. 9(4).
[35] Law, J.X., L.L. Liau, A. Saim, Y. Yang, and R. Idrus. Electrospun collagen nanofibers and their applications in skin tissue engineering. Tissue Eng Regen Med, 2017. 14(6): 699–718.
[36] Stojanov, S., and A. Berlec. Electrospun nanofibers as carriers of microorganisms. Stem cells, proteins, and nucleic acids in therapeutic and other applications. Front Bioeng Biotechnol, 2020. 8: 130.
[37] Hubbe, H., E. Mendes, and P.E. Boukany. Polymeric nanowires for diagnostic applications. Micromachines (Basel), 2019. 10(4).
[38] Rahong, S., T. Yasui, N. Kaji, and Y. Baba. Recent developments in nanowires for bio-applications from molecular to cellular levels. Lab Chip, 2016. 16(7): 1126–38.
[39] Forte, E., D. Fiorenza, E. Torino, A. Costagliola di Polidoro, C. Cavaliere, P.A. Netti, M. Salvatore, and M. Aiello. Radiolabeled PET/MRI nanoparticles for tumor imaging. J Clin Med, 2019. 9(1).
[40] Estelrich, J., M.J. Sanchez-Martin, and M.A. Busquets,. Nanoparticles in magnetic resonance imaging: from simple to dual contrast agents. Int J Nanomedicine, 2015. 10: 1727–41.
[41] Goel, S., C.G. England, F. Chen, and W. Cai. Positron emission tomography and nanotechnology: A dynamic duo for cancer theranostics. Adv Drug Deliv Rev, 2017. 113: 157–176.
[42] Kim, J., P. Chhour, J. Hsu, H.I. Litt, V.A. Ferrari, R. Popovtzer, and D.P. Cormode. Use of nanoparticle contrast agents for cell tracking with computed tomography. Bioconjug Chem, 2017. 28(6): 1581–1597.

[43] Malik, P., R. Gupt, V. Malik, and R.K. Ameta. Emerging nanomaterials for improved biosensing, Measurement: Sensors, 2021. 16: 100050.
[44] Li, X., Z. Wei, H. Lv, L. Wu, Y. Cui, H. Yao, J. Li, H. Zhang, B. Yang, and J. Jiang. Iron oxide nanoparticles promote the migration of mesenchymal stem cells to injury sites. Int J Nanomedicine, 2019. 14: 573–589.
[45] Labusca, L., D.D. Herea, and K. Mashayekhi. Stem cells as delivery vehicles for regenerative medicine-challenges and perspectives. World J Stem Cells, 2018. 10(5): 43–56.
[46] Schirrmacher, V. From chemotherapy to biological therapy: A review of novel concepts to reduce the side effects of systemic cancer treatment (Review). Int J Oncol, 2019. 54(2): 407–419.
[47] Yu, Z., T.G. Pestell, M.P. Lisanti, and R.G. Pestell. Cancer stem cells. Int J Biochem Cell Biol, 2012. 44(12): 2144–51.
[48] Ayob, A.Z., and T.S. Ramasamy. Cancer stem cells as key drivers of tumour progression. J Biomed Sci, 2018. 25(1): 20.
[49] Yang, L., P. Shi, G. Zhao, J. Xu, W. Peng, J. Zhang, G. Zhang, X. Wang, Z. Dong, F. Chen, and H. Cui. Targeting cancer stem cell pathways for cancer therapy. Signal Transduct Target Ther, 2020. 5(1): 8.
[50] Rich, J.N. Cancer stem cells: understanding tumor hierarchy and heterogeneity. Medicine (Baltimore), 2016. 95(1 Suppl 1): S2–S7.
[51] Yadav, U.P., T. Singh, P. Kumar, P. Sharma, H. Kaur, S. Sharma, S. Singh, S. Kumar, and K. Mehta. Metabolic adaptations in cancer stem cells. Front Oncol, 2020. 10: 1010.
[52] Bahmad, H.F., K. Cheaito, R.M. Chalhoub, O. Hadadeh, A. Monzer, F. Ballout, A. El-Hajj, D. Mukherji, Y.N. Liu, G. Daoud, and W. Abou-Kheir. Sphere-formation assay: three-dimensional *in vitro* culturing of prostate cancer stem/progenitor sphere-forming cells. Front Oncol, 2018. 8: 347.
[53] Vlashi, E., and F. Pajonk. Cancer stem cells, cancer cell plasticity and radiation therapy. Semin Cancer Biol, 2015. 31: 28–35.
[54] Barzegar Behrooz, A., A. Syahir, and S. Ahmad. CD133: beyond a cancer stem cell biomarker. J Drug Target, 2019. 27(3): 257–269.
[55] Patra, J.K., G. Das, L.F. Fraceto, E.V.R. Campos, M.D.P. Rodriguez-Torres, L.S. Acosta-Torres, L.A. Diaz-Torres, R. Grillo, M.K. Swamy, S. Sharma, S. Habtemariam, and H.S. Shin. Nano based drug delivery systems: recent developments and future prospects. J Nanobiotechnology, 2018. 16(1): 71.
[56] Weng, L., and J. Xie. Smart electrospun nanofibers for controlled drug release: recent advances and new perspectives. Curr Pharm Des, 2015. 21(15): 1944–59.
[57] Jarak, I., I. Silva, C. Domingues, A.I. Santos, F. Veiga, and A. Figueiras. Nanofiber carriers of therapeutic load: current trends. Int J Mol Sci, 2022. 23(15).
[58] Kim, P.H., and J.Y. Cho. Myocardial tissue engineering using electrospun nanofiber composites. BMB Rep, 2016. 49(1): 26–36.
[59] Zhang, X., Y. Meng, B. Gong, T. Wang, Y. Lu, L. Zhang, and J. Xue. Electrospun nanofibers for manipulating soft tissue regeneration. J Mater Chem B, 2022.
[60] Guo, H.D., G.H. Cui, J.J. Yang, C. Wang, J. Zhu, L.S. Zhang, J. Jiang, and S.J. Shao Sustained delivery of VEGF from designer self-assembling peptides improves cardiac function after myocardial infarction. Biochem Biophys Res Commun, 2012. 424(1): 105–11.
[61] Wu, J., Z. Zhang, J. Gu, W. Zhou, X. Liang, G. Zhou, C.C. Han, S. Xu, and Y. Liu. Mechanism of a long-term controlled drug release system based on simple blended electrospun fibers. J Control Release, 2020. 320: 337–346.
[62] Sart, S., T. Ma, and Y. Li., Preconditioning stem cells for *in vivo* delivery. Biores Open Access, 2014. 3(4): 137–49.
[63] Galderisi, U. Role of gene and stem cell therapies in the treatment of neurological disorders - Editorial. Curr Drug Targets, 2005. 6(1): 1–1.
[64] Hoang, D.M., P.T. Pham, T.Q. Bach, A.T.L. Ngo, Q.T. Nguyen, T.T.K. Phan, G.H. Nguyen, P.T.T. Le, V.T. Hoang, N.R. Forsyth, M. Heke, and L.T. Nguyen. Stem cell-based therapy for human diseases. Signal Transduct Tar, 2022. 7(1).
[65] Trounson, A., and C. McDonald. Stem cell therapies in clinical trials: progress and challenges. Cell Stem Cell, 2015. 17(1): 11–22.

9

Biomaterials use in *Ex vivo* Testing and Animal Model

Chia-Ning Shen,[1] *Wen-Yen Huang,*[2] *Sung-Jan Lin,*[2] *Nai-Chen Cheng,*[3] *Min-Huey Chen,*[4] *I-Chi Lee*[5,*] and *Yi-Chen Ethan Li*[6,*]

Tissue engineering involves the combination of scaffolds, cells, and biologically active molecules to generate functional tissues. Regenerative medicine generally is a broader concept that includes tissue engineering and research on self-healing where damaged tissues can be repaired after administration of foreign biological signals or materials. Due to easy accessibility, the paracrine effect, and multipotency, stem cell-related tissue engineering applications have been expanded into many aspects of tissue repair [1]. For example, biomaterial constructs have been employed to promote stem cellular viability to facilitate spinal cord regeneration [2] and scar-less excisional wound healing [3]. However, biomaterials or stem cell-based bioproducts may have toxic or injurious effects on biological systems. Therefore, scientists should evaluate the blood response, local response (i.e., cellular and tissue response), system response, and immune response of biomaterial and stem cell-related bioproducts in pre-clinical testing and clinical trials. In general, many different guidelines and standards for using biomaterials, regenerative medical products, and devices, from pre-clinic testing to clinical trials, have been made and governed by country-specific

[1] Genomics Research Center, Academia Sinica, Taiwan.
[2] Department of Biomedical Engineering, College of Medicine and College of Engineering, National Taiwan University, Taiwan.
[3] Department of Surgery, National Taiwan University Hospital, Taiwan.
[4] Graduate Institute of Clinical Dentistry, School of Dentistry, National Taiwan University, Taiwan.
[5] Department of Biomedical Engineering and Environmental Sciences, National Tsing Hua University, Taiwan.
[6] Department of Chemical Engineering, Feng Chia University, Taiwan.
* Corresponding authors: iclee@mx.nthu.edu.tw; yicli@fcu.edu.tw

or international organizations [4]. In general, all efficacy testing, biological effects, and biocompatibility of biomaterials or stem cell-based bioproducts occur before any clinical testing. Also, the development of the combination of biomaterials and stem cell-based culture systems can also be relevant to medical device components, biomaterials, and regenerative products that may be applied to human tissue and may be treated as biologics, reflecting the cell and tissue components. Therefore, good laboratory practices with a source for important regulatory guideline and overview that may apply to the clinical development of biomaterials, stem cell therapy, and stem cell-based culture systems is necessary for bioengineers, toxicologists, pathologists, and allied professionals. Here, we will introduce the evaluation methods from laboratory testing, *ex vivo* testing, animal models, and pre-clinical testing of biomaterials, stem cell-based system, and their derivate-relative products. In terms of standard guidelines for evaluation, there are three major international standard organizations, International Organization for Standardization (ISO), the American Society for Testing and Materials (ASTM) International standard, U.S. Food and Drug Administration (FDA), and other institutes that provide the guideline to professionals. For most laboratories, ISO-10993, ASTM F748-95, and FDA guidance document (G95-1) from three international standard organizations offer various sources of guidelines to evaluate the biocompatibility of biomaterials and stem cell-related products [5–7]. Similarly, scientists in Taiwan also can refer to CNS14393:2004 from the Bureau of Standards, Metrology, and Inspection [8]. In addition, other similar guideline documents are also described in the various standard organizations from different countries, such as the International Medical Device Regulators Forum (IMDRF), World Health Organization (WHO), European Conformity marking, Asian, African, and Latin American Regional Harmonization, and others [4]. Therefore, considering the physicochemical properties of biomedical materials, the nature, extent, and time of contact with the human body. In this chapter, we will briefly introduce the ISO-10993 guideline. This standard has been adopted nationally in Europe and the United States. It is the latest trend in the biological evaluation of biomedical materials worldwide and the complete biological evaluation standard.

Overall, the ISO-10993 guideline document is applicable to the biocompatibility of biomaterials or stem cell-based products directly or indirectly contacting the human body. Table 9-1 summarizes the content related to each part of the ISO-10993 and provides a list to scientists for testing their products. The content of ISO-10993 could be divided into three major sections (Figure 9-1). Section 1 is an overall framework to show how to select appropriate testing through the guideline document. Section 2 shows the material properties and their substantial equivalence, including parts 9 and 13–19. Section 3 indicated the importance of animal welfare and the evaluation items.

To correctly evaluate the biocompatibility of biomaterials or stem cell-relative bioproducts, the selection of tests should depend on the two important parameters, the contact way and contact time of samples on the human body. The contact way can be categorized into the non-contact, contact body surface, connect the body inside and outside, and implants; and the contact time are divided into short-term

Table 9-1. The summarized overall parts in the ISO-10993 guideline documents [5].

No.	Purposes
ISO-10993-1:2009	Evaluation and testing within a risk management process
ISO-10993-2:2006	Part 2: Animal welfare requirements
ISO-10993-3:2003	Part 3: Testing for genotoxicity, carcinogenicity, and reproduce toxicity
ISO-10993-4:2002/Amd 1:2006	Part 4: Selection of tests for interactions with blood
ISO-10993-5:2009	Part 5: Tests for *in vitro* cytotoxicity
ISO-10993-6:2007	Part 6: Test for local effects after implantation
ISO-10993-7:2008/Cor 1:2009	Part 7: Ethylene oxide sterilization residuals
ISO-10993-8:2000	Part 8: Selection and qualification of reference materials for biological tests
ISO-10993-9:2009	Part 9: Framework for identification and quantification of potential degradation products
ISO-10993-10:2010	Part 10: Tests for irritation and skin sensitization
ISO-10993-11:2006	Part 11: Tests for systemic toxicity
ISO-10993-12:2007	Part 12: Sample preparation and reference materials
ISO-10993-13:2010	Part 13: Identification and qualification of degradation products from polymeric medical devices
ISO-10993-14:2001	Part 14: Identification and qualification of degradation products from ceramics
ISO-10993-15:2000	Part 15: Identification and qualification of degradation products from metals and alloys
ISO-10993-16:2010	Part 16: Toxicokinetic study design of degradation products and leachable
ISO-10993-17:2002	Part 17: Establishment of allowable limits for leachable substances
ISO-10993-18:2005	Part 18: Chemical characterization of materials
ISO/TS-10993-19-2006	Part 19: Physicochemical, morphological, and topographical characterization of materials
ISO/TS-10993-20-2006	Part 20: Principles and methods for immunotoxicity testing of medical devices

($\leq$ 24 hours), long-term (1 day ~ 30 days), and permeant ($\geq$ 30 days). According to the two parameters, the testing items are further divided into preliminary evaluation testing and supplementary evaluation testing. The preliminary evaluation items contain the ISO-10993-4, ISO-10993-5, ISO-10993-6, ISO-10993-10, and ISO-10993-11; and the supplementary evaluation items include the ISO-10993-3, ISO-10993-9, and ISO-10993-11. Furthermore, the detailed information for preliminary and supplementary evaluation items is shown in Table 9-2 for readers to select the appropriate evaluation experiments. Here, we will make a simple example of a cytotoxicity test that the reader enables to learn how to test in laboratories. The cytotoxicity test is an *in vitro* test. In the test, we can use an elution condition medium of material to treat a mammal's fibroblast (i.e., L929). Following the ISO-10993-12 guideline, the content of materials for elution tests can be calculated

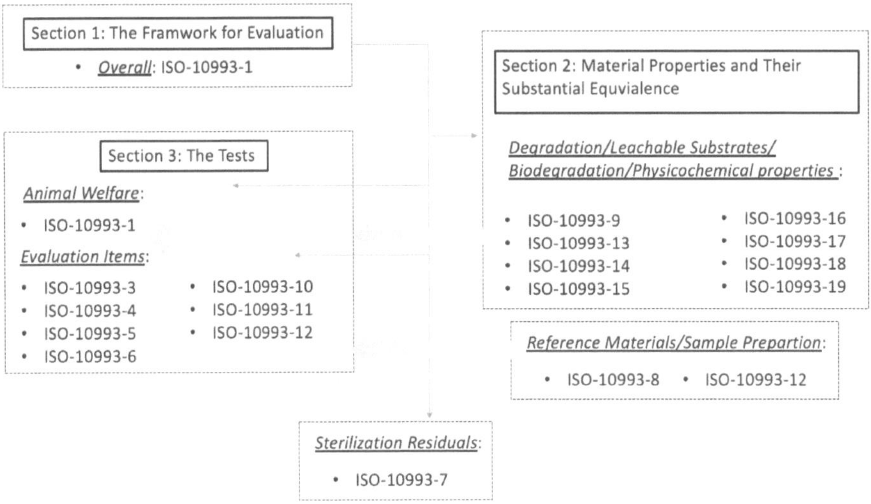

Figure 9-1. The framework of ISO-10993.

according to their surface area. For example, the content of an irregular low-density porous material is 0.1 g/mL for the elution test. Then, the material immersed in the medium is further incubated according to the needs of experimental conditions, such as 37 ± 1°C for 24 ± 2 hours, 37 ± 1°C for 72 ± 2 hours, 50 ± 2°C for 72 ± 2 hours, 70 ± 2°C for 24 ± 2 hours, or 121 ± 2°C for 1 ± 0.1 hours. Afterwards, the elution condition medium of materials is used to treat L929 cells and then qualitatively observe the cell morphology through a microscope and quantitatively measure the cell proliferation and cell apoptosis by using the test reagents such as MTT or LDH. Therefore, through biocompatibility testing, it can obtain more stability, confidence, and safety results for further *ex vivo* and other pre-clinical tests.

The *ex vivo* testings, animal studies, and pre-clinical testings of ASCs

During a pre-clinical testing process, developing new therapies for damaged tissues or diseases should be evaluated by using *in vitro* cell testing and *in vivo* animal studies [9]. However, conventional *in vitro* cell testings miss the native tissue-like complex environments, leading to the preliminary results from *in vitro* testing may affect the confidence in the effect of new therapies on *in vivo* animal models. Currently, the *ex vivo* models bring hope to provide a 3D biomimetic environment to bridge the gap between *in vitro* cell testing and *in vivo* animal studies. Therefore, *ex vivo* testings enable offering a promising to provide more confident results in pre-clinical testing. ASCs and iPSCs are two stem cells that have been widely used in regenerative medicine field. Here, we will use ASCs and iPSCs as samples to show the potential applications of stem cell-based systems and biomaterials in *ex vivo* and pre-clinical tests. *Ex vivo* testing and animal models are important translational studies for the clinical application of ASCs. In combination with certain biomaterials,

Table 9-2. The test items for biocompatibility [5, 7].

Category				Biological Effects											
Contact Way			Times	Preliminary Evaluation								Supplementary Evaluation			
				Cyto-toxicity	Sen-sitization	Ir-ritation	Acute System toxicity	Sub-system toxicity	Geno-toxicity	Implant	Hemo-compatibility	Chronic system toxicity	Carcino-genicity	Re-produce toxicity	De-gradability
Parts	Tissues	Example													
Body Surface	Skin	Electrode, Heating compress	S	●	●	●									
			L	●	●	●									
			P	●	●	●									
	Mucosa	Contact lens	S	●	●	●									
			L	●	●	●	⊕	⊕		⊕					
			P	●	●	●	⊕	●	●	⊕		⊕			
	Injured body surface	Wound dressings	S	●	●	●	⊕								
			L	●	●	●	⊕			⊕					
			P	●	●	●	⊕	●	●	⊕		⊕			
Connect body inside and outside	Non-invasive blood	Tube-adapters	S	●	●	●	●				●				
			L	●	●	●	●				●				
			P	●	●	⊕	●	●	●	⊕	●	●	●		
	Tissue Bone Dentin	Laparo-scopes	S	●	●	●	⊕								
			L	●	●	⊕	⊕	⊕	●	●					
			P	●	●	⊕	⊕	⊕	●	●		⊕	●		
	Cir-culation	Blood dialysis devices	S	●	●	●			⊕		●				
			L	●	●	●		⊕	●	⊕	●				
			P	●	●	●	●	●	●	⊕	●	●	●		

Implant	Bone Tissue	Artificial Knees	S	●	●	●	⊕								
			L	●	●	⊕	⊕	⊕	●	●					
			P	●	●	⊕	⊕	⊕	●	●		●	●		
	Blood	Artificial vessels	S	●	●	●	●			●	●				
			L	●	●	●	●	⊕		●	●	●			
			P	●	●	●	●	●	●	●	●	●	●		

S: Short-term (≤ 24 hours); L: Long-term (1 day ~ 30 days); P: Permeant (≥ 30 days)
●: ISO and FDA; ⊕: Additional test for FDA

ASCs viability can be dramatically increased *in vitro*, but the results need to be validated by *ex vivo* and *in vivo* examinations. In this paragraph, we will discuss some currently available *ex vivo* and animal models to explore the synergistic effect of ASCs/biomaterial combination for tissue engineering applications.

In the oral and maxillofacial region of ASCs-based regenerative medicine, animal studies and clinical applications have been examined for many years [10]. The previous section has mentioned that the ASCs exosome is an important component released by the ASCs paracrine, possessing multiple biological activities. Encapsulation ASCs and ASC exosomes with biomaterial have demonstrated positive results on different animal models or tissue regeneration [11]. Overall, those findings have created several ASCs-related tissue engineering strategies for future clinical application, and molecular cell regulation may play a critical role in regenerative medicine. Stem cells frequently require a specific environment or niche to fulfill their intrinsic potency. A stem cell niche is defined as a complex, multifactorial local microenvironment required to maintain stem cell biology [12]. Developing the *ex vivo* culture condition system is a viable solution for ASCs research to maintain the ASC's phenotype. *Ex vivo* organ culture of adipose tissue was designed for *in situ* mobilizations of ASCs and defining its stem cell niche [13]. In a recent study, stimulation of ASCs in fibrin-hydrogel nerve conduits was achieved by using nerve growth factor (NGF) or vascular endothelial growth factor (VEGF), and the composite was subsequently evaluated in an *ex vivo* axonal outgrowth assay to treat nerve gap injuries [14]. Although *ex vivo* tests are generally easier to conduct, they exhibit several limitations and disadvantages. Therefore, animal models are still necessary for many aspects of ASC research. Wound healing is a complex process composed of several stages of inflammation, epithelialization, neoangiogenesis, proliferation, and collagen matrix formation [15]. Poor wound healing has been associated with many medical conditions, including diabetic foot ulcers, pressure injuries, etc. ASC-based cell therapy is a promising treatment for chronic wounds, but the low survival rate of ASCs has been noted after transplantation in the wound tissue [16]. ASCs combined with biomaterials not only provide a good ecological environment for ASCs transplantation and survival, but also promotes the proliferation, differentiation, and paracrine abilities of ASCs to secrete more growth factors [17]. ASCs combined with biomaterials not only provide a good ecological environment for ASCs transplantation and survival but also promote the proliferation, differentiation, and paracrine abilities of ASCs to secrete more growth factors [16]. The application of ASC/biomaterial complex has been demonstrated to promote wound healing in animal models with excisional wounds [18, 19]. Burn wound healing is another common wound model to study the wound repair process of thermal injury. ASCs have been shown to accelerate skin re-epithelialization, angiogenesis and collagen deposition [20], and collagen deposition [21] in murine burn models. The skin flap ischemia model is a well-known animal model related to angiogenesis and wound healing [22, 23]. There are some key factors of ASCs treatment that have been direct strategies for advancing ASCs-related therapy. ASCs treatment was effective for enhancing skin flap recovery and angiogenesis after ischemia/reperfusion injury in a mouse skin flap model, and the secretory factor IL-6 was identified as a key factor [24]. Exosomes are membrane-bound extracellular vesicles (EVs) produced in endosomal compartments of most eukaryotic cells. Their main function is recognized as carrying nucleic acids, proteins, lipids, and other bioactive substances to play a role in the body's physiological and pathological processes. Exosomes derived from human ASCs were shown to enhance skin flap survival, promote neovascularization, and alleviate inflammation and apoptosis in a rat skin flap model [25]. There are three types of diabetes mellitus (DM). Type 1 diabetes (T1D) is characterized by deficient insulin production and requires daily administration of insulin. Type 2 diabetes (T2D) results from the body's ineffective use of insulin, comprising 90% of people with diabetes around the world. Type 3 diabetes (gestational diabetes) occurs during pregnancy [26]. Much research regarding DM treatment with ASCs focused on their immuno-modulation characteristics as the main factor in attenuating DM symptoms and signs. ASCs were shown to alleviate inflammation and promote tissue repair in a T2D rat model [27]. Particularly, 3D culture techniques, such as cell sheet formation, recapitulate the *in vivo* cell-to-cell and cell-to-matrix interactions more effectively than 2D cultures [28]. An ASC sheet transplantation into the subcutaneous sites could improve glucose tolerance in high-fat/sucrose diet-induced T2D mice models [29]. Since autologous cell therapy is preferred because of the lack of immune rejection after transplantation, it is of concern whether ASCs isolated from DM patients still yield potential therapeutic effects. One study revealed lower proliferation and migration capabilities in ASCs derived from diabetic patients [30]. ASCs derived from diabetic mice also displayed impaired proliferation and angiogenic potential [31]. Reduced hepatocyte growth factor (HGF) secretion was also noted in ASCs from diabetic mice. However, they still possess significant insulin-sensitizing and pancreas-protective effects [32], and they could promote wound healing [33]. Cardiovascular diseases (CVDs) are the leading causes of death worldwide, and ASCs treatment is an emerging potential strategy to decrease mortality [34]. Some reports showed cardiac ASCs display differentiation, angiogenesis, and cell regulation capacities and exhibit high differentiation potential to cardiovascular cells [35, 36]. Moreover, by neovascularization and cardiac function enhancement, ASCs sheet and cardiac patch therapies facilitated recovery from myocardial infarction in a rat myocardial ischemic model [37]. Hypertension is the major risk factor related to the CVDs and chronic kidney disease (CKD). Human ASCs extracellular vesicles (EVs) were shown to alleviate hypertension in a rat model by attenuating cardiac fibrosis and reducing pro-inflammatory cytokine and macrophage infiltration [38]. However, another report showed that ASCs-based therapy did not reduce lesion volume or functional deficits and had no effect on gliosis, neurogenesis, or vascular marker levels in hypertensive rats [39]. The musculoskeletal system includes bones, muscles, tendons, ligaments, and soft tissues, which work together to support the human body's weight and daily activities. Musculoskeletal injuries or diseases are trauma or aging-related disorders, representing a gradually emerging health problem around the world. The use of ASCs in musculoskeletal disorders presents significant therapeutic advantages, including immune modulation, cell expansion *ex vivo*, differentiating to several mesodermal cell lineages, and secretion of trophic factors [40]. Current research on ASC-based therapy in treating several musculoskeletal disease studies and their clinical applications has been reported in the fields of veterinary [41] and human medicine [42]. Cell therapy of ASCs protects against ovariectomy-induced bone loss in nude mice by promoting osteoblastic differentiation through paracrine manner [43]. *Ex vivo* treatment allows ASCs to expand more capacity to musculoskeletal tissue engineering. ASCs injection with platelet-rich plasm cure athletic horses with lameness of the superficial digital flexor tendon diagnosing by ultrasound examination analysis [44]. There are also several other aspects of ASC applications in musculoskeletal diseases, such as tendon injuries [45], bone fracture [46], cartilage defect [47], and anterior cruciate ligament tear [48]. A study further showed that ASCs inhibited glycation-mediated inflammatory cascade and rejuvenated cartilaginous tissue, thereby promoting knee-joint integrity in T1D mouse model [49]. ASCs are becoming popular for regenerative medicine applications in neurodegenerative disorders [50]. Alzheimer's disease (AD) is an irreversible neurodegenerative disease, still lacking proper clinical treatment. Intravenously or intracerebrally transplanted human ASCs improved memory impairment and neuropathology in a mouse model [51]. Exosome has been reported as a biomarker of AD illness progression, and it is considered a therapeutic target of AD patients [52]. In a study, human ADSC exosomes reduced β-amyloid pathology and apoptosis of neuronal cells derived from the transgenic mouse model of AD [53]. Circ-Epc1 as a circle-RNA was highly expressed in human ASC exosomes and improved cognition by shifting microglial M1/M2 polarization in an AD mouse model [54]. Those studies suggested the promising future development of clinical ASC-based therapy for AD. A spinal cord injury (SCI) is damage to the spinal cord that results in a loss of function, such as mobility and/or feeling. In a rat model, transplantation of ASCs into the SCI site substantially improved tissue regeneration and functional recovery [55]. Moreover, cytokine-induced neutrophil chemoattractant (CINC-1) accumulation after ADSCs intravenous infusion to activate ERK1/2 and Akt signaling pathway to promote cell proliferation and differentiation [56]. *In vivo* neural tissue engineering using ADSCs and fibrin matrix could be achieved in an SCI model in rats [57].

The *ex vivo* testings, animal studies, and pre-clinical testings of iPSCs

iPSC-derived products intended for clinical use should be used in pre-clinical studies. Immunosuppressed or immunocompromised animals should be studied to prevent human cell rejection in animal studies. Humanized animal models, particularly mice, have progressed to the point where they can reconstruct human hematopoiesis and immunity. Humanized mice have been used to recreate a variety of human disease conditions, identifying mechanisms of relapse and suggesting new therapeutic strategies. Future research should improve these models' prediction capacities and make it easier to create and deploy humanized models based on large animal species, which can better inform clinical trials. Somatic cell reprogramming was first demonstrated using mouse and human cells. The fact that the same transcription factors may reprogram non-human primate and rat cells shows that pluripotency induction pathways are similar across mammalian species. Rabbits, dogs, a range of non-human primate species, and, more recently, domestic ungulates such as pigs, cows, sheep, goats, and horses have all been used to create iPSCs. A deeper understanding of the similarities and differences between human and animal stem cells, as well as the emulation of the behavioral, cellular, and molecular manifestations seen in human disease conditions in animal models, should lead to more interpretable screening and the prediction of major complications and off-target effects of iPSC-based therapies. The goal of a pre-clinical study is to determine whether the cells can still be retained in the target tissue after being transplanted. The use of mutant cells has serious risks and it is unlikely that all mutational alterations will be avoided. The goal is to find effective strategies to minimize or prevent the effects of genetic change on the cells. This process involves carrying out studies that are designed to predict the outcome of human trials. Human cells probably will not survive in the animal host for some applications, immunosuppression protocols will not allow long-term monitoring, and immunomodulating agents will influence the disease phenotype. As a result, using autologous and homologous animal stem cell products, especially in the early stages of intervention development, may be considered. Immune responses can have a big impact on treatment efficacy and tumor development. Immune rejection is one of the key issues with iPSC-mediated replacement treatment; hence more research is needed [58].

According to published guidelines, pre-clinical testing to determine the safety, feasibility, and efficacy of iPSC-derived products for patient therapies is required. Pre-clinical research in healthy animals and diseased models should be undertaken and compared. The same cells that may be used in humans should be examined in animals, according to FDA guidelines.

Rodents have already been used to study the basic biology of iPSCs with great success, but they are unable to predict clinical efficacy. Due to physiological similarities to humans and longer life spans, larger animal species such as pigs and monkeys may be preferable for stem cell-based pre-clinical studies. Surgical and imaging techniques are required for the application of stem cells in large animals, but the use of larger animals has its own set of difficulties that should be carefully evaluated. Higher costs, more complex husbandry, fewer reagents and instruments, less understood disease mechanisms, less genomic information, a restricted number of disease models, and less ability to manipulate the genome for model development are some of the difficulties compared to rodents. Genomic instability, the immunological response, cell rejection, the capability for uncontrolled proliferation and tumorigenicity, and off-target effects are all long-term safety concerns that should be addressed in pre-clinical studies. The examination of biological activity and numerous clinically relevant outcomes will be the goal of testing the feasibility and efficacy of a treatment. Pre-clinical animal testing should also provide information on biological and behavioral effects in regard to cell transplantation timing during the course of the disease, cell delivery routes, and administration frequencies, concentrations, and doses. It's possible that no single satisfactory model exists for specific conditions. As a result, using many models will reveal potential constraints and improve the ability to develop alternate solutions. Several examples of the usage of iPSC-derived cells in animal disease models are as follow, emphasizing the need for more specific ways to compare phenotypes and therapeutic effects across species. Cell replacement therapy holds promise in the treatment of diseases such as type 1 diabetes mellitus (T1DM), which is mainly caused by the loss of islet β cells. A critical step in developing an alternative source of insulin-producing cells is reprogramming pluripotent cells to pancreatic β-like cells from a number of animal species and humans. Even when employing pancreatic β-cells as iPSC precursors, different stepwise protocols that emulate the course of pancreatic development have been employed for reprogramming, although the efficiency of the process is still very low [59]. T1DM can be reversed with human islet transplantation, which effectively restores endogenous insulin secretion in patients. However, due to the lack of a readily available source of human islets, the number of patients who can benefit from islet transplantation is currently limited. Furthermore, despite continued immunosuppression, the persistent loss of graft function and recurrence of immunological attack on the cells make islet transplantation difficult to use widely. Human pluripotent stem cells (hiPSCs) are an abundant source for generating functional cells, which include the pancreatic islet β-cells, obviating the need for donor-derived tissues. However, there are still numerous unknown factors in the translation of pluripotent stem-cell-based treatments for human transplantation and therapy, which are challenging to examine using rodent animal models. Non-human primates are genetically, anatomically, metabolically, and physiologically comparable to humans, making them an important model for evaluating novel therapies and addressing translational difficulties before clinical trials in humans. Differentiated iPSCs were able to engraft and respond to glucose stimulation by releasing insulin, alleviating hypoglycemia in streptozotocin-treated NOD/SCID mice. Rhesus monkey iPSCs derived from adult fibroblasts were also used to create pancreatic progenitor cells. The use of a TGF-β inhibitor on these cells resulted in the formation of insulin-producing cells, which restored hyperglycemia in diabetic mice treated with streptozotocin [60]. On a commonly used immunodeficient mouse model, the function of hCiPSC-islets has been studied. hCiPSC-islets survived with considerable vascularization and sustained cellular complexity after transplantation under the kidney capsule of streptozotocin (STZ)-induced diabetic mice, as evidenced by the existence of C-peptide+ cells, GCG+-like cells, and SST+-like cells 16 weeks post-transplantation (wpt). The gradual increase in human C-peptide secretion in mouse plasma revealed that the hCiPSC-islets had continued to mature after transplantation. The functional maturation of all three pancreatic endocrine cell types was validated by single-cell RNA sequencing analysis of kidney grafts at 10 weeks post-transplantation, which also demonstrated no changes in cell composition after transplantation. Long-term survival and function of transplanted hCiPSC-islets were well duplicated on multiple hCiPSC cell lines, according to these findings. The safety and efficacy of hCiPSC-islets were demonstrated in a mouse model, paving the way for study in non-human primates.

The efficacy and safety of hCiPSC-islet transplantation were tested in a non-human primate model in a study conducted by Du et al. [60]. After a single high-dose STZ injection, macaques developed diabetes, with fasting blood glucose levels of over 200 mg dl^{-1} and C-peptide values of less than 0.15 ng ml^{-1}. After overnight fasting, diabetes was induced with a single dose of STZ (90 mg kg1) administered intravenously (within 5 minutes). STZ was diluted in 0.1 M citrate buffer (pH 4.3–4.5) and given intravenously as soon as possible, followed by hydration with normal saline (40–50 ml). Immunosuppressive therapy was given nine days before transplantation to maintain the human grafts, according to a methodology based on earlier studies of islet xenotransplantation in non-human primates. Induction immunosuppression, in which both B cells and T cells are decreased, was used to create long-term immunosuppression. Tacrolimus and sirolimus were used to maintain immunosuppression, as well as biweekly belatacept treatments. hCiPSC-islets were cryopreserved as single cells after generation, then recovered and reaggregated two days before the injection. After recovery, the average viability and yield of hCiPSC-islets were 86.9%, 1.6%, 82.0%, 9.5%, respectively. The diabetic macaques received a single dosage of recovered hCiPSC-islets via intraportal infusion. The dose of hCiPSC-islets transplanted was calculated based on dosing for adult pig islets to monkey xenotransplantation. Monkeys were anesthetized with inhalable isoflurane after receiving propofol (0.5 ml/kg) intravenously. During a surgical procedure, heart rate, temperature, blood oxygenation, and blood pressure were all monitored in real time. Five percent glucose was injected to keep blood glucose levels stable. After a laparotomy, hCiPSC-islets were infused into the portal vein via a jejunal vein [61]. After hCiPSC-islet transplantation, the need for exogenous insulin was significantly reduced. All recipients had a decrease in exogenous insulin demand 1–2 weeks following hCiPSC-islet infusion, which was likely related to decreased hunger during the post-surgery recovery period and acute immunosuppression around the time of transplantation. After returning to a normal diet, the need for exogenous insulin increased. The need for exogenous insulin decreased and stabilized throughout the time when hCiPSC-islets engrafted and matured *in vivo*. Exogenous insulin need in the recipients has lowered by 31% after 15 weeks of hCiPSC-islets infusion. Postmortem histological analysis on the native pancreata of monkeys was performed to confirm that the enhanced glycemic control found in the recipient macaques was due to transplanted hCiPSC-islets. When combined with the findings in the pancreas, the data showed that hCiPSC-islet transplantation resulted in the alleviation of diabetes in the recipient macaques. Therapeutic investigations have provided good evidence that clinical islet cell transplantation can treat refractory hypoglycemia, a lifethreatening symptom in people with 'brittle diabetes. 'Recent findings showed that hPSC-islet infusion has the potential to improve glycemic control and rectify severe hypoglycemia in this specific subset of patients with labile diabetes, albeit additional animals need to be tried. Finally, the ability to be easily cryopreserved makes hCiPSC-islets a reliable, ready-to-use cell source, which is particularly significant for clinical applications and provides much-needed flexibility in human transplantation.

These findings show that pluripotent stem-cell-derived islets have a lot of potential for clinical use as a sustainable cell supply and that they could help diabetic patients. Furthermore, while no tumorigenesis was observed in any of the animals transplanted, strategies to ensure subject safety are worth further investigation because, in clinical settings, the increased transplanted cell number could increase the risk of teratoma formation, which would be especially concerning in immunocompromised patients. Further attempts to create techniques, such as establishing suitable protections and developing retrievable encapsulation devices, could be beneficial to the therapeutic application of hPSC-islets. In conclusion, the data acquired from the long-term assessment of hCiPSC-islets in a monkey model of diabetes provide useful insights for stem-cell-derived islets in diabetes clinical research [60]. The ability to create iPSCs and differentiate them into functioning cardiomyocytes, endothelium cells, and smooth muscle cells is a promising new advance in regenerative medicine. Modifications to the original processes significantly increased the low original differentiation efficiency of human cells. In mouse ischemia models, the use of diverse cell populations was investigated. Because of inadequate engraftment of the cells, injection of cardiac progenitor cells produced from iPSCs into the ischemic rodent heart resulted in functional improvement; however, the impact was mostly transient [62]. Endothelial cells from canine and porcine made from iPSCs are used to treat immunodeficient mouse models of myocardial infarction. By releasing paracrine substances, both types of cells increased heart contractility [63]. In a pig model of myocardial infarction, Templin et al. observed vascular differentiation and long-term engraftment of human iPSCs. In the pig ischemia model, the usage of human iPSC-derived cardiomyocyte sheets on temperature-sensitive polymers was investigated in an attempt to improve cell survival and engraftment. To achieve long-term therapeutic results, more technological advancements are required [64]. Age-related macular degeneration, gyrate atrophy, and certain kinds of retinitis pigmentosa can all be treated by iPSCs. The immune-privileged nature of the target tissue; the need for a small number of cells; and the ease of monitoring cell injection, potential therapeutic effects, and problems are only a few of the benefits of stem cell therapy for these disorders. Human iPSCs have been differentiated into multipotent retinal progenitor cells and RPE using protocols. Suspension culture as embryoid bodies is followed by adherent culture on an extracellular matrix in a two-step differentiation protocol for photoreceptor production from swine iPSC. Injection of cells derived from mouse iPSCs restored retina function in immunocompromised rhodopsin knock-out (Rho-/-) mice [65]. Human RPE cells injected into the subretinal area of Rpe65rd12/Rpe65rd12 mice recovered eyesight in the long run. After iodoacetic acid therapy, sub-retinal injection of differentiated iPSC resulted in cell incorporation into the retina. Future eye disease research should focus on developing strategies to aid proper transplanted cell integration, such as the use of natural and synthetic scaffolds. Human and animal iPSCs have been successfully differentiated into hepatocytes using proven techniques [66]. Gene expression profiles, secreted proteins, and metabolism show that these cells are often extremely similar to primary hepatocytes. These cells might mature *in vivo* and fulfill typical functions in rats after being engrafted into different animal models. In several cases, the cells saved the animal›s life by preventing liver failure. In immunodeficient Alb-uPA+/+;Rag2/;Il2rg/ mice, a point mutation in the 1-antitrypsin gene was repaired in human iPSCs, and generated liver cells showed normal cell function [67]. Chronic liver injury was established in NSG mice by giving them dimethylnitrosamine (DMN) for four weeks, which causes liver tissue damage that mimics cirrhosis. Intravenously injected human ESC and iPSC-derived hepatic cells (0.1×10^6 to 2×10^6 per mouse) were used to test the mice's capacity to engraft and heal the wounded liver. Approximately 89 percent of NSG mice with liver injury who received human hepatic cells survived until tissue harvesting, while 60% of mice that did not get human cells perished within the first three weeks. Neural cells produced from iPSCs have been used in experiments in numerous neurodegenerative disease models. In rhesus monkeys, Emborg et al. used brain progenitor cells produced from iPSCs to treat 1-methyl-4-phenyl-1,2,3,6-tetrahydropyridine-induced Parkinson's disease [68]. After transplantation, progenitor cells developed into neurons, astrocytes, and oligodendrocytes, and they lasted for at least six months. Reprogrammed and differentiated human iPSCs administered to rats with striatal lesions improved motor function significantly. In a rat model of the lysolecithin-induced demyelinated optic chiasm, human oligodendrocyte progenitors derived from iPSCs alleviated symptoms. After being grafted into the damaged spinal cords of rats and common marmosets, neural progenitor cells generated from murine or human iPSCs facilitated functional and electrophysiological recovery. When rodent or human iPSC-derived progenitor cells were transplanted into stroke-damaged mouse or rat brains, mixed outcomes were achieved. Tumor development and the absence of any behavioral effects were contrasted with considerable functional recovery, controlled cell proliferation, and the development of electro-physiologically active synaptic connections. The lack of defined protocols for cell preparation, modeling stroke, and testing therapy effects is one of the reasons for the disparity. Poor cell survival, statistically underpowered animal groups, biological variance, and measurement errors are all potential sources of unpredictability.

References

[1] Hassan, W.U., U. Greiser, and W. Wang. Role of adipose-derived stem cells in wound healing. Wound Repair Regen, 2014. 22(3): 313–25.

[2] Ahi, Z.B. et al. A combinatorial approach for spinal cord injury repair using multifunctional collagen-based matrices: development, characterization and impact on cell adhesion and axonal growth. Biomed Mater, 2020. 15(5): 055024.

[3] Piejko, M. et al. Adipose-derived stromal cells seeded on Integra® dermal regeneration template improve post-burn wound reconstruction. Bioengineering, 2020. 7(3): 67.

[4] Schuh, J.C.L., and K.A. Funk. Compilation of international standards and regulatory guidance documents for evaluation of biomaterials, medical devices, and 3-D printed and regenerative medicine products. Toxicol Pathol, 2019. 47(3): 344–357.

[5] https://www.iso.org/iso/iso_catalogue.htm.

[6] https://www.astm.org.

[7] https://www.fda.gov/regulatory-information/search-fda-guidance-documents/use-international-standard-iso-10993-1-biological-evaluation-medical-devices-part-1-evaluation-and.

[8] https://www.bsmi.gov.tw/wSite/mp?mp=1.

[9] Cramer, E.E.A., K. Ito, and S. Hofmann. *Ex vivo* bone models and their potential in preclinical evaluation. Curr Osteoporos Rep, 2021. 19(1): 75–87.

[10] Liu, T. et al. Advances of adipose-derived mesenchymal stem cells-based biomaterial scaffolds for oral and maxillofacial tissue engineering. Bioact Mater, 2021. 6(8): 2467–2478.

[11] Xiong, M. et al. Exosomes from adipose-derived stem cells: The emerging roles and applications in tissue regeneration of plastic and cosmetic surgery. Front Cell Dev Biol, 2020. 8(931): 574223.
[12] Becerra, J. et al. The stem cell niche should be a key issue for cell therapy in regenerative medicine. Stem Cell Rev Rep, 2011. 7(2): 248–55.
[13] Yang, Y.-I. et al. *Ex vivo* organ culture of adipose tissue for in situ mobilization of adipose-derived stem cells and defining the stem cell niche. Journal of Cellular Physiology, 2010. 224(3): 807–816.
[14] Prautsch, K.M. et al. *Ex-vivo* stimulation of adipose stem cells by growth factors and fibrin-hydrogel assisted delivery strategies for treating nerve gap-injuries. Bioengineering (Basel), 2020. 7(2).
[15] Singer, A.J., and R.A. Clark. Cutaneous wound healing. N Engl J Med, 1999. 341(10): 738–46.
[16] Cheng, N.C., S. Wang, and T.H. Young. The influence of spheroid formation of human adipose-derived stem cells on chitosan films on sternness and differentiation capabilities. Biomaterials, 2012. 33(6): 1748–1758.
[17] Li, P., and X. Guo. A review: therapeutic potential of adipose-derived stem cells in cutaneous wound healing and regeneration. Stem Cell Res Ther, 2018. 9(1): 302.
[18] Capella-Monsonís, H. et al. Extracellular matrix-based biomaterials as adipose-derived stem cell delivery vehicles in wound healing: a comparative study between a collagen scaffold and two xenografts. Stem Cell Research & Therapy, 2020. 11(1): 510.
[19] Hsu, S.H., and P.S. Hsieh. Self-assembled adult adipose-derived stem cell spheroids combined with biomaterials promote wound healing in a rat skin repair model. Wound Repair Regen, 2015. 23(1): 57–64.
[20] Zhou, X. et al. Multiple injections of autologous adipose-derived stem cells accelerate the burn wound healing process and promote blood vessel regeneration in a rat model. Stem Cells and Development, 2019. 28(21): 1463–1472.
[21] Bliley, J.M. et al. Administration of adipose-derived stem cells enhances vascularity, induces collagen deposition, and dermal adipogenesis in burn wounds. Burns, 2016. 42(6): 1212–1222.
[22] Quirinia, A., F.T. Jensen, and A. Viidik. Ischemia in wound healing. I: Design of a flap model--changes in blood flow. Scand J Plast Reconstr Surg Hand Surg, 1992. 26(1): 21–8.
[23] Quirinia, A., and A. Viidik. Ischemia in wound healing. II: Design of a flap model--biomechanical properties. Scand J Plast Reconstr Surg Hand Surg, 1992. 26(2): 133–9.
[24] Pu, C.M. et al. Adipose-derived stem cells protect skin flaps against ischemia/reperfusion injury via IL-6 expression. J Invest Dermatol, 2017. 137(6): 1353–1362.
[25] Bai, Y. et al. Adipose mesenchymal stem cell-derived exosomes stimulated by hydrogen peroxide enhanced skin flap recovery in ischemia-reperfusion injury. Biochem Biophys Res Commun, 2018. 500(2): 310–317.
[26] World Health, O., Definition, diagnosis and classification of diabetes mellitus and its complications: report of a WHO consultation. Part 1, Diagnosis and classification of diabetes mellitus. 1999, World Health Organization: Geneva.
[27] Yu, S. et al. Treatment with adipose tissue-derived mesenchymal stem cells exerts anti-diabetic effects, improves long-term complications, and attenuates inflammation in type 2 diabetic rats. Stem Cell Research & Therapy, 2019. 10(1): 333.
[28] Ylostalo, J.H. 3D Stem Cell Culture. Cells, 2020. 9(10).
[29] Cao, M. et al. Adipose-derived mesenchymal stem cells improve glucose homeostasis in high-fat diet-induced obese mice. Stem Cell Res Ther, 2015. 6(1): 208.
[30] Cheng, N.C. et al. High glucose-induced reactive oxygen species generation promotes stemness in human adipose-derived stem cells. Cytotherapy, 2016. 18(3): 371–83.
[31] El-Ftesi, S. et al. Aging and diabetes impair the neovascular potential of adipose-derived stromal cells. Plastic and Reconstructive Surgery, 2009. 123(2): 475–485.
[32] Wang, M. et al. Therapeutic effects of adipose stem cells from diabetic mice for the treatment of type 2 diabetes. Mol Ther, 2018. 26(8): 1921–1930.
[33] Sun, Y. et al. Adipose stem cells from type 2 diabetic mice exhibit therapeutic potential in wound healing. Stem Cell Res Ther, 2020. 11(1): 298.
[34] Ma, T. et al. A brief review: adipose-derived stem cells and their therapeutic potential in cardiovascular diseases. Stem Cell Research & Therapy, 2017. 8(1): 124.

[35] Nagata, H. et al. Cardiac adipose-derived stem cells exhibit high differentiation potential to cardiovascular cells in C57BL/6 Mice. Stem Cells Transl Med, 2016. 5(2): 141–51.
[36] Lambert, C. et al. Stem cells from human cardiac adipose tissue depots show different gene expression and functional capacities. Stem Cell Res Ther, 2019. 10(1): 361.
[37] Kashiyama, N. et al. Adipose-derived stem cell sheet under an elastic patch improves cardiac function in rats after myocardial infarction. J Thorac Cardiovasc Surg, 2020.
[38] Lindoso, R.S. et al. Adipose mesenchymal cells-derived EVs alleviate DOCA-salt-induced hypertension by promoting cardio-renal protection. Mol Ther Methods Clin Dev, 2020. 16: 63–77.
[39] Diekhorst, L. et al. Mesenchymal stem cells from adipose tissue do not improve functional recovery after ischemic stroke in hypertensive rats. Stroke, 2020. 51(1): 342–346.
[40] Rivera-Izquierdo, M. et al. An updated review of adipose derived-mesenchymal stem cells and their applications in musculoskeletal disorders. Expert Opin Biol Ther, 2019. 19(3): 233–248.
[41] Arnhold, S., and S. Wenisch. Adipose tissue derived mesenchymal stem cells for musculoskeletal repair in veterinary medicine. Am J Stem Cells, 2015. 4(1): 1–12.
[42] Torres-Torrillas, M. et al. Adipose-derived mesenchymal stem cells: a promising tool in the treatment of musculoskeletal diseases. International Journal of Molecular Sciences, 2019. 20(12): 3105.
[43] Cho, S.W. et al. Human adipose tissue-derived stromal cell therapy prevents bone loss in ovariectomized nude mouse. Tissue Eng Part A, 2012. 18(9-10): 1067–78.
[44] Guercio, A. et al. Mesenchymal stem cells derived from subcutaneous fat and platelet-rich plasma used in athletic horses with lameness of the superficial digital flexor tendon. Journal of Equine Veterinary Science, 2015. 35(1): 19–26.
[45] Usuelli, F.G. et al. Intratendinous adipose-derived stromal vascular fraction (SVF) injection provides a safe, efficacious treatment for Achilles tendinopathy: results of a randomized controlled clinical trial at a 6-month follow-up. Knee Surg Sports Traumatol Arthrosc, 2018. 26(7): 2000–2010.
[46] Saxer, F. et al. Implantation of stromal vascular fraction progenitors at bone fracture sites: From a rat model to a first-in-man study. Stem Cells, 2016. 34(12): 2956–2966.
[47] Song, Y. et al. Human adipose-derived mesenchymal stem cells for osteoarthritis: a pilot study with long-term follow-up and repeated injections. Regen Med, 2018. 13(3): 295–307.
[48] Alentorn-Geli, E. et al. Effects of autologous adipose-derived regenerative stem cells administered at the time of anterior cruciate ligament reconstruction on knee function and graft healing. J Orthop Surg (Hong Kong), 2019. 27(3): 2309499019867580.
[49] Dubey, N.K. et al. Adipose-derived stem cells attenuates diabetic osteoarthritis via inhibition of glycation-mediated inflammatory cascade. Aging Dis, 2019. 10(3): 483–496.
[50] Wang, Y.-h. et al. Adipose stem cell-based clinical strategy for neural regeneration: a review of current opinion. Stem Cells International, 2019. pp. 8502370.
[51] Chang, K.A. et al. The therapeutic effects of human adipose-derived stem cells in Alzheimer's disease mouse models. Neurodegener Dis, 2014. 13(2-3): 99–102.
[52] Soares Martins, T. et al. Diagnostic and therapeutic potential of exosomes in Alzheimer's disease. J Neurochem, 2021. 156(2): 162–181.
[53] Lee, M. et al. The exosome of adipose-derived stem cells reduces beta-amyloid pathology and apoptosis of neuronal cells derived from the transgenic mouse model of Alzheimer's disease. Brain Res, 2018. 1691: 87–93.
[54] Liu, H. et al. Alzheimer's Research and Therapy, 2021.
[55] Aras, Y. et al. The effects of adipose tissue-derived mesenchymal stem cell transplantation during the acute and subacute phases following spinal cord injury. Turk Neurosurg, 2016. 26(1): 127–39.
[56] Ohta, Y. et al. Intravenous infusion of adipose-derived stem/stromal cells improves functional recovery of rats with spinal cord injury. Cytotherapy, 2017. 19(7): 839–848.
[57] Chandrababu, K. et al. *In vivo* neural tissue engineering using adipose-derived mesenchymal stem cells and fibrin matrix. J Spinal Cord Med, 2021: 1–15.
[58] Martins-Taylor, K. et al. Recurrent copy number variations in human induced pluripotent stem cells. Nat Biotechnol, 2011. 29(6): 488–91.
[59] Harding, J., and O. Mirochnitchenko. Preclinical studies for induced pluripotent stem cell-based therapeutics. J Biol Chem, 2014. 289(8): 4585–93.

[60] Du, Y. et al. Human pluripotent stem-cell-derived islets ameliorate diabetes in non-human primates. Nat Med, 2022. 28(2): 272–282.
[61] Kim, J.M. et al. Long-term porcine islet graft survival in diabetic non-human primates treated with clinically available immunosuppressants. Xenotransplantation, 2021. 28(2): e12659.
[62] Mauritz, C. et al. Generation of functional murine cardiac myocytes from induced pluripotent stem cells. Circulation, 2008. 118(5): 507–17.
[63] Gu, M. et al. Microfluidic single-cell analysis shows that porcine induced pluripotent stem cell-derived endothelial cells improve myocardial function by paracrine activation. Circ Res, 2012. 111(7): 882–93.
[64] Templin, C. et al. Transplantation and tracking of human-induced pluripotent stem cells in a pig model of myocardial infarction: assessment of cell survival, engraftment, and distribution by hybrid single photon emission computed tomography/computed tomography of sodium iodide symporter transgene expression. Circulation, 2012. 126(4): 430–9.
[65] Tucker, B.A. et al. Transplantation of adult mouse iPS cell-derived photoreceptor precursors restores retinal structure and function in degenerative mice. PLoS One, 2011. 6(4): e18992.
[66] Ogiso, T. et al. Granulocyte colony-stimulating factor impairs liver regeneration in mice through the up-regulation of interleukin-1beta. J Hepatol, 2007. 47(6): 816–25.
[67] Hannan, N.R. et al. Production of hepatocyte-like cells from human pluripotent stem cells. Nat Protoc, 2013. 8(2): 430–7.
[68] Emborg, M.E. et al. Induced pluripotent stem cell-derived neural cells survive and mature in the nonhuman primate brain. Cell Rep, 2013. 3(3): 646–50.

10

Guidance & Case Study for Stem Cell Therapy

Wen-Yen Huang,[1] *Sung-Jan Lin*,[1] *Chia-Ning Shen*,[2] *Nai-Chen Cheng*,[3] *Min-Huey Chen*,[4] *I-Chi Lee*[5,*] and *Yi-Chen Ethan Li*[6,*]

Clinical transplantation of stem cells is a new medical technology, and its development process is the same as developing new drugs. There are strict regulations, rigorous preclinical studies, and clinical trials to ensure the safety and effectiveness of stem cell transplantation and the well-being of the patient. Therefore, in this chapter, we will introduce the current clinical applications of stem cells and the arrangement of related regulations. To realize the clinical application of stem cells, the following points need to be considered in the process of stem cell development: (1) the type of cells used, (2) how to obtain sufficient cell number and purity, (3) effective implant strategies, (4) immune rejection and other issues. For diseases and damaged tissues to be treated, in addition to understanding the cellular properties and functions of the stem cells to be used, we also need to consider how to obtain sufficient cell numbers and control the cells' behavior. In the *in vitro* tests, in addition to confirming that the stem cells used have their functions, human clinical trials can be carried

[1] Department of Biomedical Engineering, College of Medicine and College of Engineering, National Taiwan University, Taiwan.
[2] Genomics Research Center, Academia Sinica, Taiwan.
[3] Department of Surgery, National Taiwan University Hospital, Taiwan.
[4] Graduate Institute of Clinical Dentistry, School of Dentistry, National Taiwan University, Taiwan.
[5] Department of Biomedical Engineering and Environmental Sciences, National Tsing Hua University, Taiwan.
[6] Department of Chemical Engineering, Feng Chia University, Taiwan.
* Corresponding authors: iclee@mx.nthu.edu.tw; yicli@fcu.edu.tw

out if relevant evidence of safety, effectiveness, and functionality can be obtained in animal experiments. Generally speaking, the clinical trial will be divided into three phases according to the promulgation of Good Tissue Practice (GTP) from the Ministry of Health and Welfare in Taiwan.

The first phase is the safety test, which usually tests the safety of transplanted stem cells in normal people. The second phase comprises ten patients undergoing the safety and effectiveness testing. The final third phase is a large sample test, which requires hundreds to thousands of tests, and the validity is confirmed by statistical analysis. In human stem cell transplantation trials, the first and second phases are usually conducted simultaneously. The product can be applied for listing if the safety and effectiveness can be successfully verified in the three phases. More importantly, as a cell-transplanted medical product, its manufacturing process must comply with good manufacturing practice (GMP) to ensure the quality and safety of the transplanted stem cells. In the past 20 years, the clinical application of stem cell therapy has gradually shown the unlimited potential of stem cells. For example, nerve cells are traditionally thought to die after birth and cannot be regenerated. However, in 1996, Cheng and co-workers used intercostal nerves combined with glial repair surgery to connect the fractured spine of a rat successfully [1]. After six months, the hind limb functions of a rat were observed to regain movement significantly, disproving the theory that nerves could not regenerate. Since then, with the successful isolation of neural stem cells and a large number of studies, the role of neural stem cells in clinical applications has gradually become more important. In 2007, Lin and his colleagues injected autologous stem cells directly into the site of spinal injury and successfully restored mobility and limb function in five patients with cerebral injury [2], showing the possibility of neural stem/precursor cells (NSPCs) used for clinical applications. Moreover, in addition to hematopoietic functions, hematopoietic stem cells have also been pointed out by many studies to have the pluripotent ability to differentiate across germ layers [3]. Therefore, studies using umbilical cord blood stem cells to successfully treat brain injuries have also shown the value of stem cells in clinical treatment. According to these reports, various clinical studies and evidence show that, in addition to the importance of basic medicine, stem cells also have great application potential in clinical medicine, such as drug screening and transplantation. For these applications, many new technologies and related industries are gradually being developed, including cell isolation, culture, scale-up, differentiation induction, translational study tests and development, preclinical animal testing, and clinical testing. Every aspect of technology development for these clinical applications is very important, and product development must be customized individually for different patients. In addition, mass production is also an important factor that must be considered. Therefore, for the research and development of stem cell-related products, scientists and industry professionals must clearly understand the characteristics of various stem cells, the progress of scientific research, and the ultimate goal of the product. Basically, the first step for any product to be commercialized is to comply with the relevant regulations. In addition to complying with current medical-related regulations and ensuring the safety and efficacy of stem cell-related products, stem cell-related products, especially embryonic stem cells, have different research and application-related regulations due to different situations in each country. Therefore, before conducting technology research and development related to stem cell therapy, it is necessary to understand the relevant local regulations. The purpose of national regulations is to ensure the safety and effectiveness of stem cell medical technology used in the human body. The Center for Biologics Evaluation and Research (CBER) in the US FDA is the unit responsible for standardization and inspection. Relevant specifications can be found on the FDA website [4], refer to "Content and Review of Chemistry, Manufacturing, and Control (CMC) Information for Human Somatic Cell Therapy Investigational New Drug Applications (INDs)." Here, we will take embryonic stem cells as an example. The policies and regulations for embryo and embryonic stem cell study can be formulated to varying degrees according to the religious and historical background of each country. The overall consensus is based on the inability to conduct studies on replicators. It can be divided into about six categories: (1) The first category is to prohibit all research on human embryos, such as in Ireland, Austria, Poland, Norway, Costa Rica, Vatican. (2) The second category is that only existing embryonic stem cell lines are allowed to be studied, but human embryo research and the creation of new embryonic stem cell lines are not allowed, such as in Germany. (3) In the third category, it is allowed to use residual embryos from abortion or artificial reproduction to create new embryonic stem cell lines, such as in Canada, Israel, Japan, Australia, India, Switzerland, and Brazil. (4) In the fourth category, it is allowed to use residual embryos, and especially for research purposes, to produce embryonic stem cells by *in vitro* fertilization for research, such as in the United Kingdom, Belgium, and Singapore (subject to a case-by-case examination). (5) The fifth category permits the use of residual embryos, and especially for research purposes, the production of embryos for research utilizing somatic cell nuclear transfer into human egg cells or fertilized eggs, such as in China, the UK, Belgium, Sweden, and South Korea (requires presidential approval). In addition, Israel, India, Singapore, and Taiwan all need to be reviewed on a case-by-case examination. (6) The last category allows the use of remaining embryos, and especially for research purposes, the nuclear transfer of human somatic cells to non-human egg cells to make embryos for research, such as in China. It is worth mentioning that, in the United States, research funded by the federal government is classified into the second category, but with the support of private funds or specific state government funds, it allows researchers to study the establishment of new embryonic stem cell lines and the somatic cell nuclear transfection. Therefore, with the above regulations, scientists and industry professionals can develop products related to stem cell therapy following the guidelines of various countries. Of course, with the development of world trends, in recent years, the law and regulations of the countries mentioned above will also be revised according to the needs, so understanding the norms of modern cell therapy will help the industry to help the development and patent layout of stem cell therapy.

Cell therapy is the ultimate goal of all companies involved in developing stem cell products, and stem cells are becoming more common as a treatment option because of their ability to become any type of cell in the body. This means that stem cells can be used to treat a variety of diseases and injuries. Stem cells are now commonly used to treat leukemias and lymphomas successfully. In addition, the use of stem cells has also proven its potential in treating other types of cancer, Alzheimer's disease, Parkinson's disease, spinal cord injury, and diabetes. According to market research by SkyQuest Technology, the global stem cell therapy market is expected to grow from $6.87 billion in 2016 to $15.63 billion by 2025 [5]. This growth is driven by increased disease prevalence, increased demand for regenerative medicine, and increased investment in developing novel cell therapies. In the past decade, it can be found that the results from many studies showing most of the existing stem cell products and the cases of stem cells entering clinical trials are mainly adult stem cells, among which mesenchymal stem cells are the most. Here, we will use ADSCs as an example to introduce related cell therapy applications in medical cases.

Stromal vascular fractions (SVF) are isolated from fat tissues. There are two main ways to obtain SVF from fat tissue: enzymatic digestion and the mechanical disruption. Two major stem cell compartments in SVF are endothelial progenitor cells and ASCs. The main SVF actions are pro-angiogenic, antiapoptotic, antifibrotic, immune regulatory, anti-inflammatory, and trophic. There have been clinical trials of SVF in chronic kidney diseases, critical limb ischemia, osteoarthritis, idiopathic pulmonary fibrosis, hair diseases, and other conditions over the past years [6]. SVF is often combined with platelet-rich plasma (PRP), and the molecular signaling pool present in PRP can instruct SVF cells and protect them from the hostile microenvironments in the pathological tissues [7]. Next, ASCs can be differentiated into osteoblasts, chondroblasts, adipocytes, myocytes, and cardiomyocytes in suitable conditions. ASCs retain a high proliferation capacity *in vitro* and have the ability to undergo extensive differentiation into multiple cell lineages. Moreover, ASCs secrete a wide range of growth factors that can stimulate tissue regeneration [8]. The therapeutic value of ASCs mainly stems from their paracrine activities, including immunomodulatory, antiapoptotic, and trophic abilities [7]. There have been autologous ASC-related clinical studies for osteoarthritis, neural, sport injuries, and chronic wounds [6]. ASCs can be isolated and applied in their autologous form. In addition, stem cells derived from fat tissues can also be isolated from an individual and stored for future use in other people. This approach is relatively more efficient and cost effective comparing to the autologous approach. However, a potential immune reaction against allogenic cells is possible. Since the immunogenicity of stem cells is generally considered to be low, ASCs are generally considered suitable for allogenic applications without the need of immunosuppressive therapy [9]. Allogeneic ASCs have been used in clinical trials of multiple sclerosis, ischemia, and heart disease [6]. Furthermore, according to the trials reported in clinicaltrials.gov, the number of ASC-related trials has increased over the years. However, the investigations on humans have been done with limited size sample and conditions. Therefore, it is necessary to evaluate the safety and benefits of ASCs therapy in larger participant groups and various conditions [10]. The range of clinical applications of ASCs are broadly ranged, because the ease of cell harvest and high yield with minimal donor-site morbidity make them an ideal cell source. Platelet-rich plasma (PRP), which contains high levels of diverse growth factors that can stimulate stem cell proliferation and cell differentiation in the context of tissue regeneration, has recently been identified as a biological material that could be applied to tissue regeneration. Thus, co-transplantation of ASCs and PRP represents a novel approach for cell therapy in regenerative medicine [8]. The number of trials of ASCs is

Table 10-1. The number of clinical trials of ADSCs applications in the world.

Area or Country	Clinical trial number
North America	71
Europe	51
United States	64
East Asia	48
Japan	2
South Korea	19
Taiwan	10

increase over the years, however the number of human trials is limited in phase I and phase II (189); and lacks phase III and IV trials [14, 6, 10]. A true evaluation of efficacy and safety would require larger phase II/III studies. On clinicaltrials.gov, we found 265 ASC-related clinical trials (study period: 2010–2020). North America exhibited the highest number of trials, following by United States, Europe and the East Asia [6].

The ASC therapy conducted in different clinical trials has documented the safety of ASCs with few reports about severe adverse effects [11]. The safety and feasibility of ASC transplantation are conducted using dose escalation in different preclinical and clinical trials. After confirming the safety dose of the stem cells, an efficacy test is performed by this dose in the experimental group [12]. Possible side effects after treatment are monitored and safety events are followed for at least 12 months [12–14]. Further research is required to elucidate the application of ASCs as a safe and effective therapeutic option in the future. In the treatment of fecal incontinence, the injection of ASCs during fecal incontinence repair surgery may cause the replacement of fibrous tissue, which acts as a mechanical support to muscle tissue with contractile function [13]. Moreover, the efficacy of allogeneic expanded ASCs was maintained for up to 1 year after a single administration in treatment refractory patients with Crohn's disease with complex perianal fistulas. The short-term favorable tolerability of Cx601 also was maintained over the long-term [14]. Autologous SVF was shown to be safe and effective for reduction of pain in knee osteoarthritis (OA) patients. Autologous SVF to treat grade I to III knee OA showed statistically significant improvement in the Western Ontario and McMaster Universities Arthritis Index (WOMAC) and VAS scale three months after the intra-articular injection of SVF, which was maintained at 1 year. All patients attained full activity with decreased knee pain [15]. In addition, other studies employing autologous ASCs therapy as the treatment of knee OA also showed equal safety and effectiveness [16–18]. Even a single intra-articular injection of autologous ASCs was shown to be a safe therapeutic alternative to treat severe knee OA patients [16]. Moreover, it is likely that similar therapeutic procedures based on autologous ASCs or SVF can be extended in the future to other joints, such as the hip joint, or indications, such as intervertebral disc degeneration. A preliminary report of intradiscal implantation of SVF plus PRP in patients with degenerative disc disease demonstrated statistically significant improvements in several parameters over a six-month time period, suggesting certain clinical benefits of the SVF therapy in degenerative disc patients [19]. The use of ASCs has shown to be effective for the treatment of chronic ulcers, both individually and combined with other therapy. ASC application has been regarded as an innovative and effective approach in the treatment of chronic wounds [20]. An extracellular matrix/stromal vascular fraction gel (ECM/SVF gel) was found to exert a therapeutic effect on human chronic wounds, which is likely attributed to a favorable immunomodulatory effect, increased collagen accumulation, and improved neovascularization [21]. Moreover, centrifuged adipose tissue (CAT) may accelerate healing time in non-healing venous leg ulcers as well as reduce wound pain. The percentage of CD34+/CD45 cells in SVF seems to be a predictive biomarker of successful CAT treatment [22]. Intralesional allogeneic ASCs injection is an effective and safe treatment modality for relatively low-grade diabetic foot ulcers (DFUs) without infection. In addition to the standard treatment, satisfactory outcomes in wound healing with lower amputation and recurrence rates may be obtained with the injection of allogeneic ASCs [23]. Moreover, an allogeneic ASCs-hydrogel complex might be effective and safe to treat non-ischemic diabetic foot ulcers without infection [24]. The combination of ASCs and PRP also significantly enhanced wound closure rates when compared to standard wound care, without causing any serious complications [25]. Autologous fat grafting is increasingly used in reconstructive surgery to treat soft tissue deficiency. However, resorption rates ranging from 25% to 80% have been reported. ASC-enriched fat grafting was shown to have excellent feasibility and safety [26]. These promising results indicate that *ex vivo*-expanded ASC graft enrichment could render lipofilling a reliable alternative to major tissue augmentation, such as breast surgery, with allogeneic material or major flap surgery. Autologous SVF-assisted fat grafting promotes the survival of fat grafts with high vascular density and improves skin quality by increasing collagen content [27]. The studies comparing different facial rejuvenation approaches implied the advantages of the use of expanded ASCs or SVF-enriched fat over the use of PRP. The use of PRP led to the presence of more pronounced inflammatory infiltrates and increased vascular and nervous component reactivity. PRP may be more useful in pathological situations in which an intense angiogenesis is desirable, such as tissue ischemia, or for nervous repair [28]. For critical limb ischemia (CLI) therapy, digital subtraction angiography before and six months after ASCs implantation showed the formation of numerous vascular collateral networks across affected arteries. Hence, multiple intramuscular ASCs injections might be a safe alternative to achieve therapeutic angiogenesis in patients with CLI [29]. Clinical studies regarding the effects of human allogenic ASCs in acute respiratory distress syndrome (ARDS) showed no infusion toxicities or serious adverse events. Administration of ASCs appears to be safe and feasible for the treatment of ARDS. However, the clinical effect with the doses of ASCs used is weak, and further optimization of this strategy is required [30]. Suprachoroidal implantation of ASCs seems to be a safe and effective treatment of dry-type age-related macular degeneration (AMD) and Stargardt's macular dystrophy (SMD) [31]. ASCs within the engrafted adipose tissue have been shown to exert an anti-fibrotic effect in systemic sclerosis, and the mechanism is proposed to have an anti-fibrotic effect of transferred stem cells [32, 33]. Moreover, SVF and SVF plus PRP appeared to be safe and effective approaches for genital

Table 10-2. The current cell therapy cases of ADSCs.

Conditions	Year, Country	Type of cells	Study design	Dose, Route	Outcome Measure	Finding
Fecal incontinence	2016, South Korea [12]	allogeneic-ASCs	Randomized, Prospective dose escalation, placebo-controlled single-blinded, single-center trial 2 parallel groups phase 1	3×10^7, 6×10^7, 9×10^7 cells anal sphincter injection	Wexner score, the pressure of the anal sphincter and the score of patients' satisfaction	˙Administration of ASCs appears to be safe and feasible
Fecal incontinence (sphincter damage)	2017, Iran [13]	allogeneic-ASCs	Randomized double-blind, placebo-controlled	6×10^6 cells per 3 ml anal sphincter injection	Wexner score, endorectal sonography, electromyography (EMG) results	˙EMG and endorectal sonography: ratio of the area occupied by the muscle to total area of the lesion showed a 7.91% increase in the cell group compared with the control group
Crohn's disease and treatment-refractory, draining, complex Perianal fistulas	2018, European countries [14]	allogeneic expanded ASCs	Randomized double-blind parallel-group placebo-controlled phase 3	120×10^6 cells fistula tract injection	Fistula closure	˙The treatment appears to be safe and effective in closing external openings, compared with placebo, after 1 year

Table 10-2 contd. ...

...Table 10-2 contd.

Conditions	Year, Country	Type of cells	Study design	Dose, Route	Outcome Measure	Finding
Knee osteoarthritis	2016, France and Germany [16]	ASCs	Dose escalation safety testing phase 1	Single injection 2×10^6, 10×10^6, 50×10^6 cells intra-articular injection	Pain and function subscales of the Western Ontario and McMaster Universities Arthritis Index	˙Interestingly, patients treated with low-dose ASCs experienced significant improvements in pain levels and function compared with the baseline
Knee osteoarthritis	2016, USA [15]	SVF	Safety and feasibility study	Mean of 14.1 million viable, nucleated SVF cells (3cc) per knee intra-articular	WOMAC, VAS, range of motion, timed up-and-go (TUG), and MRI	˙3-months postoperative: statistically significant improvement in WOMAC and VAS scores (maintained at 1 year) ˙Physical therapy measurements for ROM and TUG both improved from preoperative to 3-months postoperative.
Knee osteoarthritis	2018, USA[36]	ASCs	Randomized prospective single-center parallel-group, controlled trial ASCs vs hyaluronic acid	Single injection intra-articular	WOMAC, WOMAC-A synovial fluid samples, and assess sway velocity using a force plate Analyze excess dipose tissue	˙A study protocol
Degenerative disc disease	2017, USA[19]	SVF plus PRP	Prospective, bicentric single-arm, open-label dose escalating phase 1	1 cc of SVF/PRP suspension injected into the nucleus pulpous under fluoroscopic guidance	Adverse events, range of motion, VAS, PPI, ODI, BDI, Dallas Pain Questionnaire and SF-12 scores	˙statistically significant improvements in several parameters

Conditions	Year, Country	Type of cells	Study design	Dose, Route	Outcome Measure	Finding
Chronic skin ulcer	2016, Italy [25]	ASC plus PRP	Randomized -controlled standard wound care + ASC plus PRP vs standard wound care	PRP 5 ml, 5×10^5 ASCs injection	Pictures of the lesions (area measurements) helped in the analysis of the wound closure rate	˙Healing rates: groups similar; ˙Wound closure rates: significantly different (ASC plus PRP group faster than control group)
Chronic wounds	2018, China [21]	autologous extracellular SVF	Observational Study ECM/SVF gel vs negative pressure wound therapy	Injected directly into the base and edges of the wound and the remaining gel covered the wound as a dressing	Wound healing rate, histological changes	˙Histological analysis: less lymphocyte infiltration, more collagen accumulation, and more newly formed vessels in the ECM/SVF gel group
Chronic leg ulcers	2019, Italy [22]	Centrifuged adipose tissue	Randomized-controlled phase 2	Multiple subdermal injection, depth of 1 cm, wound bed and edges	Healing time, safety, pain, complete wound healing at 24 weeks by Margolis Index wound-healing process expressed in square centimeters per week	˙Healing time, pain score : study group performed better than the control group ˙Strong reverse correlation between the percent of CD34+/CD45 non-hematopoietic cells, respectively, with the healing time and NRS
Diabetic foot ulcers	2019, South Korea [24]	Allogeneic ASCs	Randomized allogeneic ASC sheets (hydrogel complex) vs polyurethane film	1×10^6 cells/sheet weekly apply dressing change	Complete wound closure, time required for complete wound closure, the rate of wound size reduction from baseline	˙Complete wound closure (8 wk): treatment group > control group ˙Complete wound closure (12 wk): treatment group (82%) > control group (53%)

Table 10-2 contd. ...

...Table 10-2 contd.

Conditions	Year, Country	Type of cells	Study design	Dose, Route	Outcome Measure	Finding
Chronic diabetic foot ulcer	2020, Turkey [23]	allogeneic ASCs	Randomized -controlled single-blind ASCs + standard care vs standard care	6×10^6 cells, single dose dermo-epidermal junction injection	Wound characteristics, wound closure time, amputation rates, and clinical scores	˙Wound closure time (day) and number of minor amputations : study group performed better than control group ˙Postoperative physical functioning and general health (Short Form 36): significant difference between groups
Critical limb ischemia	2012, South Korea [29]	ASCs	Thromboangiitis obliterans group vs diabetes group vs healthy donors	0.5 cc ASCs (5×10^6 Cells) Multiple intramuscular	Degree of collateral vessel formation, and pain score walking distance, ABI, thermography findings, amputation, and ulcer healing	˙Five minor amputations: (amputation sites healed completely) ˙Pain rating scales and in claudication walking distance: at six months, significant improvement
Fat grafting	2013, Denmark [26]	ASCs	Randomised placebo-controlled trial 13 healthy participants enriched with ASCs vs without ASC enrichment	20×10^6 cells per mL fat injected subcutaneously (posterior part of the right and left upper arm)	MRI (compare the residual graft volumes of ASC-enriched grafts with those of control grafts)	˙Compared with the control grafts, the ASC-enriched fat grafts had significantly higher residual volumes ˙No serious adverse events were noted

Conditions	Year, Country	Type of cells	Study design	Dose, Route	Outcome Measure	Finding
ARDS	2014, China [30]	allogeneic ASCs	Randomized placebo-controlled	Single dose of 1×10^6 cells/kg of body weight peripheral intravenous infusion	Acute lung injury biomarkers	˙Clinical effect with the doses of MSCs used is weak
Systemic sclerosis (impaired hand function)	2015, France [33]	SVF	Phase 1	0.5 mL SVF was injected into each lateral side of each digit	Severity of adverse events hand disability and fibrosis, vascular manifestations, pain, and quality of life	˙Four minor adverse events ˙significant improvement in hand disability and pain, Raynaud's phenomenon, finger oedema, and quality of life
Facial Rejuvenation	2016, Italy [28]	SVF, ASCs, PRP	Randomized, SVF-enriched fat vs expanded ASCs vs fat plus PRP	ASCs (0.4 cc, 2×10^6 Cells) fat plus PRP (1mL adipose, 1mL PRP)Injection in the preauricular areas	Analyzed by optical and electron microscopy	˙The addition of PRP did not improve the regenerative effect
Facial skin rejuvenation	2019, China [27]	SVF	Randomized controlled trial SVF-assisted fat graft group vs fat graft only	1–3 × 10^7/ml of fresh SVF cells 8–12 mL (forehead), 5–10 mL (temporal), 6–10 mL (cheek), 3–6 mL (nasolabial groove), 2–4 mL(zygomatic) transplantation	Volumes of whole faces: pre-operation, immediately after surgery, and six month post-operation (3D scanner and Geomagic software) Facial skin qualities: preoperation and six months postoperation (VISIA skin detector) clinical evaluation (a visual analog scale)	˙The survival rate of SVF-enriched fat grafts was significantly higher than that of control grafts ˙The VISIA values of wrinkles and texture were significantly higher in SVF-enriched group than in thecontrol group at 6 months post-operation. ˙During long-term follow-up, the majority of patients in both groups were satisfied with the final facial esthetic results

Table 10-2 contd. ...

...Table 10-2 contd.

Conditions	Year, Country	Type of cells	Study design	Dose, Route	Outcome Measure	Finding
Age-Related Macular Degeneration	2018, Turkey [31]	ASCs plus PRP	Single-center, prospective clinical safety study phase 2	0.1cc ASCs (2×10^6) plus 1cc PRP A flap from the orbital fat was extracted what was laid on the scleral bed. The remaining space was filled with ASCs and PRP	Visual, visual field, mf-ERG	˙All of the patients experienced visual acuity, visual field, and improvement in mf-ERG recordings
Systemic sclerosis	2019, United Kingdom [32]	ASCs and PRP		Directly injected into the fibrotic oro-facial tissues	Mouth function, psychological measurements and pre and postoperative volumetric, the anti-fibrotic effect of ASCs	˙Lipotransfer may reduce dermal fibrosis through the suppression of fibroblast proliferation and key regulators of fibrogenesis
Genital Lichen sclerosuss	2020, Italy [34]	SVF and SVF plus PRP	Randomized SVF vs SVF plus PRP	SVF: 15cc, SVF plus PRP: 15cc+4cc injected intradermal (entire genital area affected by LS)	Dermatology life quality index (DLQI) before and 6 months after treatment	˙Strong safety profile ˙Both treatments allowed for a significant improvement (after 6 months) ˙Decreased efficacy in late-stage patients
Parkinson's disease	2020, USA, Nicaragua [35]	SVF		60×10^6 total nucleated cells in processed SVF implanted into the facial muscles and nose	the Oxford University Parkinson's Disease Quality of life instrument, clinical neurologic examination, and Unified Parkinson's Disease Rating Scale (captured on video)	˙Symptoms and signs may manifest quite quickly and with potential long-term stability over time

lichen sclerosus patients. Clinical results confirmed the synergic effect of SVF and PRP and supported the preferential use of the combinative therapy for early stage patients [34]. Two patients with Parkinson's disease (PD) displayed sustained clinical improvement after the implantation of autologous SVF cell preparation into the face and nasal cavity. However, the mechanism of action of this therapy is unknown. Further investigations using SVF for PD are warranted to clarify efficacy, the best route of administration, and the potential mechanism of action [35].

Next, we have organized information from the past ten years to inform readers about some companies that currently have cell and stem cell therapy-related products and technologies. From Table 10-3, the reader can see that many companies currently have cell therapy products. We can see the world's first cell therapy product, Carticel®, that the FDA has approved. This product is produced by Genzyme, a company that uses chondrocyte cell-based products for cellular repair of damaged joints. Vescell®, which is listed in Thailand, is a precursor cell developed by TheraVitae, which uses peripheral angiogenesis to repair patients with myocardial infarction. In addition, the largest umbilical cord blood collection company in the United States, ViaCell, uses umbilical cord blood stem cells to treat diseases such as heart, blood, and diabetes. Alofisel uses the donor's own ADSCs as a treatment for Chron's disease. Its development company, TriGenix, confirmed the immune-modulating function of Alofisel after testing more than 200 patients in Phase III clinical trials, which the European Commission finally approved in 2018 [37].

In addition to the listed companies, Table 10-4 also provides information about the units currently using stem cells for clinical trials. From the table, you can see that the IRCCS San Raffaele Scientific Institute currently uses autologous hematopoietic stem cells for the treatment of mucopolysaccharidosis Type I. The progress is currently in the second phase of clinical trials [39]. The Allife Medical Science and Technology Co., Ltd used iPSC-NSCs to investigate the safety and effectiveness of the treatement on Parkinson's Disease at an early phase 1 [40]. In another case, human iPSCs were induced as retinal pigment epithelium (iPSC-RPE) and cultured on a PLGA scaffolds for form a monolayer. Aferwards, the iPSC-RPE/PGLA construct was directly transplanted into the subretina of the patient with age-

Table 10-3. The companies with stem cell therapy products on the market [37, 38].

Company location	Cell types	Targets	Product status/Names
Genzyme, U.S.A.	Chondrocytes	Cartilage defects	Listed/Carticel®
CellTran, U.K.	Keratinocytes, cornea cells, melanocytes	Trauma caused by burns/diabetes	Listed /Myskin (An autologous keratinocyte-laden polymer product)
Organogenesis, U.S.A.	Keratinocytes, fibroblasts	Chronic trauma	Listed/Apligraf®
TheraVitae, Israel	Angiogenic precursor cells	Myocardial diseases	Listed/ VesCell®
ViaCell, U.S.A	Cord blood stem cells	heart, blood, and diabetes diseases	Listed/ViaCord®
TiGenix, Belgium	ADSCs	Crohn's disease	Listed/Alofisel®

Table 10-4. The current companies with ongoing stem cell therapy in clinical trail [39–42, 45, 46].

Company location	Cell types	Targets	Product status
IRCCS San Raffaele Scientific Institute, Italy	Hematopoietic stem cells	Mucopolysaccharidosis Type I	Phase 2
Allife Medical Science and Technology Co., Ltd., China	iPSC-NSC	Parkinson's disease	Early Phase 1
National Eye Institute. U.S.A.	iPSC-derived retinal pigment epithelium (iPSC-RPE)	Age-related macular degeneration	Phase 1/2
Samsung Medical Center, South Korea	MSCs	Ischemic stroke	Phase 3
Al-Azhar University Egypt	MSCs	Premature ovarian failure	Phase 1/2
Vinmec International Hospital,Vietnam	BMSCs	Autistic disorder	Phase 2

related macular degeneration for evaluating its safety and feasibility [41]. Moreover, Samsung Medical Center has been trying to use autologous mesenchymal stem cells to expand cells in an autologous serum since 2012. After the expansion, these cells are further used to treat ischemic stroke patients with severe persistent neurologic deficits. The previous period's results have confirmed that they are different from traditional methods. Comparatively, this method has a better therapeutic effect for stroke patients in the acute phase. The technique is currently in Phase 3 clinical trials for evaluation [42]. In addition to the abovementioned adult stem cell cases, some companies use human ESC-related technologies for cell therapy. One of the US-listed companies, Geron, has developed many technologies and products [43]. Due to the Geron company (with the close cooperation between the company and academic and R&D institutions), has published many well-known papers in Nature Biotechnology and Stem cells, etc. In addition, it also has a complete patent layout, which widely covers human ESC culture, differentiation, and the scope of rejection after transplantation. It also includes the medical applications of human ESCs in nerves, the heart, the liver, etc. In addition, another company, ESI BIO, has also developed a human embryonic stem cell line without xeno-contamination sources through the investment of the Singapore government and is committed to the treatment of diabetes [44]. Moreover, the company has also developed an automated culture system to scale up human ESCs. From the above cases, we can understand the importance of stem cell-based systems and products for their medical application and the potential of industrial development in the future.

References

[1] Cheng, H., Y. Cho, and L. Olson. Spinal cord repair in adult paraplegic rats: partial restoration of hind limb function. Science, 1996. 273(5274): 510–513.

[2] Ding, D.C., W.C. Shyu, M.F. Chiang, S.Z. Lin, Y.C. Chang, H.J. Wang, C.Y. Su, and H. Li. Enhancement of neuroplasticity through upregulation of beta1-integrin in human umbilical cord-derived stromal cell implanted stroke model. Neurobiol Dis, 2007. 27(3): 339–53.

[3] Schepici, G., S. Silvestro, P. Bramanti, and E. Mazzon. Traumatic brain injury and stem cells: an overview of clinical trials, the current treatments and future therapeutic approaches. Medicina (Kaunas), 2020. 56(3).
[4] https://www.fda.gov/vaccines-blood-biologics/biologics-guidances/tissue-guidances.
[5] https://www.globenewswire.com/news-release/2022/07/26/2486095/0/en/Cell-Therapy-Market-to-Generate-35-95-billion-Regenerative-Therapy-and-3D-Printing-to-Remain-in-Limelight.html.
[6] Adipose stem cells therapy clinical trials (2010–2020). https://ClinicalTrials.gov2 Jun 2021).
[7] Andia, I., N. Maffulli, and N. Burgos-Alonso. Stromal vascular fraction technologies and clinical applications. Expert Opin Biol Ther, 2019. 19(12): 1289–1305.
[8] Tobita, M., S. Tajima, and H. Mizuno. Adipose tissue-derived mesenchymal stem cells and platelet-rich plasma: stem cell transplantation methods that enhance stemness. Stem Cell Res Ther, 2015. 6: 215.
[9] Bacakova, L., J. Zarubova, M. Travnickova, J. Musilkova, J. Pajorova, P. Slepicka, N.S. Kasalkova, V. Svorcik, Z. Kolska, H. Motarjemi, and M. Molitor. Stem cells: their source, potency and use in regenerative therapies with focus on adipose-derived stem cells - a review. Biotechnol Adv, 2018. 36(4): 1111–1126.
[10] Chu, D.T., T. Nguyen Thi Phuong, N.L.B. Tien, D.K. Tran, L.B. Minh, V.V. Thanh, P. Gia Anh, V.H. Pham, and V. Thi Nga. Adipose tissue stem cells for therapy: an update on the progress of isolation, culture, storage, and clinical application. J Clin Med, 2019. 8(7).
[11] Kuriyan, A.E., T.A. Albini, J.H. Townsend, M. Rodriguez, and H.K. Pandya. Leonard RE 2nd. Parrott, M.B., P.J. Rosenfeld, H.W. Flynn Jr. and G. JL., Vision Loss after Intravitreal Injection of Autologous. N Engl J Med, 2017. 16(376(11)): 1047–1053.
[12] Park, E.J., J. Kang, and S.H. Baik. Treatment of faecal incontinence using allogeneic-adipose-derived mesenchymal stem cells: a study protocol for a pilot randomised controlled trial. BMJ Open, 2016. 6(2): e010450.
[13] Sarveazad, A., G.L. Newstead, R. Mirzaei, M.T. Joghataei, M. Bakhtiari, A. Babahajian, and B. Mahjoubi. A new method for treating fecal incontinence by implanting stem cells derived from human adipose tissue: preliminary findings of a randomized double-blind clinical trial. Stem Cell Res Ther, 2017. 8(1): 40.
[14] Panes, J., D. Garcia-Olmo, G. Van Assche, J.F. Colombel, W. Reinisch, D.C. Baumgart, A. Dignass, M. Nachury, M. Ferrante, L. Kazemi-Shirazi, J.C. Grimaud, F. de la Portilla, E. Goldin, M.P. Richard, M.C. Diez, I. Tagarro, A. Leselbaum, S. Danese, and A.C.S.G. Collaborators. Long-term efficacy and safety of stem cell therapy (Cx601) for complex perianal fistulas in patients with Crohn's disease. Gastroenterology, 2018. 154(5): 1334–1342 e4.
[15] Fodor, P.B., and S.G. Paulseth. Adipose Derived Stromal Cell (ADSC) injections for pain management of osteoarthritis in the human knee joint. Aesthet Surg J, 2016. 36(2): 229–36.
[16] Pers, Y.M., L. Rackwitz, R. Ferreira, O. Pullig, C. Delfour, F. Barry, L. Sensebe, L. Casteilla, S. Fleury, P. Bourin, D. Noel, F. Canovas, C. Cyteval, G. Lisignoli, J. Schrauth, D. Haddad, S. Domergue, U. Noeth, C. Jorgensen, and A. Consortium. Adipose mesenchymal stromal cell-based therapy for severe osteoarthritis of the knee: a phase i dose-escalation trial. Stem Cells Transl Med, 2016. 5(7): 847–56.
[17] Pers, Y.M., J. Quentin, R. Feirreira, F. Espinoza, N. Abdellaoui, N. Erkilic, M. Cren, E. Dufourcq-Lopez, O. Pullig, U. Noth, C. Jorgensen, and P. Louis-Plence. Injection of adipose-derived stromal cells in the knee of patients with severe osteoarthritis has a systemic effect and promotes an anti-inflammatory phenotype of circulating immune cells. Theranostics, 2018. 8(20): 5519–5528.
[18] Freitag, J., D. Bates, J. Wickham, K. Shah, L. Huguenin, A. Tenen, K. Paterson, and R. Boyd. Adipose-derived mesenchymal stem cell therapy in the treatment of knee osteoarthritis a randomized controlled trial. Regen Med, 2019. 14(3): 213–230.
[19] Comella, K., R. Silbert, and M. Parlo. Effects of the intradiscal implantation of stromal vascular fraction plus platelet rich plasma in patients with degenerative disc disease. J Transl Med, 2017. 15(1): 12.
[20] Li, P., and X. Guo. A review: therapeutic potential of adipose-derived stem cells in cutaneous wound healing and regeneration. Stem Cell Res Ther, 2018. 9(1): 302.

[21] Deng, C., L. Wang, J. Feng, and F. Lu. Treatment of human chronic wounds with autologous extracellular matrix/stromal vascular fraction gel: A STROBE-compliant study, Medicine (Baltimore), 2018. 97(32): e11667.
[22] Zollino, I., D. Campioni, M.G. Sibilla, M. Tessari, A.M. Malagoni, and P. Zamboni. A phase II randomized clinical trial for the treatment of recalcitrant chronic leg ulcers using centrifuged adipose tissue containing progenitor cells. Cytotherapy, 2019. 21(2): 200–211.
[23] Uzun, E., A. Guney, Z.B. Gonen, Y. Ozkul, I.H. Kafadar, M. Gunay, and M. Mutlu. Intralesional allogeneic adipose-derived stem cells application in chronic diabetic foot ulcer: Phase I/2 safety study, Foot Ankle Surg (2020).
[24] Moon, K.C., H.S. Suh, K.B. Kim, S.K. Han, K.W. Young, J.W. Lee, and M.H. Kim. Potential of allogeneic adipose-derived stem cell-hydrogel complex for treating diabetic foot ulcers. Diabetes, 2019. 68(4): 837–846.
[25] Raposio, E., N. Bertozzi, S. Bonomini, G. Bernuzzi, A. Formentini, E. Grignaffini, and P.G. M. Adipose-derived stem cells added to platelet-rich plasma for chronic skin ulcer therapy. Wounds, 2016. 28(4): 126–31.
[26] Kølle, S.-F.T., A. Fischer-Nielsen, A.B. Mathiasen, J.J. Elberg, R.S. Oliveri, P.V. Glovinski, J. Kastrup, M. Kirchhoff, B.S. Rasmussen, M.-L.M. Talman, C. Thomsen, E. Dickmeiss, and K.T. Drzewiecki. Enrichment of autologous fat grafts with *ex-vivo* expanded adipose tissue-derived stem cells for graft survival: a randomised placebo-controlled trial. The Lancet, 2013. 382(9898): 1113–1120.
[27] Yin, Y., J. Li, Q. Li, A. Zhang, and P. Jin. Autologous fat graft assisted by stromal vascular fraction improves facial skin quality: A randomized controlled trial. J Plast Reconstr Aesthet Surg, 2020. 73(6): 1166–1173.
[28] Rigotti, G., L. Charles-de-Sa, N.F. Gontijo-de-Amorim, C.M. Takiya, P.R. Amable, R. Borojevic, D. Benati, P. Bernardi, and A. Sbarbati. Expanded stem cells, stromal-vascular fraction, and platelet-rich plasma enriched fat: comparing results of different facial rejuvenation approaches in a clinical trial. Aesthet Surg J, 2016. 36(3): 261–70.
[29] Lee, H.C., S.G. An, H.W. Lee, J.S. Park, K.S. Cha, T.J. Hong, J.H. Park, S.Y. Lee, S.P. Kim, Y.D. Kim, S.W. Chung, Y.C. Bae, Y.B. Shin, J.I. Kim, and J.S. Jung. Safety and effect of adipose tissue-derived stem cell implantation in patients with critical limb ischemia: a pilot study. Circ J, 2012. 76(7): 1750–60.
[30] Zheng, G., L. Huang, H. Tong, Q. Shu, Y. Hu, M. Ge, K. Deng, L. Zhang, B. Zou, B. Cheng, and J. Xu. Treatment of acute respiratory distress syndrome with allogeneic adipose-derived mesenchymal stem cells: a randomized, placebo-controlled pilot study. Respir Res, 2014. 15: 39.
[31] Oner, A., Z.B. Gonen, D.G. Sevim, N. Smim Kahraman, and M. Unlu. Suprachoroidal adipose tissue-derived mesenchymal stem cell implantation in patients with dry-type age-related macular degeneration and stargardt's macular dystrophy: 6-month follow-up results of a phase 2 study. Cell Reprogram, 2018. 20(6): 329–336.
[32] Almadori, A., M. Griffin, C.M. Ryan, D.F. Hunt, E. Hansen, R. Kumar, D.J. Abraham, C.P. Denton, and P.E.M. Butler. Stem cell enriched lipotransfer reverses the effects of fibrosis in systemic sclerosis. PLoS One, 2019. 14(7): e0218068.
[33] Granel, B., A. Daumas, E. Jouve, J.R. Harle, P.S. Nguyen, C. Chabannon, N. Colavolpe, J.C. Reynier, R. Truillet, S. Mallet, A. Baiada, D. Casanova, L. Giraudo, L. Arnaud, J. Veran, F. Sabatier, and G. Magalon. Safety, tolerability and potential efficacy of injection of autologous adipose-derived stromal vascular fraction in the fingers of patients with systemic sclerosis: an open-label phase I trial. Ann Rheum Dis, 2015. 74(12): 2175–82.
[34] Tedesco, M., B. Bellei, V. Garelli, S. Caputo, A. Latini, M. Giuliani, C. Cota, G. Chichierchia, C. Romani, M.L. Foddai, A. Cristaudo, A. Morrone, and E. Migliano. Adipose tissue stromal vascular fraction and adipose tissue stromal vascular fraction plus platelet-rich plasma grafting: New regenerative perspectives in genital lichen sclerosus. Dermatol Ther, 2020. 33(6): e14277.
[35] Carstens, M., I. Haq, J. Martinez-Cerrato, S. Dos-Anjos, K. Bertram, and D. Correa. Sustained clinical improvement of Parkinson's disease in two patients with facially-transplanted adipose-derived stromal vascular fraction cells. J Clin Neurosci, 2020. 81: 47–51.
[36] Jones, I.A., M. Wilson, R. Togashi, B. Han, A.K. Mircheff, and C. Thomas Vangsness, Jr.. A randomized, controlled study to evaluate the efficacy of intra-articular, autologous adipose tissue

injections for the treatment of mild-to-moderate knee osteoarthritis compared to hyaluronic acid: a study protocol. BMC Musculoskelet Disord, 2018. 19(1): 383.
[37] https://www.ema.europa.eu/en/documents/product-information/alofisel-epar-product-information_en.pdf.
[38] Wilan, K.H., C.T. Scott, and S. Herrera. Chasing a cellular fountain of youth. Nat Biotechnol, 2005. 23(7): 807–15.
[39] https://clinicaltrials.gov/ct2/show/NCT04486001.
[40] https://clinicaltrials.gov/ct2/show/NCT03815071.
[41] https://clinicaltrials.gov/ct2/show/NCT04339764?term=iPSCs%2C+phase+2&phase=1&draw=2&rank=2.
[42] https://clinicaltrials.gov/ct2/show/NCT01716481?cond=stem+cell%2C+Phase+3&draw=2&rank=8.
[43] https://www.geron.com/.
[44] https://esibio.com/products/popular-brands/es-cell-international/.
[45] https://clinicaltrials.gov/ct2/show/NCT02627131?cond=Stem+cell+therapy&phase=1&draw=2&rank=4.
[46] https://clinicaltrials.gov/ct2/show/NCT02372474?cond=Stem+cell+therapy&draw=2&rank=3.

Index

2D membrane 95
3D cell/tissue culture models 175
3D *in vitro* culture system 154, 160, 163

A

Adhesion 32, 33, 36, 39, 42, 48, 49
Adipose-derived stem cells 10

B

Biocompatibility 217, 219, 220
Biodegradable and bioresorbable biomaterials 77
Biomaterials 32–36, 38–40, 43–49, 51, 53, 55, 56
Bioreactors 65–72
Biosensors 176, 187, 192–194
Bone marrow mesenchymal stem cells 8

C

Clinical transplantation 233

D

Differentiation 32–43, 45–49, 52–56
Drug controlled release 209

E

Embryonic stem cells 3, 6, 16
Ex vivo testing 216, 217, 219, 224

H

Hair follicle stem cells 16

I

Induced pluripotent stem cells 5
ISO-10993 guideline 217, 218

M

Mechanical properties 75–79, 81, 82

N

Nanofibers 208, 211, 212
Nanoparticles 206–208, 210, 211
Nanotechnology iii
Nanowires 208, 209
Natural biomaterials 74–78, 86, 87
Neural stem cells 2, 14–16

O

Oral tissue-derived stem cells 12, 13
Organ on chips (Ooc) 175, 176, 187–189, 191–194

P

Phase inversion method 108, 109
Proliferation 32–36, 38, 39, 42, 43, 45, 47, 48, 55

S

Scaffold-based culture system 116–118
Scaffold-free culture system 114–117
Scale-up 63, 64
Stem cell therapy iii, 233–236, 245, 246
Stem cells 32, 33, 36–38, 42–50, 53, 55, 56, 63–68, 72
Surface charge 80
Surface modification 96, 97, 101–103
Surface topography 78, 79

T

Thick membrane 97, 103–106, 108, 109
Thin film 97–101, 103, 104

W

Wettability 78, 81